C000292496

Bearskins, Bayonets & Body Armour

Welsh Guards 1915–2015

Bearskins, Bayonets & Body Armour

Welsh Guards 1915–2015

Trevor Royle

With a foreword by
HRH The Prince of Wales

FRONTLINE
BOOKS

The Regiment would like to express its sincere thanks to the late Dinah Staines for her generous bequest to the Welsh Guards. Her legacy has not only helped fund the production of this book, but also, through the Welsh Guards Charity, continues to benefit soldiers who have suffered life-changing injuries sustained whilst serving in the Welsh Guards. Her connection to the Regiment was through her late stepfather, Vernon Hopkins (1915–2001), who served in the Welsh Guards before and during the Second World War, and underlines the deep family ties between the Regiment and those who have been a part of it.

Bearskins, Bayonets & Body Armour
Welsh Guards, 1915–2015
First published in 2015 by Frontline Books,
an imprint of Pen & Sword Books Ltd, 47 Church Street,
Barnsley, S. Yorkshire, S70 2AS
www.frontline-books.com

ISBN: 978-1-84832-735-1

CIP data records for this title are available from the British Library

For more information on our books, please visit www.frontline-books.com, email info@frontline-books.com or write to us at the above address.

Designed and typeset by Ian Hughes, www.mousematdesign.com

Printed and bound by Gutenberg Press, Malta

FRONTISPIECE:
HM The Queen, Colonel-in-Chief Welsh Guards, and HRH The Duke of Edinburgh, Colonel Welsh Guards, leave Buckingham Palace for Horse Guards to take the salute at the Queen's Birthday Parade, 1965, painted by Terence Cuneo. This artist was famous for always including a mouse hidden somewhere in his pictures.

CLARENCE HOUSE

In this special centennial year, I am enormously proud to have been Colonel of the Welsh Guards for the last forty years. It was in fact my Great-Grandfather, King George V who, during the First World War, completed the national complement of Foot Guards Regiments by raising the Welsh Guards.

The Regiment was raised in a week and barely six months later they were in battle. Their valour and resilience in that first action, at Loos in Flanders, demonstrated how, with astonishing rapidity, they had become a formidable force – indefatigably Welsh, defined by loyalty to each other and to country, and sustained, in desperate circumstances, by their *ysbryd*, or spirit, and by humanity and humour.

Those same qualities have endured ever since, as Trevor Royle encapsulates in this regimental history, and in which the deeds of a few stand in memory to the service and sacrifice of so many.

As this book describes, the Welsh Guards have distinguished themselves in all spectrums of conflict. In World War II they fought tirelessly in France, North Africa and Italy. During national service, the men of the Welsh Guards witnessed and experienced the frictions and tensions which would later so destabilise Palestine and Egypt. They were followed by a generation who sought to assist the people of Aden, Northern Ireland and the Falklands.

The Welsh Guards' magnificent tradition of unflinching service, most recently in Iraq and Afghanistan, has brought with it a heavy and terrible cost to those men, and their families, and this should always be remembered by those who follow.

Cymru Am Byth

Contents

Company Colours:

Chapter 1
The Prince of Wales's Company
Y DDRAIG GOCH DDYRY CYCHWYN.
The red dragon gives a lead.

Chapter 2
Number 2 Company
GWYR YNYS Y CEDYRN.
The men of the island of the mighty.

Chapter 3
Number 3 Company
FY NUW, FY NGWLAD, FY MRENIN.
My God, my land, my King.

Chapter 4
Number 4 Company
GOREU ARF CALON DDEWR.
The best weapon is a brave heart.

Chapter 5
Number 5 Company
OFNA DDUW, ANRHYDEDDA'AR BRENIN.
Fear God, honour the King.

Chapter 6
Number 6 Company
NAC OFNA OND GWARTH.
Fear nothing but disgrace.

Chapter 7
Number 7 Company
Y GWIR YN `N Y BYD.
The truth against the world.

Chapter 8
Number 8 Company
OFNER NA OFNO ANGAU.
Feared be he who fears not death.

Chapter 9
Number 9 Company
A FYNNO BARCH BID GADARN.
Let him be strong who would be respected.

Chapter 10
Number 10 Company
HEB DDIAL NI DDYCHWELAF.
I will not return unavenged.

Chapter 11
Number 11 Company
HWY CLOD NA HOEDL.
Fame lasts longer than life.

Chapter 12
Number 12 Company
HEB NEFOL NERTH NID SICR SAETH.
Without heavenly help the arrow flieth uncertain.

Chapter 13
Number 13 Company
CAIS DDUW YN GAR AC NAC OFNA FAR.
Seek God as friend and fear not ill.

Chapter 14
Number 14 Company
CAS GWR NA CHARO'R WLAD A'I MACO.
Hateful is the man who loves not the land that nurtured him.

Chapter 15
Number 15 Company
RHYDDID, HEDD A LLAIYDDIANT.
Freedom, peace and prosperity.

Photographic and picture credits and copyright

The Welsh Guards would like to thank all those people who have so kindly allowed us to use various photographs, drawings, maps and paintings.

Preface and Acknowledgements

In writing this centenary history of the Welsh Guards I have tried to be guided by the advice proffered by Rudyard Kipling when he wrote the introduction to his *Irish Guards in the Great War* (2 vols, 1923). His words were simple yet profound and they got to the heart of the matter. Kipling, whose son John was killed at the Battle of Loos in September 1915, wanted to tell the story of the Regiment's wartime experiences 'soberly and with what truth is possible'. To do that he ensured that the 'point of view is the battalions', and the facts mainly follow the Regimental Diaries, supplemented by the few private letters and documents which such a war made possible, and by some tales that have 'gathered round men and their actions'. Without being presumptuous, that seemed to be as good a set of guidelines as any and in writing this book for the Welsh Guards I tried to follow them as the book slowly took shape.

Although I have had a reasonably close association with the Regiment since the 1980s, I did not serve in the Welsh Guards. That could have been a hindrance but I hope that it has also given me the degree of objectivity which is required in any balanced narrative. That being said, I have not been without essential props. The war histories written by C. H. Dudley-Ward and L. F. Ellis remain outstanding records: voluminous, comprehensive and brilliantly understated, and I have made full use of them. The same is true of the various Battalion War Diaries, and the archives in Regimental Headquarters are a veritable treasure-trove. Across the years I salute the memory of the many Welsh Guardsmen who took the trouble to record their recollections and then deposited them for use by future generations. Equally important were two recent additions to the historiography of the Regiment – the admirable Welsh Guards Collection at Oswestry, which is a model of its kind, and the equally laudable Welsh Guards Reunited website.

To keep the project on track and to give me much-needed assistance, the Regiment appointed a centenary book committee and it is not stretching a point to say that I would have been lost without their ministrations and cheerful encouragement. Chaired by Major General Robert Talbot Rice, the unflappable Regimental Lieutenant Colonel, its members were Tom Bonas, Tony Cooper, Nick Drummond, Stan Evans, Mark Jenkins, Roy Lewis, Robert Mason, Marcus Scriven, Charles Stephens and the Battalion Seconds-in-Command (Henry Bettinson and Jules Salusbury). They all read the text in draft and in its final form, as did the Commanding Officers of the past two decades and one or two other senior figures. The text has also been cleared with the Ministry of Defence. Of course, any remaining errors or misjudgements remain my responsibility alone.

The spine of this history is provided by the main narrative of fifteen chapters and the text has been enlivened by the inclusion of a number of features which illustrate different aspects of the life of the Regiment over the years. These could not have been written without the ready and cheerful assistance of several Welsh Guardsmen, who provided the necessary words and gave unstintingly of their time and experience: Micky Barnes, Tom Bonas, Alun Bowen, Tony Davies, Stan Evans,

Roy Lewis, Charles Richards, Charles Stephens, Angus Wall, John Warburton-Lee, Reddy Watt and Peter Williams.

Most of the work was undertaken at Regimental Headquarters where I enjoyed the ready support and encouragement of all the staff over a period of almost three years. By its very nature any book of this kind is a joint effort and it would be remiss of me not to acknowledge the substantial parts played by others, namely Nick Drummond and Mark Jenkins for making sure that the illustrations were collected, collated and identified and Martin Browne for carefully reading the proofs. They were all towers of strength throughout the project, as was Melanie Lewis, who came on board as the Project Officer for the Welsh Guards Centenary. Visits to the 1st Battalion at Aldershot and Hounslow kept me honest by introducing me to the ways in which the modern army has changed: I am grateful to the respective Commanding Officers, the officers and men, for hospitality, good cheer and lively conversation. Thanks, too, to Major Alun Bowen for his patience in setting up interviews and for introducing me to the mysteries of the 300 Cup (though thankfully not on the field of play).

I acknowledge with thanks the gracious permission of Her Majesty Queen Elizabeth II to make use of material from the Royal Archives relating to the formation of the Welsh Guards in 1915. I am also grateful for permission to quote from the following works: Max Arthur, *Above All, Courage: Personal Stories from the Falklands War* (Sidgwick and Jackson, 1985) and Max Arthur, *Northern Ireland: Soldiers Talking* (Sidgwick and Jackson, 1987).

Frontline Books proved to be ideal publishers – calm, even-handed and professional in all their dealings. From the outset Michael Leventhal was enthusiastic and involved and he was ably backed by his principal lieutenants – Stephen Chumbley and Ian Hughes.

I cannot end without thanking two distinguished Welsh Guardsmen who have every right to consider themselves as 'onlie begetters' of this book. About three years ago, General Sir Reddy Watt took me aside at the conclusion of the annual carol service of the charity Combat Stress (of which I was a Trustee and he the President) and asked if I might be interested in tackling the centenary history. I said I would give it some thought but a day or so later my old friend Colonel Tom Bonas, the Regimental Adjutant, pre-empted the decision when he telephoned to say that he was delighted that I had agreed. Faced by such coercive enthusiasm it would have been difficult to demur! The rest, as they say, is history.

For me it has been a huge privilege to have been invited to write this account of the first hundred years of a Regiment I much admire. I can only hope that I have done justice to its story.

Trevor Royle,
St David's Day 2014

CHAPTER ONE

Birth and First Blooding, 1915

Foundation, Battle of Loos

THE REGIMENT RAISED AS the Welsh Guards on 26 February 1915 was a child of necessity, born into a world in conflict. At the beginning of that year the First World War was only five months old, yet the opposing sides were already facing stalemate. On the Western Front in France and Flanders the war of movement had ground to a halt with the rival armies entrenched behind heavily defended barbed-wire positions, leaving the generals little option but to engage in costly battles of attrition. Later in the year new fronts would be opened in Gallipoli, Egypt and Mesopotamia as a result of the Ottoman Empire's decision to throw in its lot with Germany and Austria-Hungary. There was also fighting in east and west Africa involving British and German colonial forces. From the outset of this global conflict it was obvious that the deciding factor would be manpower: all over the world, but especially on the European fronts, men were needed in droves. Of the main combatant nations, Austria-Hungary, Germany, the Ottoman Empire, France and Russia all had huge conscript armies with hundreds of thousands of soldiers at their disposal; only Britain possessed a small volunteer army, the bulk of which had been deployed in France in the late summer of 1914.

The British Expeditionary Force (BEF), as it was

Officers of the 1st Battalion, 1915. *Rear row: (L to R)* 2Lt H. A. Evan-Thomas, 2Lt Hon P. G. F. Howard, Lt P. L. M. Battye, 2Lt H. M. Martineau, 2Lt B. C. Williams-Ellis. *Third row:* 2Lt G. A. D. Perrins, Lt Viscount Clive, Lt W. H. J. Gough, Lt R. W. Lewis, 2Lt E. T. V. Hambrough, 2Lt N. Newell, 2Lt H. Dene, 2Lt H. J. Sutton, Lt H. E. Allen, Lt J. J. P. Evans, 2Lt J. L. W. Crawshay, 2Lt J. Randolph. *Second row:* 2Lt H. Talbot Rice, Lt Lord Newborough, 2Lt E. R. Martin-Smith, 2Lt G. C. H. Insole, Lt H. E. Wethered, 2Lt Hon E. F. Morgan, Lt K. G. Menzies, 2Lt W. A. F. L. Fox-Pitt, 2Lt F. A. V. Copland-Griffiths, 2Lt G. C. H. Crawshay, 2Lt E. G. Mawby, Lt (QM) W. B. Dabell. *Front row:* Capt J. V. Taylor, Capt C. C. L. Fitzwilliams, Maj H. H. Bromfield, DSO, Capt R. G. W. Williams-Bulkeley, Capt A. P. Palmer, DSO, Lt Col W. Murray-Threipland, DSO, Maj G. C. D. Gordon, Capt O. T. D. Osmond-Williams, DSO, Capt G. W. Phillips, Capt J. H. Bradney, Capt F. W. E. Blake. *(Welsh Guards Archives)*

Welsh Guards recruits still wearing the uniforms of their previous regiments shortly after arriving at White City in 1915. *(Christina Broome Collection/Imperial War Museum)*

known, was second to none in terms of professionalism. Its fighting strength, two corps of six infantry divisions in all, consisted of around 100,000 officers and men and Brigadier General J. E. Edmonds, the Official Historian of the war, called it 'incomparably the best trained, best organized and best equipped British Army which ever went to war'. Ominously he added the caveat 'except in the matter of numbers' and that was to be the BEF's main weakness.[1] As 1914 came to an end the BEF counted the cost of the fighting at Mons, the Marne, the Aisne and Ypres and estimated that its losses amounted to 89,969 casualties, killed, wounded or missing, almost half its original number of professional Regular soldiers and time-expired volunteers; of the eighty-four infantry battalions which had taken part in the fighting to date, only nine had more than 300 fully fit men on their strength.

Fortunately the task of raising new armies had begun within days of the outbreak of the conflict in August 1914, following the appointment of Field Marshal Lord Kitchener of Khartoum as Secretary of State for War. With his tall military bearing and luxuriant moustache, he was the country's best-known soldier and the epitome of British pluck and resolve and he proved to be a popular choice. He was also associated with success: his military career had been crowned with victories from the Middle East and Sudan to the more recent Boer War and he took up his new position with characteristic energy. From the outset he disagreed with the prevailing opinion of his colleagues in the Cabinet that the war would be short-lived and largely fought at sea. He also believed that the BEF was too small to offer anything but limited support to Britain's allies, especially France, and that huge new armies would have to be raised if the country was to make any impact on the direction of the war. On 8 August he called for the first 100,000 volunteers from the 18–30 age group, the longer-term aim being to create by 1917 a force which would number 70 infantry divisions of 1.2 million men. By Kitchener's reasoning, its arrival would come at a time when Germany's resources would be over-stretched and Britain would be in a position to crush the enemy and dictate the resultant peace.

Such was Kitchener's personal standing and the strength of his political connections that he enjoyed the full backing of Prime Minister H. H. Asquith, although no one in the Cabinet really knew how the armies would be raised or if a million men would actually volunteer. Those worries were soon dispelled. Backed by a huge publicity machine, which included the press and the influential all-party Parliamentary Recruiting Committee, the national appeal for recruits was loud and insistent and the campaigners quickly showed that they meant business. Before long, Kitchener's instantly recognizable features were assisting the drive and Alfred Leete's famous recruiting poster appeared all over the country. Very few young men caught undecided outside an Army

HM King George V, Colonel-in-Chief Welsh Guards, 1915–36, painted by Oswald Birley. His Majesty signed a Royal Warrant authorizing the raising of a Welsh Regiment of Foot Guards on 26 February 1915. *(Estate of Mark Birley)*

David Lloyd George

Born in Manchester in 1863, David Lloyd George was one of the outstanding politicians of his age. A Liberal and a natural reformer, he was continuously in office from December 1903 and occupied the post of prime minister from December 1916 to October 1922. Although he was a fierce opponent of the Boer War (1899–1902), as Chancellor of the Exchequer at the outbreak of the First World War he was supportive of the war aims but frequently impatient of the direction of policy. When the coalition government was formed in May 1915, he took over the newly formed Ministry of Munitions and with it responsibility for directing much of the country's war effort.

His grandson Owen Lloyd George served in the Welsh Guards during the Second World War. In an imaginative gesture, on David Lloyd George's death in May 1945, Winston Churchill arranged for all four grandsons to be flown home for the funeral. Owen Lloyd George arrived at the family home at Llanystumdwy an hour before the funeral, having been flown there by a Polish pilot equipped only with a school atlas and no knowledge of Wales.

The family connection with the Welsh Guards continued into the twenty-first century when his great-great-grandson, Captain the Hon F. O. Lloyd-George, joined the Regiment in 2011.

Major General Sir Francis Lloyd, GCVO, KCB, DSO

As Major General commanding the Brigade of Guards and General Officer Commanding London District during the First World War, Lloyd was closely involved in the creation of the Welsh Guards in 1915. He was also responsible for the wartime defence of London, particularly from attack by Zeppelins.

Born on 12 August 1853, he was commissioned in the 33rd (Duke of Wellington's) Regiment in 1874 but transferred to the Grenadier Guards with whom he served in Sudan and South Africa during the Boer War of 1899–1902. Before the outbreak of the First World War he commanded 1st Battalion, Grenadier Guards (1903–4), 1st Guards Brigade (1904–8) and the Welsh Division (Territorial Force) (1909–13). He was promoted Lieutenant-General in 1918 and retired from the Army two years later. That same year, 1920, Lloyd was one of the leading supporters of the Regiment when it was threatened with disbandment in the defence cuts following the First World War.

His seat was at Oswestry in Shropshire and he enjoyed close family and social links with Wales.

recruiting office found it easy to ignore the great man's bristling moustache and pointed exhortation, 'Your Country Needs You!' By 12 September the first 100,000 had been recruited to form the First New Army, or K1, and by the end of the following month there were sufficient volunteers for a further twelve infantry divisions, which comprised the Second and Third New Armies (K2 and K3).

Many of those original volunteers were young Welshmen, anxious to do their bit for their country by answering Kitchener's call. By the end of 1914 the number of Welsh volunteers amounted to 58,000, a healthy contribution from a part of Britain not normally associated with any great public enthusiasm for the military. Over the years Nonconformist preachers had taught their congregations to condemn militarism, whilst in South Wales in particular there was active prejudice against the Army following its intervention in the Cambrian miners' strike and the Tonypandy riots of 1910. Recruiting in Wales could have presented a problem for the government but young Welshmen rallied to the cause with such enthusiasm that by the end of September fourteen Service Battalions had been raised for the existing Welsh line infantry regiments – the Royal Welch Fusiliers, the South Wales Borderers and the Welsh (later Welch) Regiment.

Much of the credit for this turn-around must go to

David Lloyd George, the ambitious and influential Welsh Liberal politician who was serving as Chancellor of the Exchequer in Asquith's Cabinet. As well as being a fiery radical (especially in his younger years) he was a renowned orator and in front of an audience of London Welshmen in the Queen's Hall, London, on 19 September 1914 he delivered what was generally held to have been one of the most inspired speeches of his career, one that managed to be pro-war and pro-Wales at the same time. His message was clear: the war was being fought to help 'little Belgium' and it was the task for 'little Wales' to do its duty not just by providing volunteers but by raising a sufficient number to establish a separate Welsh army corps:

> I should like to see a Welsh army in the field. I should like to see the race who faced the Normans for hundreds of years in their struggle for freedom, the race that helped to win the battle of Crecy, the race that fought for a generation against the greatest captain in Europe – I should like to see that race give a good taste of its quality in this struggle. And they are going to do it.[2]

Although Kitchener was initially dubious – he once told Asquith that 'no purely Welsh regiment is to be trusted' – he backed down and yielded to Lloyd George's demands. While a Welsh army corps was never formed, sanction was given to the creation of a Welsh infantry division which was originally numbered 43rd and came into being in December 1914 as the 38th (Welsh) Division under the command of Major General Ivor Philipps, a Liberal politician who had served earlier in the Gurkha Rifles. It was against that background that the Welsh Guards was formed.

It was not the first time that the idea of a Welsh Regiment of Foot Guards had been aired. In 1900 Queen Victoria had given permission for the formation of an Irish Regiment of Foot Guards 'to commemorate the bravery shown by the Irish regiments in the recent operations in South Africa'. Once formed, this new regiment joined the existing Foot Guards regiments in the Brigade of Guards (Grenadier, Coldstream and Scots) and its creation prompted the first proposal that Wales should have a similar regiment. The suggestion excited considerable public interest in Wales and was supported by a number of leading Welshmen and public organizations such as the Honourable Society of Cymmrodrion which had been founded in 1751 to foster Welsh culture and which was behind the creation of the National Eisteddfod Association. However, the idea never won sufficient support and was dropped when the War Office doubted that Wales would be able to find sufficient recruits for a new regiment.

Not surprisingly, the idea resurfaced in August 1914 with the outbreak of war and the call for volunteers. Once again, prominent Welshmen were involved in the campaign, notably Sir Henry Erasmus Philipps of Picton Castle, Haverfordwest, a local Militia officer who initiated a spirited correspondence in the *Western Mail* and the *South Wales Daily News*. He was supported by Walter FitzUryan Rice, 7th Lord

Colonel William Murray-Threipland, DSO, JP, DL

Born William Scott-Kerr, he adopted the surname of Murray-Threipland on 30 April 1882, following his inheritance of the estates of his first cousin, Sir Patrick Murray-Threipland, 5th Baronet. These included Fingask Castle in Perthshire, and Dale House in Caithness. He was commissioned in the Grenadier Guards in 1887 and saw action in Sudan and in South Africa during the Second Boer War.

As well as being the first Commanding Officer and Regimental Lieutenant Colonel of the Welsh Guards, he was Colonel between March 1937 and his death in June 1942. This was a unique appointment, which was normally reserved for the monarchy, members of the aristocracy or field marshals, and demonstrates the esteem with which he was held within the Regiment. His wife was Charlotte Eleanor Wyndham, co-heiress of the New House estate at Llanishen in Glamorgan.

Dynevor, a Carmarthenshire Militia officer and Conservative politician who went on to serve in the Ministry of Munitions. The government's response to these appeals was that 'the matter was under the consideration of the Army Council', but behind the scenes steps were already being taken to form the requested regiment. Crucially, the idea seems to have claimed the attention of King George V, who took a great deal of interest in Army matters and who granted weekly audiences to Kitchener in order to discuss the progress of the war. It remains unclear when the idea of forming a Welsh Regiment of Foot Guards was first proposed but the available evidence suggests that the subject was probably raised at one of the four weekly meetings in January 1915, perhaps on the 23rd, when the King's diary records 'a long talk' with Kitchener. A fortnight later, on 4 February, after a visit to Salisbury Plain to inspect a Canadian division, the King's diary mentions another 'talk with Lord K' during which the subject was almost certainly discussed. Corroboration comes from a letter written by Asquith to his close friend Venetia Stanley on 7 February, revealing that he had received a letter from Kitchener 'proposing to form a battalion of Welsh Guards'.[3] What is known is that on 6 February Kitchener summoned to his office Major General Sir Francis Lloyd, commanding London District, and gave him the blunt order that he had 'to raise a regiment of Welsh Guards'.

Lloyd was an experienced soldier of Welsh ancestry who had commanded 2nd Battalion, Grenadier Guards, during the Boer War and was steeped in the traditions of the Foot Guards regiments, but his initial response to the order was that there might be 'difficulties'. Told by Kitchener that 'if you do not like to do it someone else will', Lloyd accepted the order and promised that the new regiment would be raised immediately and would mount King's Guard at Buckingham Palace on St David's Day that year.[4] The Royal Warrant authorizing the raising of the 'Welsh Regiment of Foot Guards to be designated Welsh Guards' was signed by King George V on 26 February 1915 and the new regiment began assembling the following day at the White City, an area in west London which had housed the Franco-British exhibition of 1908 and had been taken over by the government as a training depot for troops bound for active service on the Western Front. It was an improbably tight timetable to raise a new regiment within three weeks but Lloyd had already taken the first important steps to ensure that his promise to Kitchener would be kept. His immediate

Founding Welsh Guardsmen at White City, 1915, wearing original cap-badges. Note front row figures: Lt Col W. Murray-Threipland (fourth from left); Maj G. C. D. Gordon (fourth from right); and WO1 (RSM) Stevenson (third from left). *(Welsh Guards Archives)*

task was to find a suitable officer to command both the Battalion and the Regimental Headquarters and on 11 February his choice had fallen on William Murray-Threipland, an experienced Grenadier who had served under Kitchener in the Sudan campaign of 1898 and the Boer War. Although of Scots ancestry – his family name was Scott-Kerr, of Chatto, Roxburghshire, in the Scottish Borders – Murray-Threipland, whose wife was Welsh, proved to be the ideal man for the job. According to one of his first appointments, Lieutenant J. A. D. Perrins (Seaforth Highlanders), soon to become Adjutant, 'from the time he took over command, the Welsh Guards became the ruling passion of his life'.[5]

At this point, though, alarm bells started ringing at Buckingham Palace when King George V was asked to approve Murray-Threipland's appointment – not because of any official displeasure but because the King was astonished that Kitchener and Lloyd had moved so quickly. The King discussed the problem with his private secretary, Lord Stamfordham, who wrote to Kitchener on 11 February expressing the King's grave doubts about the wisdom of rushing the new regiment into being.

> In continuation of our conversation by telephone, the King did not realise that the somewhat academic discussion as to the possibility of forming a Welsh Regiment of Guards had so rapidly developed into a fait accompli.
>
> On further consideration His Majesty had appreciated the difficulties which he foresees are likely to arise in obtaining the necessary Officers, whether Welshmen or otherwise, in time of peace.[6]

Stamfordham pointed to an earlier pre-war problem involving the Irish Guards, which had experienced 'the greatest difficulty in getting Irish officers',[7] and in another letter hoped that Kitchener had considered 'the difficulties which have forcibly suggested themselves to His Majesty'.[8] In fact, as it turned out, the King's concerns never materialized and getting the first Guardsmen proved to be relatively easy. From the outset it was agreed that Welsh soldiers in the other Foot Guards regiments would be permitted to transfer – from the Grenadiers alone, which had always recruited from Wales, there came 300 trained NCOs and other ranks, with a further 200 Welsh recruits arriving from the Guards Depot at Caterham. Amongst them was George Henry Duggan from Bargoed, who had joined the Grenadiers in the previous year but transferred to the Welsh Guards on 27 February 1915 with the regimental number 429. He survived the fighting on the Western Front and was placed on the Reserve in 1919.[9] Both his sons, Hugh and John, later joined the Regiment, thus establishing the notion that this was a Welsh family regiment. Strenuous efforts were made elsewhere and appeals were made to the young men of Wales to join this new Welsh regiment. The matter was formalized on 22 February when Kitchener issued a circular to the Chief

Brigadier the Hon. A. G. A. Hore-Ruthven, VC, later 1st Earl of Gowrie

Born at Windsor in 1872, Hore-Ruthven was educated at Eton and entered the tea trade before being commissioned in the Highland Light Infantry in 1892. Following the Sudan campaign he was decorated with the Victoria Cross for his courage in saving a wounded officer under heavy fire at Gedaref. He served subsequently in the Queen's Own Cameron Highlanders and the King's Dragoon Guards and between 1908 and 1910 was military secretary to Lord Dudley, the Governor-General of Australia. He joined 1st Battalion Welsh Guards, as second-in-command shortly after its formation and commanded the 1st Battalion between May 1919 and December 1920. As Lord Gowrie he was Colonel between June 1942 and July 1953.

After leaving the Army in 1928, he served in Australia, first as Governor of South Australia, then as Governor of New South Wales, before being appointed Governor-General of Australia in 1936, an office he held until 1945. He died on 2 May 1955.

Battalion staff officers, 1915. *(L to R)* Capt R. G. W. Williams-Bulkeley, Lt (QM) W. B. Dabell, Lt Col W. Murray-Threipland, DSO, Major and Adjutant G. C. D. Gordon, Lieutenant Viscount Clive, Lieutenant P. L. M. Battye. *(Welsh Guards Archives)*

Welsh Guardsmen at Chelsea Barracks in 1915, shortly after receiving their new uniforms. *(Christina Broome Collection/Imperial War Museum)*

of the Imperial General Staff (CIGS), informing him that volunteers were required for the Welsh Guards 'from all Regular, Reserve or Territorial Battalions of Welsh Regiments in the United Kingdom', that they must be of Welsh parentage on one side at least, or be domiciled in Wales or Monmouthshire and must possess Welsh surnames. All volunteers had to be at least five feet seven inches in height.[10] Even after the first volunteers arrived the pressure was maintained, especially in Wales. On Saturday, 17 April, a 'military international' was played at Cardiff Arms Park between a Wales XV and the Barbarians which resulted in a 26–10 win for the latter and the enlistment of 183 Guardsmen, with many more joining up in the week that

Numbers as names: Welshmen with the same names

In Welsh regiments it has long been the practice for soldiers with a 'common' name, such as Davies, Evans, Jones, Hughes, Roberts, Thomas or Williams, to be referred to by the last two or three digits of their regimental numbers. This immediately identifies the individual. This system has evolved over many hundreds of years and is peculiar to Welsh regiments. No English, Scottish or Irish regiment has anything similar.

It is a practice that works extremely well at platoon and company level, and is used as a very good bonding and team tool. It has long been the envy of other regiments, when referring to the close relationship of officers and other ranks in Guards Regiments, well known and envied for their discipline whether in the field or on parade. Certainly this system has been practised in the Welsh Guards since its formation. It is common practice, not casualness, for officers to refer to junior ranks by their 'last two'; indeed, a Guardsman suddenly referred to as Williams/Davies/Jones would wonder what he had done wrong. And Jones 64 is always Jones Sixty-four, not Jones Six Four.

One story says it all. In Caterham, three new recruits joining Support Company from basic training were marched before the Company Sergeant Major. He pointed to the first and said, 'What's your name?' to which the Guardsman replied, 'Jones, Sir.' Leaning sharply across his desk the CSM said, 'Don't they teach you lot anything at the Depot these days? What's your last two?' 'Twenty-four, Sir', came the reply.

He quickly moved on to the second and said and 'Who are you?' 'Davies, Sir.' Turning to the sergeant he said, 'Jesus Christ, another one. What's your last two?' 'Twenty-two, Sir,' to which the CSM said, 'Well, that's two North Walians.' Looking at the third, he said, 'And who are you?' 'Radmilovic Eighty-eight, Sir.' There has only ever been one Radmilovic in the Regiment since 1915 but he was known as 'Eighty-eight' throughout his Army service.

Regimental Sergeant Major William Stevenson, MBE, DCM, MM

The first Regimental Sergeant Major of 1st Battalion Welsh Guards, and possessor of the regimental number '1', William Stevenson originally joined the Scots Guards in 1901 and served with them on the Western Front in the opening months of the war. Known throughout the Regiment as 'Stevo', he had been wounded at Gheluvelt during the Battle of Ypres in 1914 and was not sufficiently recovered to cross over to France with the Battalion on 17 August 1915. Having recuperated, he re-joined 1st Battalion Welsh Guards, at Boyelles in July 1918 and resumed his duties as Regimental Sergeant Major. Stevenson remained in that position until June 1928 when he was commissioned. He retired from the Army in the rank of major in August 1937 and died on 14 January 1961 aged seventy-eight.

Painted by Kenneth Green. Observant readers will note that RSM Stevenson is incorrectly dressed: only nine buttons are on display, rather than the ten that are required of a Welsh Guardsman! Stevenson was RSM 1WG 1915 and 1918–28, QM 1WG 1928–37. *(Estate of Kenneth Green)*

followed. It was the first match of its kind and helped to cement the link between the Welsh Guards and the game of rugby union.[11]

Murray-Threipland was also successful in persuading a number of younger Welsh officers in other regiments to transfer – J. W. L. Crawshay (Welsh Regiment), W. A. F. L. Fox-Pitt (Cheshire Regiment), for example – but, as he admitted to his wife, 'seniors are the problem'. Appeals to older potential officers in other Foot Guards regiments often fell on stony ground as some felt that they would jeopardize their careers while others were disinclined to accept Murray-Threipland's invitation because they did not feel comfortable with the thought of joining a Welsh regiment. That was the case with George Montgomerie, a personal friend of Murray-Threipland in the Grenadier Guards, who responded bluntly that he would not transfer for the simple reason that he was 'not a Welshman ... and take no interest in Welshmen so I should probably be a failure'.[12] Even so, some excellent candidates did appear and Murray-Threipland was fortunate in key appointments such as Captain G. C. D. Gordon (Scots Guards) as Adjutant and Major A. G. A. Hore-Ruthven, VC (King's Dragoon Guards), as Second-in-Command. The other crucially important appointment was the Regimental Sergeant Major and here the fledgling regiment was helped enormously by the selection of William Stevenson (Scots Guards) who was given the regimental number '1'. He retained it until 1920 when eight-digit Army numbers were instituted to replace regimental numbers.[13] Known throughout the Regiment as 'Stevo' he proved to be a

Regimental Sergeant Major Walter Bland, MC

A native of Church Stretton in Shropshire, Walter Bland originally joined the Grenadier Guards but on the Regiment's foundation transferred to the Welsh Guards in the rank of Drill Sergeant with the regimental number '2'. In the absence of William Stevenson he became Regimental Sergeant Major in October 1915 and held the appointment until May 1918.

great regimental stalwart and a fair-minded disciplinarian who left his mark on all who met him. A young officer, J. C. Windsor-Lewis, later Brigadier and Regimental Lieutenant Colonel between 1951 and 1954, remembered him as 'fierce and he looked terrifying, but he was just a man and intensely loyal and nothing was too much for him when the welfare of the Regiment was at stake'.[14] Earlier in the war Stevenson had been badly wounded during the fighting at Ypres and was therefore unable to accompany the Battalion when it first went to France. It was not until May 1918 that he was passed fit for duty and for much of the conflict the Regimental Sergeant Major was Walter Bland, a former Grenadier, who had the regimental number '2'.

The speed with which these appointments were made enabled Lloyd to keep his promise to Kitchener: on St David's Day 1915 the Welsh Guards mounted King's Guard at Buckingham Palace with Murray-Threipland acting as captain of the guard, the first time in the history of the Brigade of Guards that a commanding officer had gone on duty in that capacity. This was not so much a case of the Colonel stealing the limelight; rather, as Murray-Threipland explained to his wife, the preferred officer lacked the necessary experience in ceremonial duties and 'That won't do!' The guard mounting proved to be a great success and that night it was celebrated by a dinner party which included Lloyd, Murray-Threipland's brother General Scott-Kerr, Kitchener and the field marshal's future biographer Sir George Arthur. Although Murray-Threipland told his wife that 'these big binges don't suit me', he also said with some pride, 'Weren't the Guard smart, everyone is full of it.'[15]

Welsh Guards uniforms and insignia

The uniform was settled at the time that the Regiment was formed. According to Dudley-Ward's history, 'Both officers and men wore the leek as cap badge, which national emblem was repeated on the button designed by Mr Seymour Lucas, RA. The peacetime forage-cap of officers and men was to have a black band; the tunic was to have buttons in groups of five; the collar badge to be the leek, repeated on the men's shoulders. The bearskin cap would be the same as in the other regiments of Guards, but would have a distinctive plume of green and white.'

The choice of the leek reflected its traditional position as the main national emblem for Wales. Despite the more recent claims of the daffodil to fulfil the same function – supported by, amongst others, David Lloyd George – it was felt that the leek had a stronger association, especially with the military. During the Hundred Years War, Welsh archers wore green and white colours, whilst in William Shakespeare's play, *Henry V*, when Fluellen observes that Henry is wearing a leek, the Prince replies: 'I wear it for a memorable honour, for I am Welsh, you know, good countryman.'

The other great emblem of Wales is the Red Dragon which became a Royal Badge in 1807 and was officially recognized by HM The Queen in 1959 . It is also widely used as the Welsh national flag with a green and white background. The motto 'Y Ddraig Goch Ddyry Cychwyn' ('The Red Dragon Gives a Lead') is taken from the poetry of Deio ab Ieuan Du (1450–80) and is also used by The Prince of Wales's Company.

This cartoon appeared in the *Western Mail* on 4 March 1915, celebrating the formation of the Welsh Guards. HM King George V approved the Leek as the badge, and *Cymru am Byth* as the motto for the newly formed regiment. The choice of the leek as the badge had caused fierce debate. The daffodil and the dragon had been rejected as options. The cartoon depicts a leek walking away from a slaughtered bull (which represents the dragon and daffodil). The caption underneath reads as follows: Un Coup de Grace. Welsh leek: 'Of course, it could only end in one way!' The original of this cartoon was presented to RHQ in memory of Colonel D. G. Davies-Scourfield, MC, by his family. *(Western Mail)*

The Regimental Band

The Band of the Welsh Guards was formed in the same year as the Regiment and initially consisted of forty-four musicians and a Warrant officer, Andrew Harris, who became the first bandmaster. Harris was later commissioned and became the Senior Director of Music of the Brigade of Guards until he retired in 1937. The Band's first instruments were presented by the City of Cardiff and it accompanied the Regiment for the first time when it mounted King's Guard on St David's Day, 1916. The same evening the Band gave its first concert at the London Opera House. The occasion was such a success that the Band's reputation for musical excellence was immediately established.

When the 2nd Battalion Welsh Guards was formed in 1939, the number of bandsmen was increased to sixty, in line with the total of the other Foot Guards regiments. During both world wars the Band gave regular performances and concerts to members of the Regiment, other services and civilian audiences. A highlight was its performance at the 'Farewell to Armour' Parade of the Guards Armoured Division in 1945.

The bandmaster between 1948 and 1962 was the talented composer and arranger, Leslie Statham. Writing under the name of Arnold Steck, he is best known for his composition 'Drum Majorette' which was the original theme tune for BBC Television's long-running flagship football programme, *Match of the Day*. This was used from 1964 until 1971. Lieutenant Colonel Peter Hannam, BEM, MBE, became the first Welsh Guards Director of Music to be promoted to this rank. At the time of his retirement in 1993, he was the last remaining National Serviceman still serving in Army music.

After the Options for Change Defence Review in 1990, the number of British Army bands was reduced from sixty-nine to twenty-nine and a new Corps of Army Music (CAMUS) was established. Based at Kneller Hall in Twickenham, which was already home to the Royal Military School of Music, musicians from all of the existing regimental bands were transferred to the new formation. In 2006, the total number of bands was further reduced to twenty-three. Today, the Band of the Welsh Guards is comprised of musicians from the Corps of Army Music, but continues to be based at Wellington Barracks. Throughout its history, the Band of the Welsh Guards has been much in demand both at home and overseas. Its primary role is to support state and ceremonial events such as the Sovereign's Birthday Parade and the Changing the Guard ceremony at Buckingham Palace. As well as providing a marching band and concert band, the Band has a number of smaller ensembles including a dance band, fanfare trumpeters, a salon orchestra and a brass quintet.

Early photograph of members of the Band of the Welsh Guards with Director of Music, Major A. Harris, LRAM, exact date unknown. *(Welsh Guards Archives)*

Recruits at White City, 1915. The uncomfortable-looking beds were known as 'biscuits'. *(Welsh Guards Archives)*

The next step was approving matters of uniform and regimental insignia and these too received speedy attention. On 2 March, King George V approved the leek as the Regiment's cap badge from a design by John Seymour Lucas, RA, the distinguished historical artist and costume designer. Other uniform details were also settled at this time: the forage cap for officers and men was to have a black band; the tunic was to have buttons in groups of five; the collar badge was to be the leek; and the bearskin cap would have the distinctive white-green-white plume. It was also agreed that the lead infantry company would be known as Prince of Wales Company. It contained the taller Guardsmen in the Regiment and it soon acquired a nickname – the 'Jam Boys', on account of a belief that its officers paid for an extra ration of jam as additional sustenance for their taller men. (Another derivation is that as there was always a shortage of jam it always went to the taller men in Prince of Wales Company.) In later years, on 23 June 1980, HRH The Prince of Wales, announced that he would like the Company to be called The Prince of Wales's Company and this was duly done.[16] Also later in the Regiment's life, during the Second World War, No. 3 Company,

The first drummer boy to enlist, Drummer Charles Baden-Sware. To his right is WO2 (CSM) D. Cossey 16 from Skewen, and to his left is CSgt (CQMS) Jenkins 16 from Cardiff. *(Welsh Guards Archives)*

Colonel the Lord Harlech

George Ralph Charles Ormsby-Gore was commissioned in the Coldstream Guards in 1875 and later served in the Shropshire Yeomanry, becoming its Honorary Colonel in 1908. Elected Conservative MP for Oswestry in 1901, he succeeded his father as Lord Harlech three years later and entered the House of Lords. He was Regimental Lieutenant Colonel of the Welsh Guards from June 1915 to October 1917. He died in May 1938, aged sixty-three.

The family seat was Brogynton Hall near Oswestry.

containing shorter men, acquired its own nickname – the 'Little Iron Men'.

All the while, the training of the nascent battalion continued apace in the somewhat bizarre surroundings of the disused funfair at the White City where the Regiment's first historian, Major Dudley-Ward, remembered recruits being instructed 'on the very ground where the populace used to "wiggle-woggle" and "water-chute" to the strains of brass bands and under the glow of a hundred thousand coloured lights'.[17] Other steps included an inspection by the Lord Mayor of Cardiff, whose city corporation provided funds for the purchase of instruments for the Regimental Band which came into being in October under the direction of Bandmaster Andrew Harris, LRAM (South Lancashire Regiment). Slowly but surely the new regiment and its 1st Battalion were taking shape and on 18 March Murray-Threipland was able to report that its total strength was 763 all ranks. This was followed on 21 April by a letter from Lloyd to Stamfordham asking that the King be informed that the Welsh Guards were over-establishment and explaining that he had authorized the creation of a reserve battalion to provide drafts and reinforcements under the command of Lieutenant Colonel J. B. Stacey Clitheroe (Scots Guards). This was followed by the appointment of Lord Harlech as Regimental Lieutenant Colonel in succession to Murray-Threipland, who retained command of the 1st Battalion, by then training at Sandown Park near Esher.

Only one formality remained: the presentation of the Regiment's Colours and this took place on 3 August when the new regiment was engaged on public duties in London. Before the ceremony, Stamfordham wrote to Murray-Threipland stating that the King 'is very glad that a Hymn will be sung; but also asks if the Men could sing "The Land of My Fathers" in Welsh'.[18] As a result, in the gardens of Buckingham Palace, after the Colours had been consecrated by the Bishop of St Asaph, the Welsh Guards Choir sang '*Ton-y-botel*', 'Aberystwyth' and '*Hen Wlad Fy Nhadau*'. Then the two senior subalterns, H. E. Allen and H. E. Wethered, knelt before the King to receive the King's Colour and the Regimental Colour. Following the presentation the ceremony closed with the Battalion marching past amidst a huge downpour. Four days later, on 7 August, the following announcement appeared in the *London Gazette*: 'His Majesty the King has been graciously pleased to confer on the Welsh Guards the honour of becoming Colonel-in-Chief of the Regiment.' A week later, 1st Battalion, Welsh Guards, was off to war, travelling from Waterloo Station to Southampton in three special trains with 29 officers and 1,108 other ranks. Following the Channel crossing, it landed at Le Havre in the early morning of 18 August. The final destination was GHQ St-Omer where the Guards Division was being formed in July 1915 under the command of Major General the Earl of Cavan; in this new formation, 1st Battalion Welsh Guards was placed in 3rd Guards Brigade under the command of Brigadier General H. J. Heyworth whose first advice to his men was brutally simple: 'The only way for us to win this war is to go on killing Germans like vermin.'

Intensive training followed the Battalion's arrival: battlefield tactics, use of machine guns (of which there were initially only four), selection of snipers, grenade training, wiring, trench-digging, loading transport, march discipline and shooting practice. The strategic situation in mid-1915 was dominated by the stalemate on the Western Front where the soldiers had to come to terms with the reality of trench warfare. The conditions on the British side were basic and occasionally unsanitary but they offered safety to their inhabitants with a complicated system of underground shelters, support and communication trenches protected by breastworks and barbed wire. Between them and the German line lay no-man's land, a space of open ground which could be as wide as 300 yards or as narrow as twenty-five. From the air it looked orderly and secure but the creation of the trench system also dominated the tactics used by both sides and would scarcely change until the return to more open warfare in the last months of the war. In 1915, the dilemma facing British and French planners was how to break the German trench line by attacking key points which would force the enemy to fall back on their lines of communication and in so doing return some fluidity to the fighting. Lines of advance had to be chosen and in January the Allies had agreed to mount offensives against both sides of the German salient which ran from Flanders to Verdun. In this spring offensive the British and the French would attack in Flanders and Artois, the French alone in the Champagne. For the British this involved them in battles at Neuve Chapelle, Aubers Ridge, Festubert and later in the year at Loos.

Loos has been called many things by the soldiers who fought in it and by the historians who have picked over its bones, but most are agreed that the best description is that it was both an unnecessary and an unwanted battle. In strategic terms it was meaningless as the end result did little to help the French offensive in Artois and Champagne, the main reason why Kitchener insisted that the battle should take place. From the outset, the planning for the battle was a story of improvisation and optimism that the weight of the French attack would lead to an early breakthrough, thereby taking some of the strain off the attacking divisions. Under the direction of General Foch, the British First Army (General Sir Douglas Haig) would attack from its lines in the north along a wide front which stretched from a position known as the Hohenzollern Redoubt to the town of Loos, while the French Tenth Army would attack the German defensive positions on Vimy Ridge. Further south, in Champagne, the French Second and Fourth Armies would strike a convergent blow aimed at destroying German lines of communication at Sedan and Douai. Marshal Joseph Joffre, the French Commander-in-Chief, hoped that the massive assault would lead to a collapse in the German centre and drive the enemy back to the Ardennes, but he was already at loggerheads with his allies.

'Position and hold must be firm enough to support the weapon.' Rifle training at White City, 1915. *(Welsh Guards Archives)*

Welsh Guards uniforms 1915–18. *(From illustrations by Charles C. Stadden and Eric J. Collings, commissioned by the Regiment)*

1915 Captain – Guard Order, London. The new uniform for Welsh Guards officers was based on the same barathea service dress worn by the other Foot Guards regiments. The jacket has unpleated breast pockets (unlike those of line regiments). It is fastened with five equally spaced buttons, not two groups of five as on the full dress tunic. This detail is unique to the Regiment. By 1915, the blue top of the forage cap had been replaced by a khaki one. There are no collar badges. The badges of rank were originally Bath stars not Garter stars, which were adopted later. Badges of rank were worn on the epaulettes, not on the cuffs. As a dismounted officer, the captain wears plus fours. The steel guard of his sword bears the regimental badge instead of the royal cypher used by non-Guards regiments.

1918 Lieutenant – Fighting Order, Flanders. The British Army adopted the 'Brodie' steel helmet in 1915. Here it has a cloth cover to reduce shine and break up its outline. This officer's rank badges are now Garter stars, although they were not officially authorized until the 1920s. Instead of a sword, attached to his Sam Browne belt is a holster for a .38 Webley Mk V pistol and field glasses case. A rolled Burberry raincoat is carried on a shoulder sling and he carries an ashplant stick. The new type of box respirator (gas mask) is shown here in the 'alert' position. From 1917, it was usual for officers to wear other ranks' pattern dress and equipment when in the line. Badges of rank distinguished them as officers.

1916 Private – Transport Driver, France. The Welsh Guards Transport Section was part of battalion headquarters and commanded by a sergeant. This driver wears the soft winter Service Dress cap, known as the 'Gor blimey', which was issued from February 1915. Bedford cord pantaloons were worn instead of service dress trousers and jack spurs were attached to the ankle boots. Other equipment was carried on the GS wagons or limbers. Rifles were mounted in clips attached to the vehicle footboards. Men leading pack animals wore full Marching Order with slung rifles.

1915 Lance Corporal – Marching Order, France. The 1902 Pattern khaki drab service dress was the standard uniform throughout the British Army. On the stiffened round cap, the brass leek cap badge is worn overlapping the brown chinstrap. The tunic is fastened with five brass buttons and has brass shoulder titles above the initials 'WG'. When the 1908 Pattern web equipment was worn with the valise or large pack, it was called Marching Order. The overall weight carried was about 30kg and represented a considerable burden, especially when soaking wet. The rifle is the .303 Short Magazine Lee Enfield (SMLE) Mk1 (1903 Pattern). A small satchel contains the smoke helmet, an early respirator of limited value.

1917 Private – Fighting Order, France. Known as 'Battle Order', this order of dress substituted the pack with the smaller haversack. Although heavy equipment was left behind with the transport, extra bandoliers or ammunition, grenades, rockets, flares, entrenching tools and other equipment still contributed to a heavy combat load. By 1917, the tunic has black and white worsted shoulder titles above cloth battalion numerals. 'Overseas' chevrons are worn on the right cuff, a blue one for each year of service abroad, except for 1914 which is red. Gold-braid wound stripes and trade badges are worn on the left. This private is armed with a .303 Lewis Machine Gun. Although it frequently jammed, it had a rapid rate of fire and could be used against aircraft. When the Lewis Gun entered service in September 1915, brigade machine-gun companies were formed and the existing Vickers machine guns were transferred to them. In March 1917, the four companies serving the Guards Division became known as the Machine Gun Guards.

Even before the battle began, Haig was forced to extend his line and was concerned that the First Army's objectives included the obstacle of the huge industrial complex within the coalmining triangle of Loos–Lens–Liévin. As he noted in June after visiting the surrounding countryside and riding the ridge north of Ablain, the country was 'covered with coal pits and houses ... this all renders the problem of an attack in this area very difficult'.[19] He was also worried about the lack of available artillery and ammunition to neutralize the German positions in advance of the main infantry assault and he questioned 'the suitability of New Army Divisions for this duty on their first landing'. There was also a muddle over the command and use of the reserve divisions that were to be ordered forward to exploit any breakthrough – these would remain under the strategic command of General Sir John French at GHQ but Haig reckoned that the congested nature of the ground precluded any rapid deployment.

As it turned out, though, when the attack began on 25 September, the first hours were promising enough. It was a mistake to use gas for the first time – in some sectors it drifted listlessly towards the German lines on the soft south-westerly wind – but the first assault divisions managed to reach their objectives. Interest for 1st Battalion Welsh Guards held in XI Corps reserve with the rest of the Guards Division, centred on the performance of 15th (Scottish) Division whose objectives in the first wave were the Lens Road and Loos Road redoubts, two formidable salients in the German line. Having taken them, the division was supposed to capture Loos and the nearby Hill 70, the one commanding point of any height in the area. Led by their pipers, the Scottish battalions quickly took all their targets and by mid-morning were like a 'bank holiday crowd' on top of Hill 70.

Then things started to go wrong. During the assault, 15th (Scottish) Division had become enmeshed with 47th (London) Division and a gap began to open up between them and the 1st Division to the north. This was exploited by the Germans and by the early evening 15th Division was holding an uneven line in front of Hill 70. 1st Battalion Welsh Guards was to win the Regiment's first battle honour. By the time that they and the rest of 3rd Guards Brigade had arrived at Vermelles, two miles short of Loos, on 26 September, having marched overnight from St-Omer, the central problem in the British tactics had begun to emerge. The Guards Division, along with the inexperienced 21st and 24th Divisions (both New Army), had been held in reserve far too far behind the battle to be effective. Instead of waiting behind the British lines in open country, they had been forced to move up to Loos along cluttered roads that were choked with columns of transport and wounded men. Moreover, when they reached the battlefield they found that their role had changed. The expected breakthrough was no longer possible; now 1st Battalion Welsh Guards was required to support 4th Battalion Grenadier Guards, in the attack to retake Hill 70. If that assault faltered it would be left to the Welsh Guardsmen to consolidate the line.

By the time the brigade reached Vermelles the pattern of the battle had become clear – the ruined village was awash with wounded and stragglers leaving the line, while in the near distance could be heard the

Inter-company milling, White City, 1915. *(Welsh Guards Archives)*

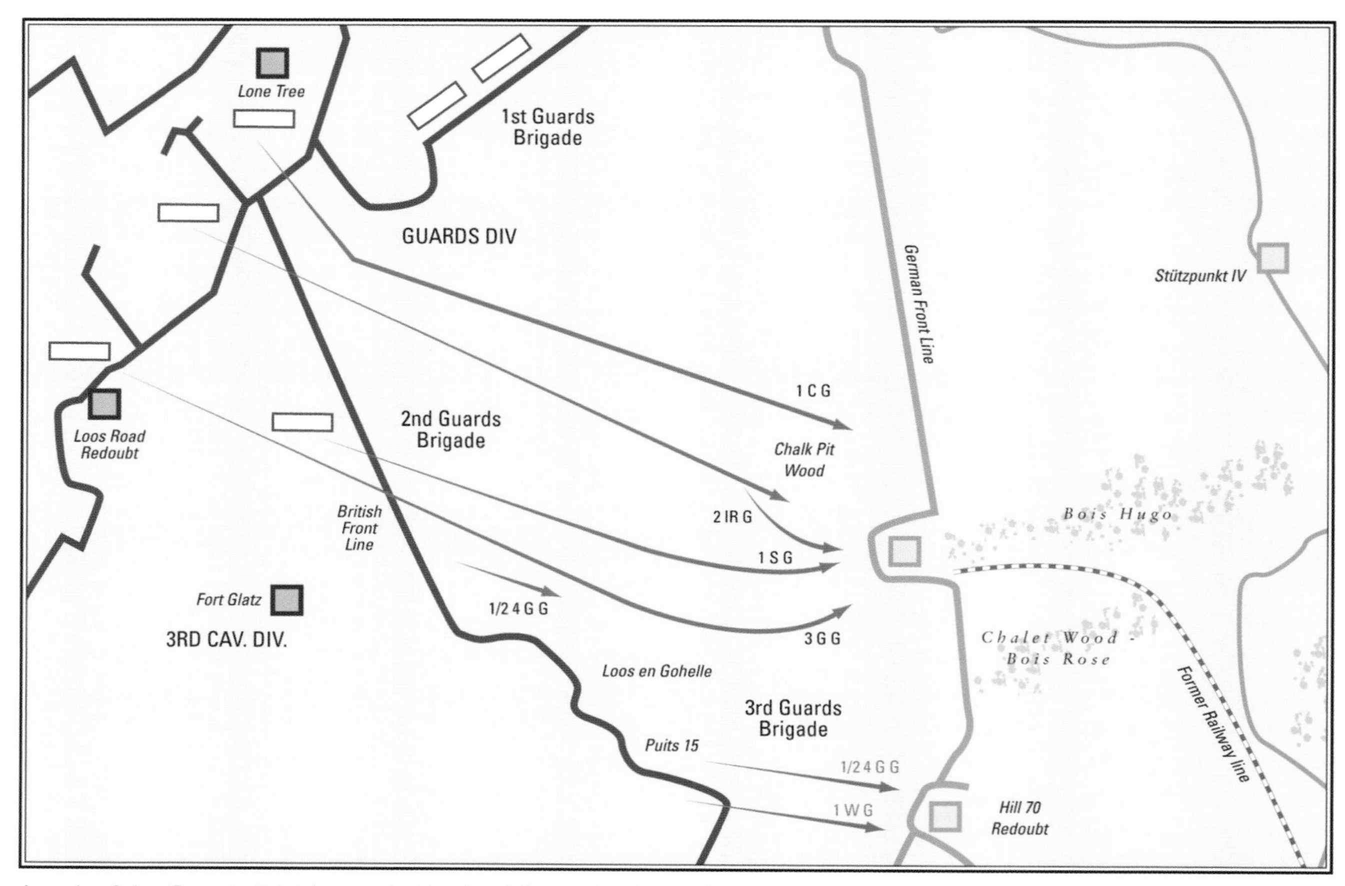

Attack of the Guards Division at the Battle of Loos, 27 September 1915; 1st Battalion attack towards Hill 70.

ominous 'short sharp thunderclap of guns'. At 14.30 Murray-Threipland received orders to march his Battalion towards the line to commence the attack on Hill 70. To do this he had to move up towards Loos across a two-mile stretch of open ground in full view of the enemy; it was then that the men had to show their mettle. Like many another soldier facing combat for the first time, Private R. Smith from Pentre (the

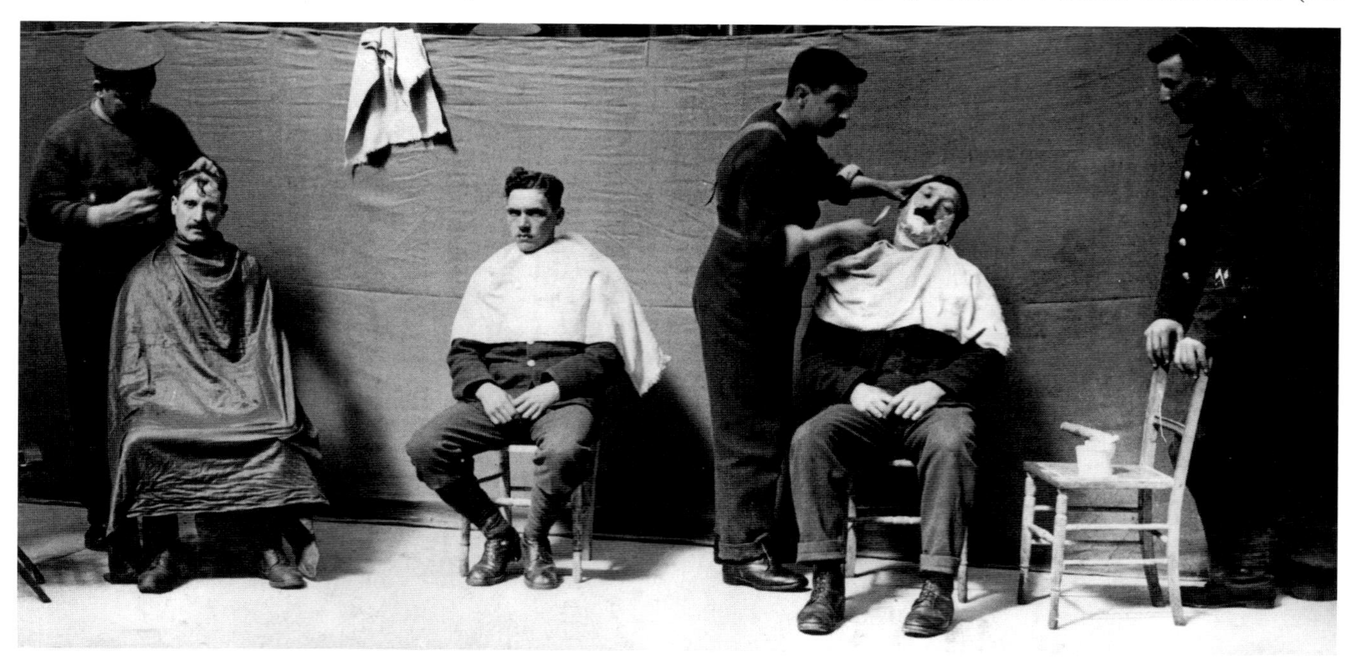

Haircuts at White City, 1915. *(Welsh Guards Archives)*

rank of Guardsman did not come into use until 1918) took comfort in the close-order drills which had been drummed into him during his basic training:

> It was then that we knew what war was, but every man was ready. A finer sight you would never want to see. On we went, shells and bullets and shrapnel falling all around us but not a man wavered: you would have thought we were on parade at Wellington Barracks.[20]

Having reached Loos, the Battalion assembled in the main street to await further instructions and at that point the men had their first experience of a German gas attack. With an 'ecstasy of fumbling' the cumbersome HP hypo helmets were hurriedly put on to meet the new emergency. 'There was the battalion standing about anyhow,' remembered Captain Humphrey Dene, 'and making noises like frogs and penny tin trumpets as they spat and blew down the tubes of their helmets.'[21] In the midst of the chaos Murray-Threipland made his way to the Grenadiers' headquarters to plan the attack: it was agreed that Prince of Wales Company would attack on the right, with No. 2 Company (Williams-Bulkeley) in support, while No. 3 Company (Phillips) would support 4th Grenadier Guards. The attack was timed for 18.02, a quarter of an hour before dark.

Many thoughts ran through the minds of the Welsh Guardsmen who were about to press home their attack in the face of the expected German counter-barrage. A curious blend of fear, apprehension and excitement dominated everything, but there was, too, an overwhelming desire to do well. Lieutenant Perrins had served previously in the Seaforth Highlanders and it had been made clear to him by other officers in the Guards Division that the Welsh Guards were very much 'a poor relation', an untried battalion without battle honours or traditions. His main fear was that the Welsh Guards might let themselves down in the heat of battle and he was concerned 'that we should not turn out to be the family skeleton'.[22]

Private David Britton from Swansea felt much the same way. He was in the machine-gun section which was the last to make the assault on Hill 70. While unloading the equipment from his machine-gun limber, his section came under heavy fire. 'I was a bit flurried,' he confessed later, 'but soon got over it. We had to advance along a communication trench where there were most elaborately fitted out dug-outs. I saw a dead German with his head and leg blown off.'[23]

St David (Dewi Sant)

The patron saint of Wales and the Bishop of Menevia (later St David's) in the sixth century, David was born in roughly AD 500 and spent most of his life in Wales. Thanks to the writing of Rhigyfarch whose hagiography *Buchedd Dewi* was written in the late eleventh century there are many stories surrounding David's life, although modern historians are sceptical about most of the claims. Many miracles are associated with his name, the most important happening during the Synod of Llanddewibrefi when the ground rose up beneath his feet so that all could see and listen to him. It is claimed that David lived until he was 100 and that he died on 1 March AD 589, with his body being buried in the cathedral at St David's in Pembrokeshire. The date quickly became associated with David as a feast day and by the late Middle Ages had taken on a patriotic complexion. Also associated with the saint was the leek, which is regarded as one of the national emblems of Wales and forms the Regiment's cap badge. In 2000 the National Assembly of Wales voted unanimously to make St David's Day a public holiday.

From the outset St David's Day has been central to the Regiment's history. It was on that day in 1915 that the Regiment first mounted Guard at Buckingham Palace – as promised by Major General Sir Francis Lloyd – and ever since it has been celebrated as a Regimental Day with a parade and the presentation of representational leeks to all serving Guardsmen.

The Monastic Rule of St David prescribed that monks had to pull the plough themselves, without draught animals, must drink only water and eat only bread with salt and herbs, and spend the evenings in prayer, reading and writing. No personal possessions were allowed; even to say 'my book' was considered an offence. David lived a simple life and practised asceticism, teaching his followers to refrain from eating meat and drinking beer. Some Welsh Guardsmen have difficulty with these injunctions, but they do at least wear his symbol, the leek, with pride.

The Patron Saint of Wales, St David. *(Aidan Hart Icons)*

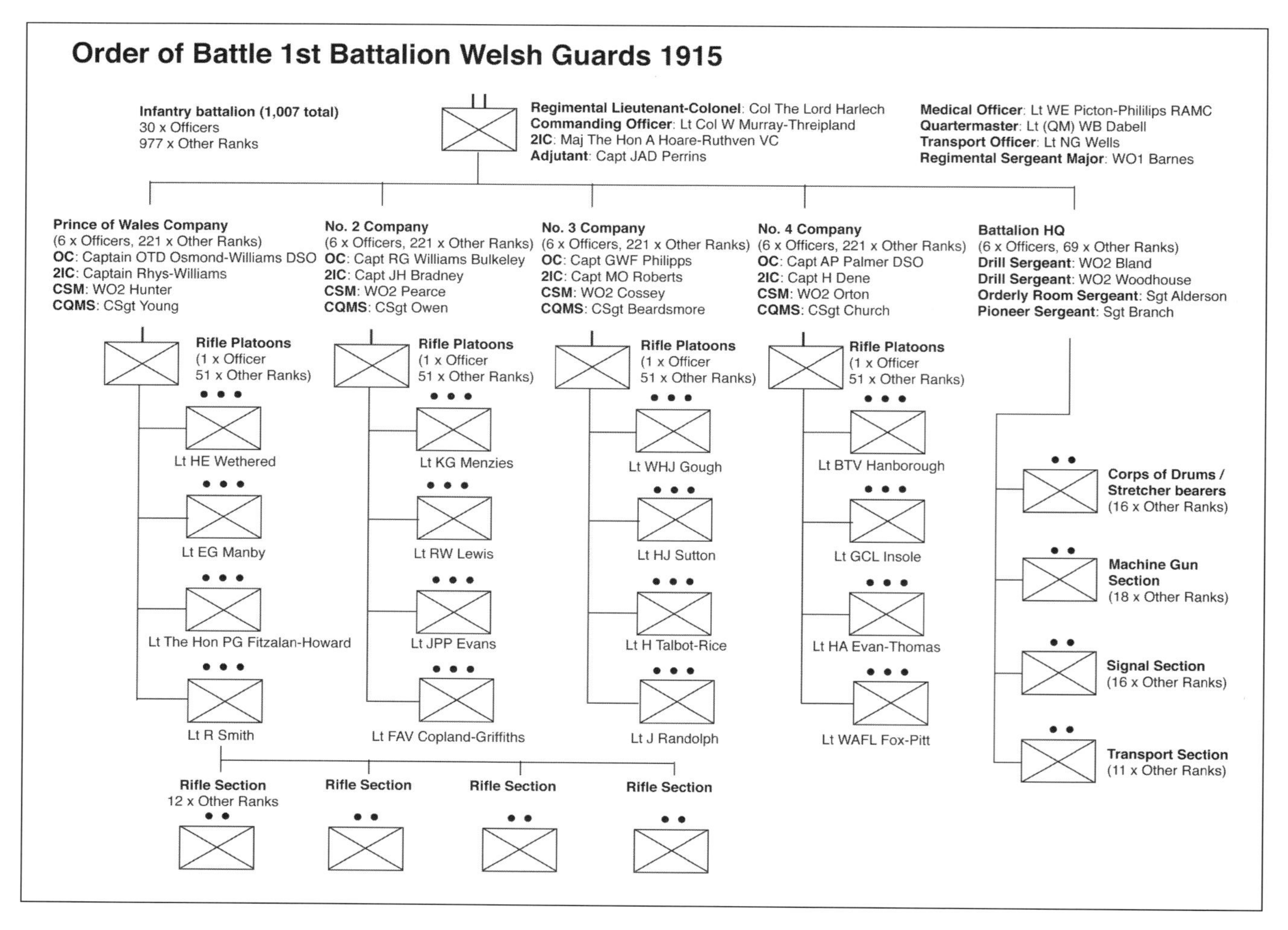

Worse sights were to follow as the Battalion reached the upper slopes of the hill. In the early evening darkness the leading sections of Prince of Wales Company arrived on the summit where they were subjected to a barrage of artillery and machine-gun fire which Lieutenant F. A. V. Copland-Griffiths thought was more like a tornado or monsoon downpour than a regular bombardment. Forced to dig in, the Battalion began to take heavy casualties. Captain O. T. D. Esmond Williams, whom many regarded as a potential future Commanding Officer, was mortally wounded and the remainder of Prince of Wales Company were scattered by the ferocity and suddenness of the German riposte.

Lying in the darkness under the crest of a hill which gave them some protection but which also denied them any line of fire, the Welsh Guards could only take cover and wait. The Battalion had been blooded in battle but it was at some cost: when the men retired from the line the following morning and regrouped at Loos the losses were five officers killed and five wounded and 162 NCOs and men killed, wounded or missing. 'I can't see any of the glory of war that people talk about,' wrote Private G. A. Cooksley to his parents in Cardiff a few days later. 'It's all perfectly hellish.' Private A. C. Morgan of Radnor experienced similar feelings once the battle was over. Having told his parents that he was proud of his Regiment's conduct, he admitted that Hill 70 had been a terrifying experience: 'I don't want to see such a sight again, all the same. The dead, dying and wounded were awful to witness.'[24]

Just as sportsmen who represent their countries in a team game such as rugby football always say that their first international match passes in a blur, so too did many Welsh Guardsmen admit that their experience of combat during the Battle of Loos was a mass of noise and confusion. 'It was the most exciting time and one I will never forget,' recalled the machine-gunner David Britton who eventually reached the top of Hill 70. 'Our casualties were heavy but we had a splendid name which will always live in a Welsh Guardsman's memory.' The Welsh Guards had pressed home their attack in darkness and under the trying conditions of a heavy artillery bombardment, yet they had reached their objective and in so doing had helped to steady the

British front line. For the Commanding Officer that was enough and on 28 September he wrote a glowing report to Lord Harlech praising the men's fortitude and courage under pressure:

> Nothing I can say would represent what I feel at the behaviour of the battalion and the leading of every officer during this trying time. There was no disorder and every platoon appeared to me to be in its place and advancing quietly and in excellent order.[25]

That thought was echoed by Private H. T. Walton of Birmingham: 'I have never witnessed a finer sight. The men swept across the shrapnel-swept ground as if they were on parade.' Others like Lieutenant Perrins were less sure that it had been such a momentous victory. Although he was pleased that the Regiment had proved its worth – 'after Loos there was no more talk of the Welsh Guards being inferior in any way' – he was contemptuous of the bungling that had led to so many deaths. 'Loos was yet another instance of the incompetence of the higher command being redeemed by the tenacity and gallantry of the common soldier.'[26]

That judgement might seem harsh but it was not far off the mark. At the end of the month the French attack in Champagne was brought to a standstill, and, coupled with the failure at Vimy, the Allied autumn offensive achieved little in return for huge losses. Bowing to the demands of the French, Haig kept the battle going in the British sector until 16 October by which time the British casualties at Loos and the subsidiary attacks amounted to 2,466 officers and 59,247 other ranks, killed, wounded or missing. And yet it could have been otherwise. Had Sir John French positioned the reserve XI Corps nearer the front, the early British breakthrough could have been exploited; faulty lines of communication and overcrowding in the reserve area prevented that from happening. There is also a question mark over the deployment of the experienced Guards Division, which should have been used in the first stages and not kept in reserve. Had it taken part in the initial assault, the outcome of the battle could have been very different.

The 1st Battalion mounts King's Guard for the first time, St David's Day, 1915. Officers at the front include, *L to R*, Major G. C. G. Gordon, Lt Col W. Murray-Threipland, Lt R. G. W. Williams-Bulkeley, Lt Viscount Clive, 2Lt P. L. M. Battye. The sergeant of the guard was WO2 (CSM) Woodhouse 297, formerly Coldstream Guards, and the drill sergeant (pictured to the rear of the Guard) was WO2 Bland, formerly Grenadier Guards. *(Welsh Guards Archives)*

On 3 October 1st Battalion Welsh Guards proceeded to Vermelles where they relieved 9th Highland Light Infantry. By then the Guards Division had replaced 28th Division to the left of the line opposite a heavily defended German position known as the Hohenzollern Redoubt, which was overshadowed by Fosse 8, a huge slag heap that has long since disappeared. Desultory but heavy fighting continued throughout the month and few Welsh Guardsmen ever forgot the hellish conditions of having to exist in trench systems which were often improvised: No. 2 Company found itself on the West Face, the sides of which, according to Dudley-Ward's account, 'were composed of dead men, equipment and a little loose earth'. Most of the Guardsmen had an instinctive dislike of handling dead bodies and had to be coaxed into clearing up the trenches, Having taken over command of No. 4 Company, Claude Insole overheard the following whispered conversation between two of his Guardsmen:

> 'You heard what the officer said, Dai – we are to throw the man out.'
>
> Inaudible mumbles.
>
> 'Come on Dai – you take the man's legs and I will take his shoulders. Now then …'
>
> 'Oh damn! Ianto, the man has no legs. What shall we do?'
>
> Inaudible conversation.
>
> 'The officer said so. Come on now, take hold of him anywhere and let us throw him out!' [27]

Relief came on 26 October when the Battalion handed over to 6th Battalion Queen's Royal Regiment, and moved out to billets in the village of Allouagne. This allowed the Battalion to regroup. Hugh Allen, who had arrived from the Royal Fusiliers, took over command of Prince of Wales Company, Dick Williams-Bulkeley remained in command of No. 2 Company, Herbert Aldridge arrived with a draft from London to command No. 3 Company, while Claude Insole took over temporary command of No. 4 Company from Captain A. Palmer, DSO, a remarkable mining engineer who had served earlier with the South African Constabulary. Leave was granted and intervals were spent in the front line below Aubers Ridge, scene of fierce fighting earlier in the year. At the end of 1915 the Battalion moved to Laventie, a reasonably quiet sector.

With the Guards Division now in XIV Corps, 1st Battalion Welsh Guards moved on to the Ypres Salient in March 1916. Dudley-Ward's diary gives a telling description of the town and its iconic buildings following their gradual destruction by enemy artillery fire:

> It is beyond description! No earthquake has ever devastated a town in such a fashion. The cathedral struggles valiantly to retain the title of 'ruin'. One side of the tower stands to a good height, with a corner turret almost perfect, and the entrance has an untouched Christ over the door. But in other places the ruined walls are half their original height, and in some are battered down into large mounds of stone. Of the inside nothing remains – just broken pillars lying about and piles of powdered stone and smashed chairs.[28]

The Battalion served in the Ypres Salient from March to July 1916 and a pattern developed whereby it spent six days in the front line, followed by a further six days in brigade reserve and six days

Major (QM) W. B. Dabell, MBE, MC

William Bates Dabell was born in Nottinghamshire in 1873 and joined the Grenadier Guards in January 1892. By the outbreak of the First World War he had reached the rank of WO1, serving as Superintending Clerk in the Regimental Orderly Room. When the Welsh Guards was formed he transferred to the 1st Battalion as Quartermaster where his 'knowledge, helpfulness and good nature at all times' were noted as being 'remarkable'. During the fighting on the Western Front he was awarded the Military Cross and his abilities at ferreting out much needed supplies were second to none. He left the Army on 22 June 1928 but served at London District throughout the Second World War. He died in July 1957 and was buried in Brighton. Brigadier J. C. Windsor-Lewis, who knew him well, described him as 'a tower of strength' and 'an enthusiastic follower of racing'.

further back in divisional reserve where there were facilities for relaxation such as film shows and concert parties. There were also opportunities for training, but life was often reduced to repairing the trench systems, a task which had to be completed during hours of darkness. There was always a danger of men becoming stale or unfit – during one route march of fourteen miles twenty-three men had to fall out having collapsed from exhaustion. To counter those effects patrols were mounted against the German lines to gather intelligence. One such raid by No. 4 Company against a German position at Mortaldje on 1 July developed into a minor battle, complete with artillery support. The position, a ruined *estaminet*, was being used by the Germans as a machine-gun post and had to be retaken. This was done but at some cost to the Battalion – ninety-six Welsh Guardsmen were killed or wounded during the raid. At the end of July the Battalion moved with the rest of the division to the Somme sector where a fierce battle had been raging since the beginning of the month.

Founder members of the regiment training at White City, 1915. *(Welsh Guards Archives)*

The 1st Battalion moving off from Vermelles for the attack on Hill 70, Battle of Loos, 27 September 1915. *(Imperial War Museum)*

CHAPTER TWO

The Western Front, 1916–1918

Somme, Ypres, Cambrai, Hindenburg Line

IN THE AFTERMATH OF the Battle of Loos there were several changes in the Army's high command. General Sir John French was replaced as Commander-in-Chief of the BEF by General Sir Douglas Haig. At the same time General Sir William Robertson became Chief of the Imperial General Staff (CIGS) – the senior officer responsible for directing operations on all fronts – and command of a new Fourth Army was given to General Sir Henry Rawlinson who shared Haig's reservations about the misuse of the reserves at Loos. Those three generals would be the guiding figures behind the major British campaign of 1916, a large offensive along the River Somme to the north of the French lines. At the same time the French Sixth Army would launch an attack south of the river to guard the right flank of the British assault.

Haig's military philosophy and his approach to the strategic situation on the Western Front were clear-cut. Looking at the trench system which separated the rival armies, he argued that far from being permanent or an insuperable obstacle, it was the key to victory. Once the Allies had built up large enough armies backed by overwhelming firepower, they could attack and

Welsh Guardsmen in a reserve trench, Guillemont, September 1916. *(Imperial War Museum)*

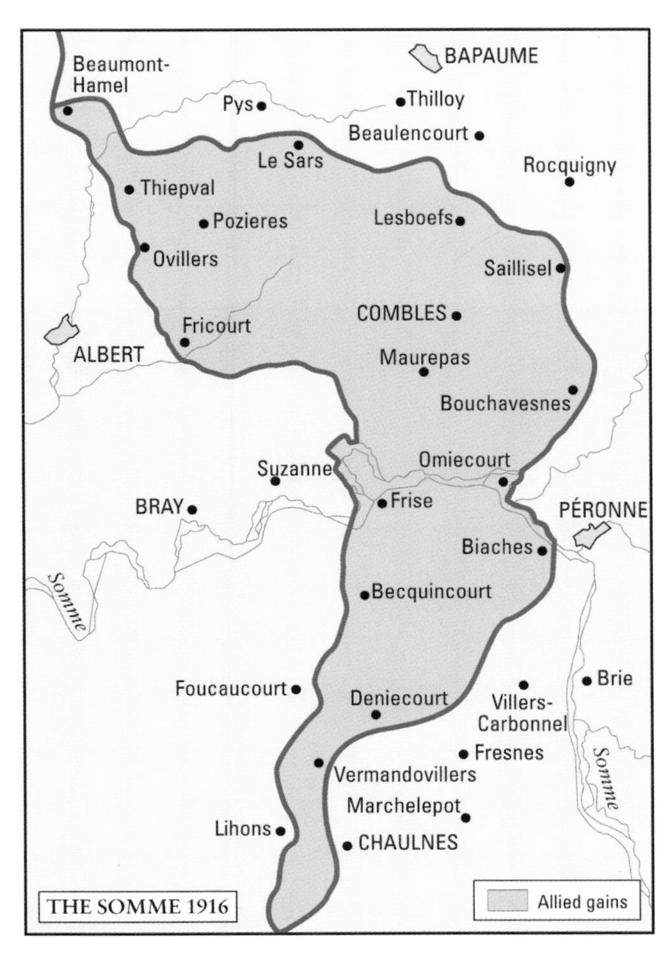

The area of the Battle of the Somme, fought between 1 July 1916 and 18 November 1916, showing the extent of British gains.

destroy the German positions with complete confidence. Successful infantry and artillery assaults would then allow cavalry to exploit the breakthrough by sweeping into open country to turn the German system of defence and ultimately defeat the enemy by removing his ability to resist. It was against that background that Haig contemplated the planning for the major offensive battle of 1916.

The tactics were deceptively simple. Haig aimed to attack the German lines using the maximum force at his disposal, to break the defences and then to move forward to take possession of the area to the rear. To do this the British would attack with Rawlinson's Fourth Army, numbering nineteen divisions, which would, in the words of the Tactical Notes produced for the battle, 'push forward at a steady pace in successive lines, each line adding fresh impetus to the preceding line'. Following an enormous week-long bombardment involving the firing of a million shells along a twenty-five-mile front, the Germans would be in no condition to resist; the British infantry would simply brush the opposition aside as they took possession of the German lines. Behind the front lines, from Albert to Amiens, the British created a huge rear area with new roads, ammunition dumps and encampments in preparation for the push which would win the war.

For all the careful planning which went into the battle, Haig's optimism had little connection with the reality of the situation. The Somme had been a quiet sector for most of the war and the Germans had used the long periods of relative inaction to good effect by creating a formidable defensive alignment in the firm chalk downlands. Some of their dug-outs were thirty feet deep and had been constructed to withstand the heaviest bombardments. The lie of the land also favoured the Germans. A first line incorporated several fortified villages such as Thiepval and Fricourt and a second defensive line had been constructed behind the ridges of the higher ground stretching from Pozières to Combles. Protected by barbed wire, these

The Corps of Drums leading the Battalion, Battle of the Somme, 1916. *(Welsh Guards Archives)*

Lt the Hon P. G. Fitzalan-Howard shaving outside his dugout. *(Harrop Collection HQ WG)*

would provide stern defences, yet Rawlinson likened the topography to the familiar surroundings of Salisbury Plain and reassured the King's private secretary, Clive Wigram, that the Somme was 'capital country in which to undertake an offensive when we get a sufficiency of artillery for the observation is excellent and we ought to be able to avoid the heavy losses which infantry have suffered on previous occasions.'[1] Alas for those fond hopes, the Somme was to be remembered not for the expected breakthrough but as the killing ground of the British Army – no other battlefield of the First World War created more casualties per square yard. The opening day of the battle, 1 July 1916, produced the bloodiest day for the infantry regiments which took part in the initial attack. From the eleven divisions which began the assault, 57,470 men became casualties – 21,392 killed or missing, 35,493 wounded and 585 taken prisoner.

Despite the huge losses, which were not immediately apparent at GHQ, Haig decided that the 'correct course' was to press ahead with attacks on an enemy who, he believed, had been 'severely shaken'. In making that estimate he was relying largely on guess-work but he was not far wrong in backing his hunch. Although the Germans had beaten off the attack on their right, they had to rush reinforcements to the front and as a result were forced to cancel all further attacks against Verdun in the French sector. The next major British attacks began on 14 July with the intention of straightening the line. The attacks were centred on the German defensive positions between Ginchy and Martinpuich at Delville Wood, High Wood and Bazentin-le-Petit Wood and they achieved substantial results, largely due to the effective use of artillery. Smaller-scale attacks continued throughout August and the next major assault was planned for the middle of September.

Lt Charles Romer-Williams transferred to the Welsh Guards from the 4th Dragoon Guards in July 1917. He served on the Western Front between December 1917 and January 1919. His campaign record states that he saw action during the Retreat from Mons (wounded in action 24 August 1914), Aisne, First Battle of Ypres, Neuve Chapelle, Second Battle of Ypres (gassed 24 May 1915), Loos, Somme and Arras. He took a pack of hounds with him to France, which he kept behind the lines. Horses and hunting were his life. He narrowly avoided death on two occasions: first, when a bullet went through his cap, and killed an NCO standing behind him; secondly, when a bullet was deflected by a large silver brandy flask which he carried in his pocket. Romer-Williams was killed in the 1920s after becoming caught up in a gun battle between rival Mafia gangs in Chicago. *(Painted by Warrington Mann)*

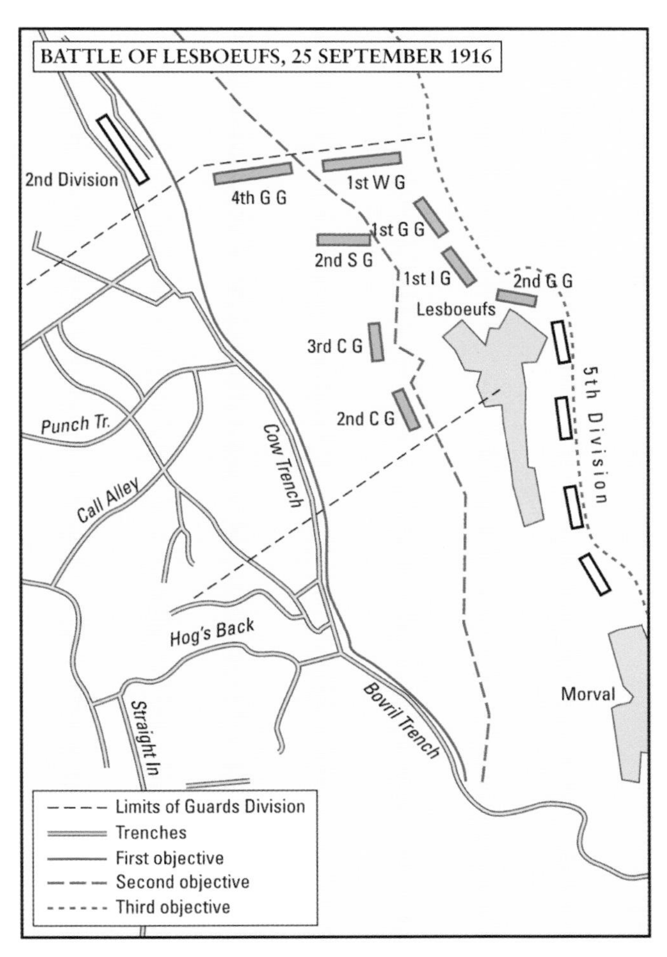

Battle of Lesbouefs, showing position on 25 September 1916.

Designated the Battle of Flers-Courcelette, the objective now was to assault the German 'Switch Line' between Martinpuich and Bouleaux before breaking into the line at Flers and occupying a line of villages parallel to the Bapuame–Péronne Road – Gueudecourt, Lesboeufs and Morval. The Guards Division together with 5th, 6th, 16th, 20th and 56th Divisions formed XIV Corps. By then 1st Battalion Welsh Guards had moved up to Mericourt and were quickly involved in the preliminary attack on the small hamlet of Ginchy standing on high ground surrounded by woods and scene of some of the most bitter fighting on this sector of the Somme battlefield. It had defied all attempts by the British to take it and on 9 September the task was handed to 16th (Irish) Division, with 56th Division supporting on the right. The attack began in late afternoon with 48th Brigade approaching from the south-west only to be halted by fierce German machine-gun fire. Attacking in support from the south, 47th Brigade fared little better in appallingly wet conditions and failing light, but in fighting 'characterized by dash, turmoil and heavy casualties', Ginchy had been cleared by the evening. It was at that point in the battle that 1st Battalion Welsh Guards were committed to relieve 48th Brigade. The order was for Prince of Wales Company and No. 2 Company to take over the line to the north of the village while No. 3 Company would move to the left towards Delville Wood. At the same time 4th Battalion Grenadier Guards would take over the line towards the east held by 47th Brigade.

As happened so often during this phase of the battle, the plan quickly unravelled. In the darkness Prince of Wales Company found itself in the wrong alignment, facing north-west instead of north-east; and enemy activity soon revealed that Ginchy had not been completely secured. At dawn the Germans counter-attacked through the mist-shrouded woods that shielded the village and soon all three companies were involved in fierce close-quarter fighting. One

Observation Post Calcutta, Western Front, just 200 yards from the enemy's trenches, 29 May 1918. *(Harrop Collection HQ WG)*

Capt R. W. Lewis outside his Company Headquarters, May 1918. *(Harrop Collection HQ WG)*

example will stand for many: during the fighting Private Williams 56 was seen despatching several Germans with his bayonet until it became jammed in a victim's body. He continued fighting using his fists until he was overcome by superior numbers. Later, the Adjutant, Captain J. A. D. Perrins, described it as 'a classic soldier's battle where every man had to fight on his own', and in the same document he rightly claimed that 'Ginchy should always be held in the greatest honour by the Regiment.'[2] It was also a battle won at great cost: the Battalion sustained 205 casualties.

After the fighting at Ginchy the Battalion moved into quarters near Fricourt to refit and receive a draft of 180 reinforcements. New orders were also received for the next stage which would see the Guards Division attack the village of Lesboeufs on 15 September along a twelve-mile front south-west of Bapaume. Twelve British divisions took part in the assault, which was made memorable by the first use of a revolutionary new weapon known as the tank that had been developed under conditions of great secrecy at Hatfield earlier in the year. Thirty-six of these lumbering leviathans were operated officially as 'machine-gun carriers' by the Heavy Section, Machine Gun Corps. Their presence on the battlefield dictated the tactics, with hundred-yard lanes being cleared ahead of them by the creeping barrage which preceded the attack. While the tanks caused consternation in the German front lines, the impact was reduced by several machines breaking down and others getting lost, with the result that the attacking divisions took heavy casualties from German machine-gun fire, particularly in the sector assigned to the Guards Division. Even so, these 'curious objects' had caused considerable astonishment and formed a substantial part of Murray-Threipland's post-battle report to Lord Harlech.

> They were said to go through or over anything: direct through wire, over trenches, nothing stopped them: completely armoured, said to weigh 30 tons and were of the male and female order.
>
> The male carried 2 six-pounders, the female several machine guns, and they were to go in advance of infantry and break-up any strong points and knock-out machine guns. As we were

Welsh Guardsmen at Guillemont, Western Front, 1916. *(Welsh Guards Archives)*

Not a type of body armour, but a tea urn being brought up to the front line, Western Front. *(Harrop Collection HQ WG)*

> in reserve I did not see their work; some are said to have been a success, others not. I saw them next day lying in different parts of the field of battle, some still with a kick in them, but others out of action, where they, at any rate, were useful landmarks.[3]

Lack of support meant that the Guards Division had to dig in well short of their objective. The attack began again the following day, 16 September, in heavy rain, with 1st Battalion Welsh Guards on the left of the line and 4th Grenadier Guards on the right. Once again the attack faltered and the day ended with heavy losses and meagre gains. Casualties in 1st Battalion Welsh Guards were 144 killed, wounded and missing. Despite the losses and the fact that so little ground had been taken, Haig was determined to continue the assault and on 25 September the Battalion was in action again, leading 3rd Guards Brigade in a renewed attempt to take the line running north from Lesboeufs and Morval (the latter being the name for this phase of the Somme battle). For once the creeping barrage was well co-ordinated and by 15.00 hours the Battalion had achieved all its objectives, apart from an enemy stronghold on the left flank whose garrison eventually surrendered following an attack by a single tank. This was the last formal action undertaken by 1st Welsh Guards during the Battle of the Somme and the Battalion went into quarters in the Carnoy valley from where the men took their turn in the divisional line. Fighting continued sporadically as winter began to take a grip on a landscape already shattered by five months of continuous combat. In his *History* Dudley-Ward left a telling description of the conditions faced by the Battalion while they were working on road-building operations between Carnoy and Montauban.

> It was no longer a road – it looked like the bed of a mountain torrent. Horses splashing through water two inches deep would fall into a hole with water up to their bellies; hundreds of men along the road would be engaged, dodging in and out amongst the traffic, trying to fill up the holes by throwing large stones into them whenever the opportunity served; and the traffic never stopped, and the loose stones thrown into the holes were as quickly ground into mud and washed away.[4]

Major C. H. Dudley-Ward, DSO, MC, and the First World War History

The writing of the Regiment's first history commenced immediately after the conflict and it was published by John Murray of Albemarle Street, one of London's oldest and most distinguished literary publishers. It carried an introduction by Lieutenant General Sir Francis Lloyd, GCVO, KCB, DSO, and its author was Major Charles Humble Dudley-Ward, DSO, MC (1879–1945), who had joined the Regiment in October 1915 from the London Regiment (Queen's Westminster Rifles) and fought with the 1st Battalion throughout the war. His private diary, on which much of the history is based, is held in the Imperial War Museum. A noted military historian in his own right, Dudley-Ward also wrote histories of the 53rd Welsh Division and the 74th Yeomanry Division, as well as editing the *Regimental Records* of the Royal Welch Fusiliers. He served as Regimental Adjutant between September 1939 and January 1944.

In such atrocious conditions men quickly became soaked and filthy and were unable to keep warm. Trench foot became a problem, with sixteen cases presenting on a single day. The mud was no respecter of rank: when Major General Sir Frances Lloyd visited the Battalion in December he jumped out of his car and promptly plunged into eighteen inches of mud and slush. A good indication of the severity of the conditions can be found in the experience of the Regimental Band which arrived before Christmas to provide some cheer for the Battalion. This it did with some vim but the men were so exhausted and the conditions were so vile that they had to be played back to their billets in slow time. To counteract the dreadful weather the Quartermaster (Captain W. B. Dabell MBE, MC) and his staff worked wonders in getting cookers up to the front line so that the men could receive hot food and, as Captain Arthur Gibbs, commanding Prince of Wales Company, told his father, there were hot baths and changes of clothing every fortnight. In this case they were situated in a disused factory behind the lines.

> The men wash in big wooden tubs, which are just large enough to sit down in. Before they have their bath they leave their dirty underclothes upstairs and get clean ones in exchange. Thousands of shirts etc. are being washed all the time by the girls who were formerly mill hands. There is a space partitioned off for officers, with 5 baths in it, proper baths this time. There is any amount of hot water and the whole thing is very well run.[5]

A picture taken by Lt the Hon P. G. Fitzalan-Howard showing a typical communication trench on the Western Front, which joined the front line with the rear area. *(Harrop Collection HQ WG)*

To the sorrow of everyone in the Battalion, Murray-Threipland succumbed to illness as a result of his long

Captain J. A. D. Perrins, MC

John Allan Dyson Perrins was born in Malvern on 20 November 1890 and was commissioned in the Seaforth Highlanders in 1910. With that regiment's 2nd Battalion he served in France when the British Expeditionary Force deployed on the outbreak of the First World War and saw action in the opening battles at Mons and on the Marne. He joined the Welsh Guards on their formation and was appointed the second Adjutant of the 1st Battalion, serving with it in France until August 1917 when he transferred to a staff appointment. After the cessation of hostilities he was demobilized but returned to the Colours on the outbreak of the Second World War. After serving for a short time with 2nd Battalion Welsh Guards he worked in various staff jobs, including aide de camp to GOC 48th Division and GOC XII Corps. In July 1940 he was discharged and returned to his civilian occupation as Managing Director of Lea and Perrins, the family firm, to undertake essential food-production work. In March 1958 he retired to live in southern Africa, first in Southern Rhodesia (now Zimbabwe) and then in South Africa. He died on 11 August 1972 at Knysna, Cape Province. His son Neil Perrins joined the Welsh Guards in 1939 and was taken prisoner while serving with the 2nd Battalion at Boulogne in June 1940.

Lieutenant Colonel G. C. D. Gordon, DSO

The first Adjutant of the 1st Battalion, Douglas Gordon was born in London 8 April 1883 and after education at Eton was commissioned in The Gordon Highlanders in 1900. In the following year he transferred to the Scots Guards and served with them in the Second Boer War. In 1913 he transferred to the Special Reserve and was recalled to the Colours on the outbreak of the First World War, serving with 3rd Battalion Scots Guards in France. One of the founding officers of the Welsh Guards, he commanded the 1st Battalion in 1916–17 and retired from the Army in June 1919. He died on 3 October 1930.

exposure to the harsh conditions and was ordered home. He was succeeded as Commanding Officer by Major Gordon, another founding officer, described by Dudley-Ward as 'a disciplinarian [who] kept the standard of smartness in the Battalion at a high pitch'.[6] His second-in-command was Major Humphrey Dene, known as 'Broncho', who had a deserved reputation as a thrusting if occasionally impetuous fighting soldier.

Another Welsh Guardsman who did not survive the winter of 1916–17 was Private David Jones 01, nicknamed '*Dei Llwyn Cwbl*', the son of Robert and Margaret Jones of Llwyn Cwbl farm, Llangwm, Denbighshire. He had been wounded during the last fighting on the Somme undertaken by the Battalion on the Gueudecourt front in mid-December and died of

Lieutenant Colonel Humphrey Dene, DSO

Nicknamed 'Broncho' because of his outstanding horsemanship, Humphrey Dene was one of the founding officers of the Welsh Guards and commanded the 1st Battalion in 1918. Born on 21 March 1877 at Bideford, North Devon, he was educated at Eastman's Royal Naval Academy at Southsea but decided against a career in the Royal Navy. After emigrating to South Africa he served in the South African Constabulary between 1901 and 1906, when he commanded a troop during a Zulu uprising. Having joined the Reserve of Officers in the British Colonial Service in West Africa, he returned to the UK at the outbreak of the First World War and served with XX Hussars in France in 1914. He retired from the Army in September 1919 and died at Nottingham on 21 January 1948.

(Illustration courtesy of the Welsh Guards Archives)

Lt the Hon P. G. Fitzalan-Howard, No. 3 Company, observing from a forward position. *(Harrop Collection HQ WG)*

Lt L. F. Ellis while commanding No. 3 Company, Reserve Lines, Western Front, February 1918. *(Harrop Collection HQ WG)*

Prince of Wales Company officers having lunch, Western Front. (*L to R*) Capt Goetz, MC, 2Lt C. F. Fleming, 2Lt H. A. Spence Thomas, 1918. *(Harrop Collection HQ WG)*

Another view of a forward trench – although probably not that far forward, as the risk of being shot by snipers meant that few soldiers risked sticking their heads above the parapet in daytime. *(Harrop Collection HQ WG)*

Welsh Guardsmen resting, Boyelles, August 1918. *(Imperial War Museum)*

Sgt Robert Bye during the action at Langemarck for which he was awarded the Victoria Cross, Third Battle of Ypres, 31 July 1917. Painted by David Rowlands for the Sergeants' Mess.

his wounds on 7 January 1917 aged twenty-six. A talented harpist, he received his training from Nansi Richards '*Telynores Maldwyn*', the doyenne of traditional Welsh harpists, who claimed that she herself had learned to play the triple harp from the gypsies who lived on her father's farm near Oswestry. After Jones's death a plaque was unveiled in his memory at Capel Gellioedd and an elegy was written in his honour by his friend and fellow soldier Private David Ellis of Penyfed, Ty Nant, Corwen:

Brwd alaw ei bêr Delyn – ddistawodd
ys tywyll ei fwthyn.
Hyd erwau gloes – drwy y glyn
Aeth o ymdaith a'i emyn.

The fervent song of his sweet harp is silent;
his cottage is in darkness;
through the aching acres – through the valley
he went, leaving his sojourn and his hymn.[7]

Relief from the miseries of the Somme front came on 25 March 1917 when the Battalion was moved to the vicinity of Péronne-en-Malentois (near Lille) where the men went under canvas and, to the relief of everyone, the weather began to improve. The two months of relative relaxation provided a great tonic and the mood within the Battalion began to change as everyone rediscovered their old form and gradually forgot about the cold, the wet and the all-pervasive mud. A gardening competition was held and judged by the brigade commander, himself a keen horticulturalist. Apart from muttering 'damned Welsh thieves!' he managed to keep a straight face when confronted by some of his choice specimens in the winning flower garden. Work, too, was relatively pleasurable – assisting Canadian engineers to rebuild railway lines destroyed by German artillery fire. Many of them had worked for the Canadian Pacific Railway and they got on famously with the Battalion when the officers entertained them to dinner in their mess, a converted barn. As the Adjutant recalled later, the occasion produced 'a glorious joke'.

> Towards the end of dinner the Corporal-in-Waiting, complete with stick and book, came in and went through the drill of asking the CO's leave to speak to Captain R. E. C. [Luxmoore] Ball. Turning smartly he went down the length of the table and went through the ritual with Ball who kept him waiting a bit and then rather testily said, 'Yes, what is it?' Corporal: 'Sir, I have to report that your tent is on fire.' Ball: 'Put the bloody thing out.' Exit Corporal. Our guests shouted in delight, but the real joke was that they thought that it was a carefully rehearsed act put on for their benefit. They just couldn't believe that it was the real thing.[8]

In every respect Péronne provided a time out of life; it was a picturesque place with a long and venerable history and possessing a great deal of charm. Being close to the French border, it had seen its fair share of fighting and had been badly pillaged by the Germans in 1914. Fortunately the local museum curator had managed to save several artefacts, one of which caught the attention of the Welsh Guards – an old chamber pot with an eye painted on the bottom, together with an inscription in French slang which translated as: 'I see your little cock.' Everyone was greatly amused and the Commanding Officer immediately vowed to present it to the Major General. (There was more to this gesture than simple ribaldry: the badge of the Guards Division was 'the all-seeing eye'.)

Suitably refreshed, the Battalion left Péronne on 31 May for St-Omer to prepare for a new offensive that had been planned for that summer – the Third Battle of Ypres, also known as Passchendaele, which lasted from 31 July to 10 November. The plan was to break out of the Ypres salient and occupy the high ground to the east, prior to an attack on the ports of Ostend and Zeebrugge from the rear. Although the tactics for the battle had been planned meticulously and had been practised with every attention to detail, the expected 'breakthrough' did not materialize, largely as a result of the unexpectedly robust German defences and the onset of heavy and prolonged rain. Even so, for the Welsh Guards the initial assault got off to the very best of starts when 1st Guard Brigade was able to capture the lightly defended German trenches on the other side of the Yser Canal, thereby removing the need for a potentially difficult assault crossing. Before the

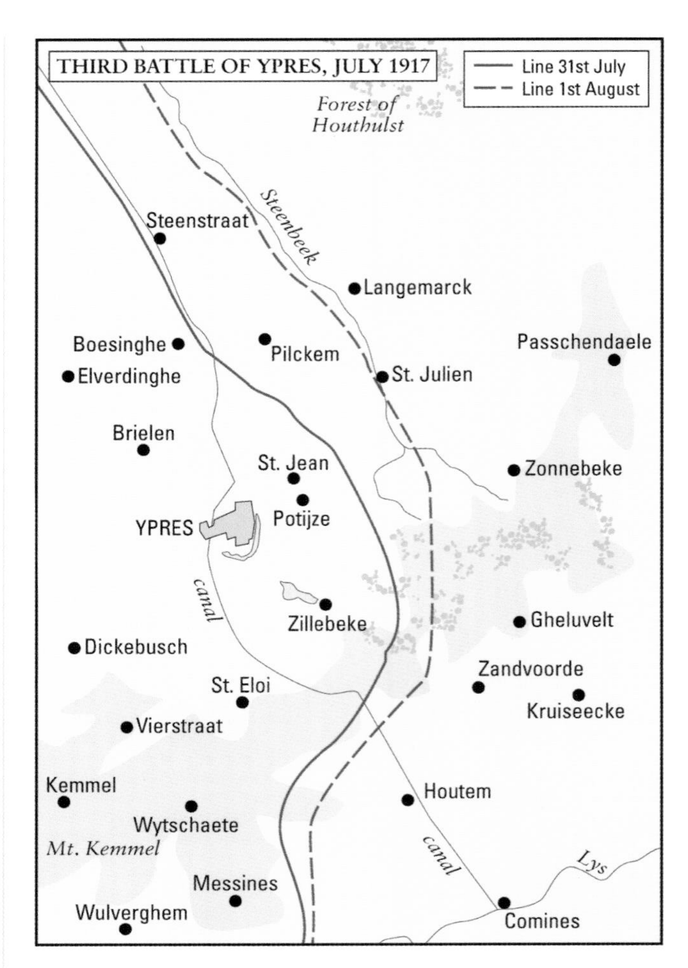

The Third Battle of Ypres, also known as Passchendaele, showing British gains by 10 November 1917.

attack began on 31 July there was a huge barrage which was so powerful that one young officer believed that 'it could have been easily heard in England'.[9]

Beneath a creeping barrage which moved forward at twenty-five yards a minute, the Battalion reached its first objective at Wood 15 where it was halted by machine guns in a blockhouse which had managed to escape the bombardment. It was the first time that anyone had seen this kind of obstacle on the Western Front, but there was one man who was equal to the task of dealing with it – Sergeant Robert James Bye from Pontypridd, who had worked at the Deep Dyffryn Colliery, Mountain Ash, before joining the Welsh Guards on 3 April 1915. Seeing that the leading waves were under heavy fire from the enemy blockhouse, Bye crawled forward and rushed at the position from the rear, putting it out of action with hand grenades. He then rejoined his company and went forward to a second, then a third, blockhouse and gave them the same treatment. During the action, over seventy Germans were killed, wounded or captured and Bye

Welsh Guards Uniforms 1915–18. *(From illustrations by Charles C. Stadden and Eric J. Collings commissioned by the Regiment)*

1917 Musician – Regimental Band. Even during the war full dress was retained by the King's Life Guard in Whitehall, and also by the bands of the Foot Guards on public duties in London. Note the old bandsman's sword with its monogrammed hilt, and its black leather scabbard with a trefoil shape. A white music pouch is worn to the rear.

1915 Private – Guard Order. Sentries removed their packs but first and last reliefs were posted in full guard order.

1916 Bandmaster – Regimental Band. The Band's instruments were paid for by the City of Cardiff. In 1916 it undertook a four-month tour with the Guards Division on the Western Front. Major Andrew Harris, MVO, LRAM, the first Bandmaster, composed the regimental quick march, 'Rising of the Lark'.

1916 Sergeant – Walking Out Order, France. When the battalion was out of the line, its training programme alternated with the three Rs: rest, renovation and recreation. Equipment was cleaned, trousers pressed and clothing cleaned. Training included musketry, drill, PE, exercises and route marches. The Battalion Choir gave a number of well-received performances during these moments of respite.

1918 Sergeant – Walking-Out Order, Cologne. Shortly after the Armistice of November 1918, the Guards Division was ordered to winter in Cologne as part of the Allied Army of Occupation. During this time the Battalion helped suppress a Marxist rebellion, as well as performing ceremonial duties at British Army headquarters. The Battalion returned to the UK on 11 March 1919. The Guards Division now ceased to exist, and the reserve battalion was absorbed into the 1st Battalion at Ranelagh Club, Barnes. In November 1918 HM King George V ordered that, in recognition of wartime service, privates of the Brigade of Guards would be known as Guardsmen.

was rightly recommended for the award of the Victoria Cross. The citation appeared in the *London Gazette* on 6 September and three weeks later Bye was invested with his VC by King George V at Buckingham Palace.[10]

Bye's example showed how to deal with these seemingly impregnable positions and other Guardsmen followed his example. The first objectives had been taken early in the morning but, as it turned out, that was the sum of the Battalion's success during Third Ypres. By the time that the fighting came to an end in the late autumn the British line had been extended to Poelcapelle, Passchendaele and Polygon Wood, but it had come at a fearful cost. An estimated 310,000 men had been killed or wounded – around 70,000 had drowned in the waterlogged shell holes and lagoons – and Haig was widely criticized for keeping the battle going when it was clear that there was no chance of success. Casualties in 1st Battalion Welsh Guards were 451 killed, wounded or missing.

After its involvement in the battle, the Battalion was quartered at Serques from 20 October until 9 November, and there time was given over to training in open warfare. Inspections were also the order of the day and there was an opportunity to hold two race meetings featuring some questionable mounts. The War Diary also noted that Serques was home to 'one of the prettiest

Sergeant Robert James Bye, VC

Robert Bye was born on 12 December 1889 at 13 Maritime Street, Graig, Pontypridd, the son of Martin and Sarah Jane Bye. Shortly after his birth the family moved to 21 Woodfield Street, Penrhiwceiber, near Aberdare, and Robert was educated at the local school. After leaving school he worked at Deep Dyffryn Colliery, Mountain Ash, also near Aberdare. On 14 October 1912, at Pontypridd, he married Mabel Lloyd from Penrhiwceiber, with whom he had two sons and two daughters. Shortly after the formation of the Welsh Guards he enlisted in the 1st Battalion on 3 April 1915. His advance through the ranks was rapid. He was promoted to lance corporal in March 1916, corporal in September of that year, and then to sergeant in April 1917. He was awarded the Victoria Cross for 'most conspicuous bravery' at the Yser Canal in Belgium on 31 July 1917 during the Third Battle of Ypres.

Bye, who moved to Nottinghamshire after the war to work as a coal miner, also served in the Second World War as a sergeant major in the Sherwood Foresters, guarding prisoners of war, until ill health (arising from his pit work) forced him to leave the Army. He died on 23 August 1962. His grandson served in the Welsh Guards in the 1980s.

Extract from the *London Gazette,* dated 6 September 1917:

'No. 939 Sjt. Robert Bye, Welsh Guards (Penrhiwceiber, Glamorgan).

On 31st July 1917 Sergeant Robert Bye displayed the utmost courage and devotion to duty during an attack on the enemy's position. Seeing that the leading waves were being troubled by two enemy block-houses, he, on his own initiative, rushed at one of them and put the Garrison out of action. He then rejoined his company, and went forward to the assault of the second objective.

When the troops had gone forward to the attack on the third objective, a party was detailed to clear up a line of block-houses which had been passed.

Sergeant Bye volunteered to take charge of this party, accomplished his object, and took many prisoners. He subsequently returned to the third objective, capturing a number of prisoners, thus rendering invaluable assistance to the assaulting Companies.

He displayed throughout the most remarkable initiative.'

Sergeant Robert Bye being presented with the Victoria Cross by HM King George V. *(Welsh Guards Archives)*

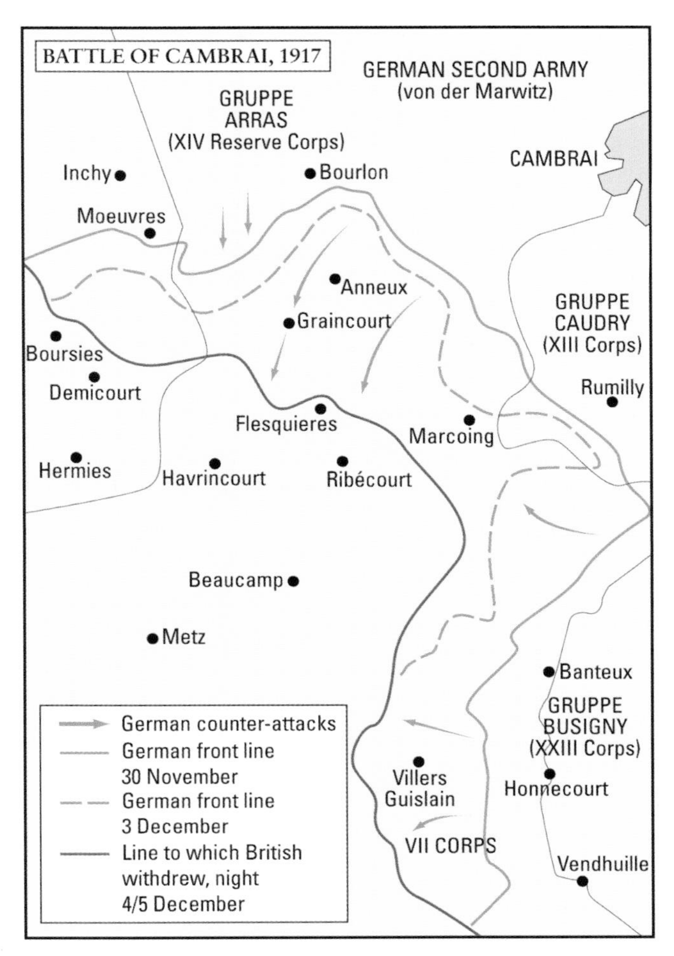

The Battle of Cambrai, showing limit of British advance, 3 December 1917.

girls in France' who dispensed coffee free of charge in her mother's *estaminet* in the town. A month later, on 17 November, the Guards Division began a move towards the front at Cambrai, which was to be the site of the next British objective. An important railhead and supply point for the Hindenburg Line, the town was heavily defended by two lines of fortifications, with barbed-wire belts, concrete blockhouses and underground works. Against them new artillery–infantry tactics had been developed for the initial attacks, which would then be reinforced by over 400 tanks to clear barbed-wire defences and to support the assault troops.

When the attack began on 20 November, it achieved complete surprise – over a thousand artillery pieces had taken part in the creeping barrage, which had been enhanced by the use of better maps and calibration of individual guns to improve accuracy. Cambrai was also the first battle in which the guns did not have to register on their targets prior to the attack, thus compromising surprise. By midday a salient eight miles wide and five miles deep had been created. This was the moment to bring up reserves to exploit the success, but in the confusion this did not happen. Dudley-Ward's diary entry neatly sums up the muddle and how it affected the Battalion: 'the Welsh Guards were the rear battalion, and spent long periods either waiting by the side of the muddy track or doubling wildly after disappearing men in front of them'.[11] Inevitably the indecision had a bad effect on morale within the Battalion and this was compounded by the emergence of an undercurrent of bad feeling amongst the officers. Not only were several absent, having been sent on courses before the battle, but Dene thought that too many were exhausted or (his word) 'gaga'. On the eve of battle Dudley-Ward had been displeased to be replaced as commander of No. 4 Company by Captain Philip Dickens, who had had only one experience of fighting in the line and had been wounded in the process. When Dudley-Ward confided his concerns to his friend Broncho Dene, he received the far from reassuring reply:

> If anything happened to DG [G. C. D. Gordon, the Commanding Officer] 'these are my orders,' said Broncho, 'you are to take charge of the battalion till I arrive and pay no attention to any paper seniority.' Nice position![12]

At that stage, 22–23 November, the Guards Division was fighting alongside 51st (Highland) Division which had managed to get a foothold in the villages of Flesquières and Fontaine-Notre-Dame, but the Germans had begun counter-attacking and it took time for the line to be stabilized. Throughout the days that followed, confusion reigned: the Battalion War Diary records the fluctuating and frequently contradictory orders which passed through headquarters until the end of the month when the Battalion was withdrawn into reserve at Trescault.[13] The waiting did not last long: on 30 November came orders that 1st Battalion Welsh Guards were to secure the Gonnelieu ridge and the high ground to the south-east of their position, which was being threatened by the German counter-attack. They would be supported by 4th Battalion Grenadier Guards on the left and 2nd Guards Brigade on the right, but as there was no time for reconnaissance in daylight, the attacking forces had no inkling of what was waiting for them once they reached the high ground.

Attacking uphill on a moonless night, the leading companies (No. 3 Company on the right, No. 4 Company on the left with No. 2 Company forming a second wave) approached the crest only to find it had

Weapons and Equipment used during the First World War

The British Army of 1914 was one of the best trained and equipped armies ever sent to war by the United Kingdom. It was the first army to introduce a form of camouflage dress when 'khaki drab' uniforms were issued to the Corps of Guides on the North-West Frontier in 1848. The term 'khaki' comes from the Urdu word for 'dust'. In 1915, Welsh Guardsman wore the 1902 Pattern Service Dress. This was made from a thick woollen serge fabric, dyed khaki green. The tunic had a Prussian collar and was fastened with five brass buttons. It had two pleated breast pockets (one of which was used to carry the soldier's AB64 Pay Book), two waist pockets, and a small internal pocket where a field dressing was kept. Shoulder straps were fastened with brass buttons and carried brass regimental insignia. Badges of rank were sewn onto the upper tunic sleeves, while trade badges and long service and good conduct stripes were placed on the lower sleeves.

A stiffened peak cap made from the same khaki material was worn as headdress. It had a leather strap secured with two brass buttons. The leek cap badge was made of brass. By 1916, the realities of trench warfare made head protection mandatory and the Brodie steel helmet replaced the cap. Brown ammunition boots with hobnail soles were the standard issue footwear and were worn with long puttees to support the ankle. Boots were made of reversed hide with reinforced steel toecaps and heel plates. Similar boots are still worn by Guardsmen today.

The British Army was the first European army to replace leather belts and pouches with 'webbing', a strong material made from woven cotton. Developed by the Mills Equipment Company, the1908 Pattern equipment comprised a 3" belt, left and right ammunition pouches (which held 75 rounds each), left and right braces, a bayonet frog, an entrenching tool carrier, a water-bottle and carrier, a small haversack and a large pack. The haversack was used to carry mess tins, a knife, fork and spoon set, washing and shaving kit, 'housewife' (sewing and repair kit) and rations. The soldier's greatcoat and blanket were carried in the large pack.

A Guardsman's primary weapon was the .303 Short Magazine Lee-Enfield (SMLE) rifle. This had a 10-round magazine that made it one of the fastest bolt-action rifles of the period. A well-trained soldier could fire between 20 and 30 rounds per minute at ranges of up to 600 metres. With various versions of this rifle used between 1895 and 1982, no other weapon has remained longer in British Army service.

The Mills Bomb (hand grenade) was a defensive weapon used to repel attacks. With a grooved cast iron construction to help it fragment, it looked like a pineapple and was lethal within a radius of 30 metres. A rifle grenade, the Hales Rifle Bomb, was also used.

The .303 Vickers Machine Gun was a belt-fed, water-cooled support weapon. Based on the original Maxim design with various improvements, it was extremely reliable and durable, but heavy. The rate of fire was between 450 and 500 rpm and it had a direct range of over 2,000 metres. It was used with devastating effect and remained in service until 1968. The .303 Lewis Machine Gun was lighter and more portable than the Vickers and was thus issued in larger numbers. Instead of being belt-fed, it had a drum magazine that held 47 rounds. Infantry battalions initially had two, but this was increased to sixteen by 1918.

obscured a slight depression behind the village of Gouzeaucourt where the Germans had sited their machine guns. These immediately opened up with a 'perfect hurricane' of fire and very quickly 'the ground was thick with dead and wounded'. Then, as dawn approached, bringing fog to add to the smoke and glare of star shells, wounded men started streaming down the hill, adding to the confusion. Writing not in the Battalion War Diary but in his personal diary, Dudley-Ward recorded a scene which he wished he had not had to witness.

> It was staggering in its effect. I at once thought we had struck a strong Bosch attack and yelled to the men in reserve to line the bank and prepare to fire. Broncho [Dene] dashed off to one side yelling too, and then they were on us and we saw it was our own men. Never have I seen such a thing. They were sobbing and cursing.[14]

As he watched 'the whole attack crumbling to nothing', Dudley-Ward ordered the fleeing men to halt and to turn and face the enemy but it was difficult to make himself heard. The hammering of the machine guns was so loud that Gordon and Dene were forced to shout in each other's ears. The appearance of a tank

Officers of the 1st Battalion Wormhoudt, Western Front, May 1916. *Rear Rank:* De Wiart, Crawford-Wood, de Satgé; *Second rank:* Rowlatt, H. Talbot Rice, J. Crawshay, A. Gibbs, Kearton, Pugh, Thursby-Pelham; *First rank:* J. J. Evans, C Insole, Copland-Griffiths, Lord Clive, B. Hambrough, Perrins, Dabell; *Seated:* Aldridge, Price, Murray-Thriepland, H. Dene, R. Bull; *Seated in front:* C. Dudley-Ward, P. Batttye. *(Welsh Guards Archives)*

and some well-directed defensive artillery fire helped to restore some order and in mid-morning the advance resumed as the Germans began to pull back. Of the 370 Welsh Guardsmen who had begun the attack, 248 had been killed or wounded within the first three minutes. It had been yet another instance, common throughout the war, of a battalion carrying out an attack even though the lack of support and enemy superiority were already known. As Lance Corporal Charles Evans of Swansea put it later: 'our colonel stuck to his orders, that he had to attack, which he shouldn't have done'.[15] On 3 December the Battalion was relieved by 2nd Battalion Irish Guards and entrained for Etricourt before marching into reserve at Gouy-en-Artois, south-west of Arras. The new year of 1918 found the Battalion in the line north of Arras at Gavrelle and Fampoux where the condition of the trenches was variable, often half-full of water, with the result that the mud was the consistency of porridge. Between then and March they were in the line six times in ten weeks, each tour lasting four days at a time. At this time the Guards Division transferred from XVII Corps to VI Corps in Third Army.

All the while the strategic situation was changing. Russia had collapsed in the spring of 1917 and this was followed by the entry of the USA into the war on the side of the Allies. Towards the end of the year, as Russia descended into revolution, the German high command came to the conclusion that they had to regain the initiative before American forces were able to deploy in strength in Europe and at a time when the British and French armies had been badly weakened by the previous year's offensives. In February 1918 a decision had been taken to reduce the size of all British divisions from thirteen to ten battalions, thereby substantially weakening the size and shape of the forces available to Haig. In the Guards Division this meant shedding 4th Battalion Grenadier Guards, 3rd Battalion Coldstream Guards and 2nd Battalion Irish Guards, which together formed a new 4th Guards Brigade in the 31st Division.

The German plan was conceived at a conference in Mons in November and it took advantage of the fact that the Russian surrender had released substantial numbers of German troops which could be used on the Western Front against the weakened British and French armies. Not only would this give the German commander, General Erich Ludendorff, a numerical superiority, but many of the formations were tried and tested infantry regiments – Prussians, Guards and Swabians – representing the cream of the German Army. Ludendorff's strategy was brutally simple: his armies would drive a wedge between the British and French armies, striking through the old Somme battlefield between Arras and La Fère, before turning to destroy the British Third and Fifth Armies on the left of the Allied line. Three German armies would be used in the assault. The Second Army (von der Marwitz) and the Seventeenth Army (Below) would take the offensive across the Somme battlefield before driving north to wrap-up the British, while the Eighteenth Army (Hutier) provided flank support to the south in the St-Quentin sector. Codenamed 'Michael' (Germany's patron saint), the offensive plan called for a massive rolling 'hurricane' artillery barrage, followed by a rapid and aggressive advance by the infantry, who, in Ludendorff's words, would 'punch a hole' in the British defences and lay the foundations for defeating the enemy in Flanders. Strongpoints would be bypassed to be dealt with later by the mopping-up troops.

The plans, advanced in great secrecy, included an intensive training programme for the assault formations, which were composed of lightly equipped but heavily armed 'shock' or 'storm' troops. Their orders were to press on quickly and assertively, to take ground without thinking too much about the safety of their flanks and above all to maintain the momentum of the assault, regardless of casualties. It was a bold policy and in adopting it, Ludendorff had high hopes that it would succeed in pushing the British back towards the Channel and opening the way to Paris. In addition to enjoying numerical superiority, he also knew that the British lines were unbalanced in their deployment as a result of manpower shortages. The Germans would be attacking with forty-three divisions and a total of 2,508 heavy artillery pieces; as the *Official History* noted later, 'never before had the British line been held with so few men and so few guns to the mile'.[16]

The storm broke in the early hours of the morning of 21 March when the German artillery produced a

Welsh Guardsmen in a reserve trench, Guillemont, Western Front, September 1916. *(Welsh Guards Archives)*

huge bombardment which lasted for five hours and which left the defenders badly shaken and disoriented. Gas and smoke shells added to the confusion, which was increased by an early morning mist, leaving commanders with no exact idea of where and when the infantry attack had originated. At the time, 1st Battalion Welsh Guards had just come out of the line on the Arras front, but on 23 March they relieved 2nd Battalion Scots Guards in trenches to the east of the village of Boiry-Becquerelle. On the second day of the Michael offensive, the Third Army's battle lines remained intact but in the south the Fifth Army had come under such intense pressure that it was unable to respond to the speed and intensity of the German attack – Hutier had made startling progress, advancing up to twelve miles and prompting fears that a huge split was about to be opened up in the Allied lines. So serious was the situation that the Commanding Officer received a warning that 1st Battalion Welsh Guards, on the divisional right, might have to withdraw five miles to a new defensive position, but this was changed soon afterwards to a more limited withdrawal of 2,000 yards.

Now involved in fighting an organized defensive battle, the Guards Division dug in and refused to budge; with accurate artillery fire to support them, the men grew in confidence as they saw the enemy falling to their small-arms and machine-gun fire. During this phase the Battalion held a front 2,000 yards long to the right of Boyelles and one Welsh Guards officer, Captain Claude Insole, remembered that his men 'fairly cheered with delight when they saw the Huns coming, and some stood on the parapet and shouted to them to come on'.[17] Dudley-Ward's diary noted that 'the execution was very great' and this was confirmed on 21 March when Sergeant O. F. Waddington from Pontypridd took out a fighting patrol and found eighteen enemy dead in the vicinity and many more a few hundred yards away.[18] Casualties in the Battalion were six killed and twelve wounded. By the end of the month the fighting had died down on the Arras front and the Michael offensive was finally called off on 5 April. The German break-in battle had succeeded in capturing a large salient but this had proved difficult to hold and the expected breakthrough to split the Allies failed to materialize. There had also been heavy German casualties – some quarter of a million killed, missing or wounded – and morale within the assault formations had been shattered by their failure to produce a decisive blow in the so-called *Kaiserschlacht* (Kaiser's Battle) which was supposed to win the war.

A week later, the Guards Division went into Third Army reserve near Fosseux and Barly south-west of Arras where the battalions were able to regroup and refit. Dudley-Ward's diary gives a colourful portrayal of the 'hand-to-mouth' conditions endured by the men following a month of sustained fighting:

Prisoners of War 1915–1919

In the early fighting in 1915 and 1916, fifteen Welsh Guardsmen were taken prisoner by the Germans and held in prisoner-of-war camps in Germany. During the war, under the direction of Lord Harlech, Mrs Murray-Threipland organized the despatch of parcels of comforts to these men. It was agreed that an officer's wife would assume responsibility for each prisoner of war and that the parcels would be sent by the Royal Savoy Association for the Relief of British Prisoners of War. The soldier chosen by Mrs Murray-Threipland was Private Fowler of Cadoxton near Neath, who had been one of the Commanding Officer's orderlies. When he complained in May 1917 that the parcels were not reaching him, Mrs Murray-Threipland successfully took up his cause with the Secretary of the Savoy and the supply was resumed. Coxton's mother later wrote a letter of thanks, saying that only a family regiment would have shown such kindness to her son.

Pte Cannon *(left)* and Pte Butcher, soldier-servants to Maj Goetz and Capt Harrop respectively, Canal de Nord, 1918. *(Welsh Guards Archives)*

> I have men now who are ragged about the trousers to the point of indecency, and many have their bare toes sticking through their boots. I wash on an average of every other day, but the men are worse off, and only get a shower-bath and a change of clothing once a fortnight. These small things count, as a clean man is always more refreshed than a dirty one.[19]

For the next four months the Battalion's routine followed a familiar pattern of tours on the front line, followed by stretches in reserve for training, refitting and sporting activities with other regiments. The one problem was getting reinforcements, as the Battalion was consistently under-strength – not helped by a plague of boils, which caused a number of Guardsmen to be sent to hospital for treatment. Not that their fate won the victims much sympathy. When Sergeant Manuel, of No. 4 Company, returned to the line after an appreciable absence, he found CQMS Trott from Cardiff gesturing to him with an ammunition clip in his hand and explaining that he had something interesting to show his friend. 'This', he said, 'is what we use in the line to kill Germans – we call it a "round". Want to keep it as a souvenir?'[20]

Officers were also in short supply: on 29 May 1918 2nd Lieutenant Alexander Stanier arrived at the 1st Battalion in charge of a detail of three officers who were all in their forties. He was only nineteen and looked so young that when he turned up at Battalion headquarters some wit shouted out: 'Good Heavens! Are we reduced to receiving the infant Samuel?' From that moment Stanier was always known as 'Sammy'.[21] That summer 1st Battalion Welsh Guard was joined by US infantrymen of 1st Battalion, 320th Infantry Regiment, who arrived for instruction at their camp at Blairville. An infantry battalion in the US 80th Blue Ridge (Mountain) Division, this unit recruited in Virginia and Pennsylvania and had landed in France in May and June 1918 as part of the American Expeditionary Force (AEF) which had been despatched in the wake of the US declaration of war on 6 April 1917. Although the Americans were keen to learn, they were also naïve and badly trained. To the bemusement of the Guardsmen they all carried large packs with a huge roll on the bottom which prevented them sitting on the fire-step as British soldiers would have done. When asked what each man was carrying, they explained to Stanier that it was a tent and that 'it's sure handy on the plains of Texas!' Later Stanier saw them ripping down timber from the revetments, arguing that they needed the wood to light their cooking fires. Nevertheless the Yankees (as they were christened by the Welsh Guardsmen) were welcome reinforcements and Dudley-Ward, now back commanding No. 4 Company, was moved to record: 'They make one laugh, they are so green but they are so devilish anxious not to be caught napping they are positively dangerous to anyone going round the line – every sentry you visit receives you with a bayonet in your face.'[22]

By that point, although few could discern it at the

Brigadier Sir Alexander Stanier, Bt, DSO, MC

Born in Shropshire in January 1899, Stanier was known as Alex to his civilian friends but his many military friends always called him Sammy. Commissioned in the Welsh Guards in 1918, he served with the 1st Battalion in the last months of the First World War. A keen horseman, he encouraged fellow officers to participate in all kinds of equestrian activities and was especially active in that respect during his time as Adjutant of the 1st Battalion in the inter-war years. At the outbreak of the Second World War he was Commanding Officer of the 2nd Battalion and led them during the operations in Boulogne in June 1940. Although he was injured during a training exercise and lost the use of his left eye, he commanded 231st Brigade during the D-Day invasion in June 1944 and was responsible for the liberation of Arromanches. Later, a memorial was unveiled in the village of Asnelles to commemorate the action. The town also renamed the principal square in his honour. In 1945 Stanier became the first post-war Regimental Lieutenant Colonel and later retained strong links with the Regiment. He retired from the Army in 1948 and died aged ninety-six, in January 1995. Shortly before his death he was awarded the freedom of Arromanches which was accepted by his son Billy, also a Welsh Guards officer, on 6 June 1995.

Colour party, Cologne, January 1919. (*L to R*) Sgt Grant, DCM; Lt Paton, MC; CSM Pearce, DCM; Lt A. B. G. Stanier, MC; Sgt Pates. *(Welsh Guards Archives)*

time, the war was entering its final stage, the so-called 'Hundred Days', which saw the advantage swinging inexorably towards the Allies. In the middle of September the AEF demonstrated its mettle when the US First Army attacked the German salient at St-Mihiel, which they had held since 1914 and which was an obstacle for any French attack in Lorraine. Compared to many other battles fought on the Western Front, it was a minor offensive. It lasted two days and the American success was as much due to the German decision to withdraw as to the skill of the AEF; but a victory is a victory and it gave great heart to those who took part in it. This was followed by Foch's offensive in the Meuse-Argonne where French and US forces pushed northwards towards Sedan and Mezières. At the same time the British Second Army recaptured the Messines Ridge and the Third Army advanced towards Maubeuge, while further to the north the Belgian Army Group commanded by King Albert pushed out of Ypres and over the Passchendaele Ridge towards Roulers.

During this period, 1st Battalion Welsh Guards was involved in the attack towards St-Léger on 23 and 24 August, which demonstrated how British tactics had improved, with the Allies advancing steadily under an accurate and powerful artillery barrage. Even so, the Battalion lost 144 casualties, killed, wounded or missing. Ahead lay the vastness of the Hindenburg Line, the redoubt that was the Germans' last stronghold. The next target was the Canal du Nord

Capt A. Gibbs, MC *(left)* with his brother, Lt B. Gibbs, at Blairville, August 1918. *(Welsh Guards Archives)*

and the attack began on 2 September with the Battalion, now under the command of Lieutenant Colonel R. E. C. Luxmoore-Ball, advancing steadily across open and desolate terrain which Dudley-Ward characterized as 'good country for any sort of manoeuvres, as it is criss-crossed with trenches, though they are over-grown and look like ditches'. Within three days the whole Guards Division had advanced five miles. A fortnight later the assault on the Canal du Nord began with 1st Battalion Welsh Guards and 1st Battalion Grenadier Guards crossing Lock 7 in heavy rain in the early morning of 27 September. The first thing most Guardsmen heard was the stentorian voice of Regimental Sergeant Major Stevenson telling everyone to get up – it was 04.30 hours – followed by 'good solid army oaths' and the demand: 'Where is my valet? Does he want me to bring him a cup of tea?'[23]

By the end of the day the crossing had been secured, allowing the Battalion to regroup – no easy matter as the strength had fallen to 399 effectives, with its fighting strength from four to three companies. With the war now entering its last phase, the Battalion joined the advance towards the River Selle to capture the high ground behind it and then to continue the advance towards the River Harpies, which was secured on 22 October. This brought the Guardsmen into new country untouched by the violence of war and for the first time they were able to get billets in buildings which had not been completely wrecked. The last hostile contact with the Germans came on 4 November when No. 3 Company, under Captain L. F. Ellis, encountered a group of German soldiers in a railway cutting near Bavay, south-east of Valenciennes. Finding himself unarmed, Ellis advanced, waving his walking stick; fortunately he was being followed by Lance Corporal E. W. Gordon, from Bridgend, and Lance Sergeant W. M. Jones, from Tonypandy, who shot eleven Germans before the rest fled. This allowed the company to secure the cutting and when Prince of Wales Company, came up the German positions in a nearby farm were cleared. It was a good moment; Dudley-Ward noted with commendable understatement that 'it could almost be said that the men "enjoyed" the fighting'.[24] A week later the Battalion moved to the small and uninspiring town of Douzies near Maubeuge; on 11 November the Battalion War Diary recorded the bald entry: 'News reached the battalion that an Armistice had been signed to take effect at 11.00 hours today.'[25]

Almost four years earlier when King George V had presented the Welsh Guards with its first stand of Colours, he reminded the Regiment that:

> … these Colours bear no names of Battles fought or of Victories won, your noble deeds in the coming days will be inscribed upon them. In committing these Colours to your care I know that you will look up to them and prove yourselves true sons of loyal and gallant Wales and worthy of the glorious traditions of the Brigade of Guards.

At the war's end the Colours showed that the Regiment had indeed kept that faith. In addition to Loos, the battle honours awarded refer to three phases fought during the Battle of the Somme – Ginchy, Flers-Courcelette and Morval; to the Pilckem and Poelcappelle actions of Third Ypres; and to the Canal du Nord and Sambre battles fought during the closing stages of the war.

Regimental Colours

There are two Colours:

The Royal or First (usually called the Queen's) Colour has the following heraldic description: 'Gules [crimson]. In the centre a dragon passant Or, underneath a scroll with the motto "Cymru am Byth". The whole ensigned with the Imperial Crown.'

The Second (usually called the Regimental) Colour: 'The Union. In the centre a Company Badge ensigned with the Imperial Crown. The Company Badges are Borne in rotation. The number below the Badge is the number of the Company to which the Badge belongs. The number in the dexter canton is the number of the Battalion.'

The first pair were presented by HM King George V at Buckingham Palace on 3 August 1915 and were laid up at Llandaff Cathedral on 19 November 1925. During the Second World War the cathedral was bombed and the Colours were moved to Regimental Headquarters for safe-keeping. They were returned to Llandaff Cathedral on 3 May 1959.

The second pair were presented by HM King George V at Windsor Castle on 23 June 1925 and were laid up at St David's Cathedral on 10 June 1950. The third pair were presented by HM King George VI at Buckingham Palace on 25 May 1949 and were laid up in the Guards Chapel on 7 July 1965. The fourth pair were presented by HM Queen Elizabeth II at Buckingham Palace on 5 May 1965 and were laid up in St Mary's Church, Swansea, on 16 September 1982. The fifth pair were presented by HM Queen Elizabeth II at Windsor Castle on 14 May 1981 and were laid up in All Saint's Church, Pirbright, on 25 October 1990. The sixth pair were presented by HM Queen Elizabeth II at Buckingham Palace on 30 May 1990 and were laid up in Bangor Cathedral on 27 June 2007. The seventh pair were presented by HM Queen Elizabeth II at Windsor on 4 May 2006.

The 2nd Battalion Colours were presented by HM King George VI at the Tower of London on 14 February 1940 and were returned for safe-keeping to Windsor Castle on 16 June 1947 when the Battalion went into suspended animation. The King's Colour was transferred to the Guards Chapel on 26 May 1956; the Regimental Colour was transferred to the Guards Depot Church on 23 May 1959 and then to the Guards Chapel on 4 June 1965.

No Colours were presented to the 3rd Battalion during its short existence (1941–6).

The Queen's Colour, 1st Battalion Welsh Guards.

The Regimental Colour, 1st Battalion Welsh Guards.

CHAPTER THREE

Peacetime Soldiering, 1919–1939

Egypt, London, Outbreak of War

NO ONE EVER FORGOT the moment that the guns fell silent along the Western Front. For the men in the Battalion relief was mixed with the knowledge that kit had to be cleaned for a parade the following day and, while celebrations were muted, there was too an air of quiet satisfaction. Sammy Stanier had been sent home, having been wounded in the arm a short time earlier, and was out shooting at the family home in Oxfordshire on 11 November when he heard the church clock striking eleven o'clock. His return had been preceded by a War Office telegram announcing that he had been badly wounded in the upper leg – an error caused by the fact that an equally badly wounded Guardsman had fallen on top of him, soaking his greatcoat with blood. Seeing this, the stretcher party had assumed the worst.

When the war ended, the Battalion found itself at Maubeuge where it prepared for the march into Germany. This turned out to be 'very tiresome' for all concerned: there was no transport for the men and the roads were in poor condition, badly rutted and made worse by the winter rains. The Battalion marched in column of fours and these had to be rotated every hour or so as the men in the middle had difficulty keeping step in the ruts caused by the retreating German transport. The Commanding Officer, the solitary figure on horseback, ordered transport wagons to be emptied if he saw horses faltering or becoming over-tired. As many of the wagons contained Guardsmen's kit, they lost many souvenirs which they had picked up in France. Eventually, after marching for 200 miles, the Battalion reached Cologne in December. Colour parties were sent back to Wellington Barracks to collect the Colours of the ten Foot Guards battalions and these were then paraded in front of the local population, who for the most part

Anti-aircraft shooting practice (note aircraft targets) with the .303 Lewis gun, Pirbright, 1923. *(Welsh Guards Archives)*

Lieutenant Colonel R. E. C. Luxmoore-Ball, DSO, DCM

On joining the 1st Battalion at Vermelles on 2 October 1915, Richard Edmund Coryndon Luxmoore-Ball was described in the Regimental War History as 'a gigantic ex-Welsh Fusilier', who had come straight from the successful South-West African campaign and was burning to fight. The short time he had spent in England had been mostly in the orderly room asking to go out to France.' He was wounded in action on 16 June 1916 and invalided home on account of those wounds. He commanded the 1st Battalion from 8 September 1918 to 19 March 1919. Mentioned in Despatches three times, he was also awarded the DSO, DCM and Croix de Guerre. He left the Army in 1919 and joined the Colonial Service as a Political Officer in the former German colony of Tanganyika (later Tanzania). On the outbreak of the Second World War he served in 9th Battalion, West Yorkshire Regiment, before transferring to 9th Battalion, Royal Sussex Regiment, as second-in-command to the future Field Marshal Sir Gerald Templer. Luxmoore-Ball was born in 1884 and died in 1941 while en route to the Middle East.

WO2 (DSgt) Dunkley, WO1 (RSM) Stevenson, DCM, MM, Capt (Adjutant) Sir Alexander Stanier, Bt. DSO, MC, WO2 (DSgt) Roberts, Tower of London, 1922. *(Welsh Guards Archives)*

were respectful. Those who were not, noted Stanier, were given some rough and ready treatment from the Sergeant Major's pace stick.

On returning to England on 12 March 1919 the Battalion was based first at Wellington Barracks and from August 1920 at Warley in Essex where, according to Stanier, 'We began to learn all the traditions and customs used by the Brigade of Guards in a peacetime setting.'[1] There was much to digest. On 22 March 1919, the Guards Division had marched through London and the Welsh Guards were represented by the merged 1st Battalion, the 2nd (Reserve) Battalion having disappeared four days earlier. At the head of the Regiment rode Colonel William Murray-Threipland, Regimental Lieutenant Colonel in his second appointment (12 October 1917–13 December 1920), together with Lieutenant Colonel G. C. D. Gordon, DSO, and Major Humphrey Dene, DSO, who had both served as Commanding Officers during the conflict. The Battalion itself was under the command of Lieutenant Colonel R. E. C. Luxmoore-Ball, DSO, DCM, and the Colours were carried by 2nd Lieutenants A. B. G. Stanier and R. R. D. Paton. That summer, on 19 July, the Regiment also took part in the Victory Parade through London when the Colours were carried by Captain H. Talbot Rice and Lieutenant F. A. V. Copland-Griffiths, MC, escorted by a platoon. For many of the officers and men who had not served in a pre-war Foot Guards regiment this was a novel experience, but almost all of the time-expired men simply wanted to be demobilized as quickly as possible and get back home.

One initiative which came to nothing was a request made to London District on 9 April 1919 by the Regimental Lieutenant Colonel offering his and the 1st Battalion's services for the North Russia Relief Force that was being assembled. This was part of an ill-starred attempt by Britain, France and the USA to intervene with ground and naval forces in northern Russia in the aftermath of the Bolshevik revolution. The aim was to support pro-Tsarist 'White Russian' forces and all told fourteen British infantry battalions were despatched to the area from September 1918 onwards. However, by the beginning of 1919 the intervention had become unpopular at home and the last infantry battalions were withdrawn that summer. Murray-Threipland's offer was rejected by telegram on 29 April.[2]

As had happened throughout Britain's history the conclusion of hostilities in November 1918 brought a

Inspection of the Regimental Band by the Regimental Lieutenant Colonel, Wellington Barracks, 1920s. *(Welsh Guards Archives)*

rapid reduction in defence expenditure. Following the post-war election which was won by the wartime coalition, David Lloyd George's government introduced a Ten Year Rule to govern defence spending: expenditure on the armed forces was planned on the assumption that there would be no major war for ten years and this rule was extended annually until 1932. Equipment was not renewed and under the Geddes Axe of 1920 (named after Sir Auckland Geddes, chairman of the Committee on National Expenditure), manpower levels were reduced to make further savings. By the end of 1920, the year of the initial cuts, the Army's numbers had fallen from 3.5 million to 370,000 and between 1923 and 1932 the Army's budget contracted from £43.5 million to £36 million. Following the deconstruction of the huge volunteer and conscript army, which was rapidly disbanded, the post-war Regular Army returned to its position as a small professional force and horizons narrowed as regiments went back to a way of life that all professional soldiers recognized and understood. In most respects it was a case of 'business as usual' as they took up again the familiar patterns and routines of peacetime soldiering. These changes had an effect on the way the Army viewed itself: wartime formations disappeared and the Army became a smaller institution as infantry regiments lost battalions and cavalry regiments underwent a series of amalgamations. For ambitious soldiers the outcome was disheartening as a bottleneck in promotion prospects led to complacency and to a comatose condition that discouraged radical thinking and put a stop to reform. Pacifism, arising largely from the huge death toll in the war, was also a disincentive to increased defence expenditure.

The Welsh Guards soon discovered that it would not be immune from the post-war changes. As a regiment raised in time of conflict to meet an urgent wartime need for recruits, its position was always going to be precarious and this was especially true at a time when hard-headed economies had to be made. In common with all other regiments returning to the United Kingdom, the first task for the Welsh Guards was to concentrate on demobilizing the war-service men, a process which had already begun in Germany. It was an unsettling period as priority was given to skilled men such as miners, while many others with long service had to wait their turn. As a result, the strength of the 1st Battalion was gradually whittled down to around 400 men and it would take time for new recruits to pass through the system to replace them.

The first news of impending change came in a story published in the *Daily Mail* on 4 June 1920. It was written by the newspaper's military correspondent, Valentine Williams, formerly Irish Guards, who announced that he had been given sight of a sensitive War Office document recommending that the Irish Guards and the Welsh Guards should be disbanded as individual Foot Guards regiments. Furthermore, the recommendation contained a proposal that the Foot Guards should be reduced to the three senior regiments, each consisting of three battalions, and that the Scots Guards should revert to an earlier title of the Third Guards. To ameliorate the feelings of the two

Guardsman Jones, Prince of Wales Company, turned out in review order, ready for inspection by HRH The Prince of Wales, Colonel, Welsh Guards, 1928. *(Welsh Guards Archives)*

disbanded regiments, the Irish Guards would become the 3rd Battalion of the new Third Guards (at the time the Scots Guards only contained two battalions), while the Welsh Guards would be reduced to company strength within the Grenadier Guards. Economy was the single reason: the War Office argued that the new arrangement would save £100,000 a year. Money would also be saved by closing down the Regimental Headquarters and establishments of the two disbanded regiments. In an attempt to explain the situation the War Office issued a statement confirming that a discussion paper existed but that it had not 'directed attention solely or especially to the Brigade of Guards but has considered that force as part of a general review for its object the promotion of every possible economy consistent with the maintenance of a thoroughly efficient army of sufficient strength for the requirements of Country and Empire'.

However, the following day *The Times* added a different version when its military correspondent suggested that the change was being driven by the hostility of the Grenadiers as they 'have never viewed the younger regiments with favour'.[3] This conclusion

HRH The Prince of Wales inspects the Corps of Drums, Wellington Barracks, 1934. *(Christina Broom Collection/Guards Museum)*

Marching Party, Coronation of HM King George VI, 1937. *(Welsh Guards Archives)*

was given added impetus when it was further revealed that the paper seen by Williams had prompted two separate reports on the future structure of the Foot Guards regiments. The first, a majority report, supported the proposed changes and had been signed by the Regimental Lieutenant Colonels of the regiments involved – Field Marshal the Duke of Connaught (Grenadier Guards), Lieutenant General Sir Arthur Codrington (Coldstream Guards) and Field Marshal Lord Methuen (Scots Guards). It was also supported by the Major General, Sir George Jeffreys (Grenadier Guards). A second, minority, report opposing the plan, had been signed by Field Marshal Lord French (Irish Guards) and Colonel William Murray-Threipland (Welsh Guards). Both reports had then been submitted to King George V.

Not unnaturally the news attracted immediate protests in Wales and friends of the Regiment stepped forward to voice their support. Now retired as Major General, Sir Francis Lloyd wrote to the editor of *The Times* describing the decision as 'not only a crime, but a blunder of the worst description' and that there was 'every reason for maintaining this great regiment in its entirety'.[4] With feelings running high, Welsh Members of Parliament put aside party differences and demanded a meeting with Lloyd George to put forward the national case for retaining the Regiment. They also demanded the release of accurate establishment and recruiting figures to prove that the Welsh

Corps of Drums, 1927. These two drummers later became WO1 (RSM) A. K. Baker, MBE *(left)*, and CSgt Williams, BEM. *(Welsh Guards Archives)*

Inspection of the Battalion by HRH The Prince of Wales, Wellington Barracks, London, 1934. *(Christina Broom Collection/Guards Museum)*

Guards were up to strength – one of the main points in the majority report was an argument that the Irish Guards and the Welsh Guards had experienced low recruiting and retention figures since the end of the war. This proved to be the Regiment's salvation. When the figures were eventually released by Secretary of State for War Winston Churchill, they revealed that the Welsh Guards' strength was 728 with a nominal establishment of 1,101, while the strength of the three battalions of the Grenadier Guards was 2,317 with a nominal establishment of 3,202. In other words, there was little to choose between the two regiments. Pressed further by Lieutenant Colonel Sir John Hope (Conservative, Midlothian North and Peebles), who had served with 9th King's Royal Rifle Corps during the war, Churchill admitted that during the previous six months 500 recruits had passed through the Foot Guards training depot at Caterham, 200 of whom were Welshmen bound for the Welsh Guards.

The uproar caused by the revelations took the government by surprise. Not only did it come from Wales; the Irish were incensed too, as were the Scots, who did not take kindly to the fact that their national Foot Guards regiment would revert to a numbered title it had last held in 1831. As the *Daily Telegraph* remarked in a leader on 7 June, 'a re-organization which affronts the pride of Scotland, Ireland and Wales is hardly to be pronounced a happy invention, and to offend national and local patriotism, at this moment in the history of the Empire is a policy of which it is difficult to speak with moderation'.[5] The government was in a difficult position. The proposed changes to the

Lewis gun training, Pirbright, 1923. *(Welsh Guards Archives)*

Prince of Wales Company camp on Chobham Common, Surrey, August 1923. *(Welsh Guards Archives)*

Colonel R. E. K. Leatham, DSO

Universally known by his nickname 'Chicot', Robert Edward Kennard Leatham was born in Gloucestershire on 23 February 1885, educated at Eton and commissioned in the Grenadier Guards in 1904. In the First World War he was severely wounded during the First Battle of Ypres and took no further part in the fighting. In 1920 he transferred to the Welsh Guards and took over command of the 1st Battalion four years later. He was Regimental Lieutenant Colonel from October 1928 until September 1934 when he retired from the Army. At the outbreak of the Second World War, he returned to the post and it was said of him that 'he worked untiringly, and the quality and quantity of the officers he recruited were such that the achievements of the three battalions of the Regiment in the Second World War were, in part, due to his work.' He retired in 1942 and died on 11 May 1948.

'Chicot' Leatham was a quintessential Welsh Guardsman with strong family connections to the Regiment. His son Michael served in the Regiment during the Second World War; two nephews, Robert Pomeroy and Tom Pomeroy (sons of his sister Mary), also served in the Regiment; his step-daughter Jean married Sir William Vivian Makins who commanded the 2nd and 3rd Battalions (his brother Paul also served) and Charles Anthony la Trobe Leatham, the son of his second cousin Admiral Sir Ralph Leatham, joined the Welsh Guards in 1936, was wounded at Fondouk in 1943, commanded the 1st Battalion 1955–8, and was Regimental Lieutenant Colonel 1961–4.

Portrait painted by Alex Talbot Rice, after William Orpen. *(Alex Talbot Rice)*

Foot Guards regiments were clearly unpopular, especially in the Celtic countries, yet Churchill was a known supporter of the majority report. (According to his friend Lord Chandos, a Grenadier, Churchill would say of the Brigade of Guards: 'Star, Thistle and Grenade! They should be the only Guardsmen.'[6]) Nevertheless something had to give. Following a further barrage of protests, including a strongly worded statement from the Independent MP Brigadier General Sir Owen Thomas, who called the proposal 'crass stupidity',[7] the government decided to yield. A week later, on 15 June, Churchill rose in the House of Commons to announce that: 'The subject has now been considered by the Army Council. There is no intention of disbanding the Irish or Welsh Guards so long as they are able to maintain their recruiting in such a manner as to preserve the national character of the Regiments.'[8] There the matter rested, although the subject of disbanding the Welsh Guards rumbled on beneath the surface to reappear on several other occasions during the 1920s and 1930s whenever defence cuts were being debated – for example, on 4 February 1930 when Peter Freeman, Labour MP for Breconshire and Radnorshire, asked the Secretary for War 'whether it is his intention to take any steps to vary the strength, organization, or composition of the Welsh Guards?' The answer was 'No, sir' but the doubts were never far away. In March 1926 a meeting of the Welsh Guards Comrades Association had been shaken by the emergence of a rumour, reported in the *Western Mail*, that the Welsh Guards was in danger of being converted to become a machine-gun battalion.[9]

The Battalion cheering HRH The Prince of Wales, Windsor Castle, St David's Day 1933. *(Welsh Guards Archives)*

Convoy escort through Poplar during the General Strike, May 1926. *(Welsh Guards Archives)*

Once the excitements of the immediate post-war period had died down, the Welsh Guards returned to the orderliness of regimental soldiering in London District with its familiar round of public duties and sporting activities. On 3 June 1919, HM King George V appointed his eldest son, HRH The Prince of Wales, as Colonel of the Regiment, thereby beginning a relationship which both sides treasured and from which both benefited. Within the Regiment, Lieutenant Colonel A. G. A. Hore-Ruthven, VC, CB, CMG, DSO, had taken over command of the 1st Battalion in May 1919 with Captain W. A. F. L. Fox-Pitt as Adjutant and the redoubtable William Stevenson, MBE, DCM, MM, returned to his post as Regimental Sergeant Major. The Quartermaster was Captain W. B. Dabell, MBE, MC, another regimental stalwart, who had held the post since the Regiment's foundation. At this time too, the Regiment was joined by an officer who was to exert a profound influence throughout the next two decades – R. E. K. Leatham, Grenadier Guards, the incomparable 'Chicot' Leatham, who went on to command the Battalion between August 1924 and September 1928 and to serve two terms as Regimental Lieutenant Colonel, the first between October 1928 and September 1934 and the second when he came out of retirement from September 1939 to February 1942.[10] It was said of Leatham that 'his experience of soldiering in the Brigade, his judgement of character, charm, personality and fine appearance … made him outstanding not only in the Regiment and the Brigade of Guards, but in the whole Army'.[11]

During this period the Regiment was presented with new Colours at Windsor Castle on 23 June 1925, with *The Times* reporting the following day that the cheers for HM King George V 'were given with enthusiasm and could be heard a mile away'.[12] Later in the year the old Colours were laid up in Llandaff Cathedral.

By the admission of many of who served in the

Gargling parade HM Tower of London, 1937. *(Christina Broom Collection/Guards Museum)*

Marching into barracks at Aldershot after changing stations, October 1931. *(Welsh Guards Archives)*

The 1st Battalion marches past HRH The Prince of Wales at Chelsea Barracks en route for Egypt, 1929. *(Welsh Guards Archives)*

Aldershot Military Tattoo, 1937. Maj H. C. L. Dimsdale to the front. *(Welsh Guards Archives)*

Regiment during the period following the First World War, the conditions were not very different from what they had known in earlier years. As one historian of the British Army put it, 'the old, never really extinguished, conception of soldiering re-asserted itself – a gentleman's occupation that married well with social and sporting life in the countryside – smartness[13] on parade and stiff regimental etiquette and custom'. This manifested itself in many ways; some were useful, others less so. Officers were encouraged to engage in field sports and were not supposed to be seen in the Company Office after midday; this practice continued into the following decade. R. E. W. Sale joined the 1st Battalion in 1933 and he told his son Roddy (also destined to be an officer in the Regiment) that all officers 'worked mostly in the morning inspecting barrack rooms dressed in frock coat'. At each bed they might deal with the men's blankets by piercing the bed block with their sword and shaking it to see if illicit items had been hidden inside. If anything fell out the offending Guardsman paid for a new blanket; if nothing was found the officer had to pay.[14] As Guardsmen often used the space beneath the mattress to conceal civilian suits – a forbidden practice – it could be doubly expensive as the officer's sword invariably did damage. Asked if this were not a bloody-minded and unfair thing to do, an old soldier of the period responded: 'Well, he [the officer] was certainly a hard man, very hard indeed. But he was fair: after all, he did give us the chance to own up first.'[15]

The Transport Platoon, Egypt, 1930. *(Welsh Guards Archives)*

With ceremonial duties being paramount in the Regiment, appearances were important. After the Battalion's return to London District, the Commanding Officer was rebuked by the Major General who had seen 'two men going about like poets with long hair'.[16] As a result, Orders on 22 March 1920 included the injunction: 'In future men will have their hair cut once a fortnight, instead of once a month as at present. This order however does not form an excuse for long hair for any individual whose hair requires cutting more often.' Three months later, Orders on 23 July warned that 'soldiers are not allowed to loiter in the Park or join crowds collected around politicians and other speakers in Hyde Park'. And for cyclists there was the order that, when passing an officer, they should turn their head smartly towards him but should never drop their hands to the side.

The intricacies of a Guardsman's uniform also had to be drummed into recruits until it became second nature. Some idea of the difficulties facing Guardsmen in an age when fabrics were mainly coarse, natural and unyielding can be found in a short memoir written by Harold William Humphries, MBE (later Lieutenant Colonel), from Buckinghamshire, who joined the Regiment in August 1933 after enlisting in Oxford, where he was told quite untruthfully that at five feet ten inches, he was too short to join the Grenadiers or the Coldstream, his preferred choices of regiment. Having been attested in the Welsh Guards, he was sent to the depot at Caterham but even after joining the Battalion he found the niceties of a Guardsman's uniform difficult to master. Greatcoats and capes had to be correctly folded and it took time, patience and considerable dexterity to make sure that it all passed muster on parade.

> Pressing uniforms was a problem as there were no ironing boards or irons as there are these days. The only way we could press our trousers etc was to dampen them and then place them under the bottom blanket and sleep on them. In those days we wore puttees which were three-inch wide khaki bandages which were wound round our legs starting at the top of the boot and ending just below the knee so that the end was on the outside of the leg. Trousers had to have a crease in them which pulled down to cover the top of the puttees just below the knee. Puttees were worn until battledress was introduced during the war! One of the other things we had to sleep on was the blue-grey

Field training, August 1937. Here the battalion escorts the fictitious 'Mugwump of Eastland' on a friendly visit to a rival potentate's territory. *(Welsh Guards Archives)*

> overcoat which had to have knife-edged pleats, where the coat was gathered into the belt for parades, guards etc.! As a result, later on in the Battalion I slept on my blue-grey overcoat, the jacket of my service dress (to get knife-edged pleats in the sleeves), blue full-dress trousers and khaki service dress trousers – quite a mound! In order to get the knife-edge pleats we used to 'soap' the inside of the pleat or crease.[17]

By the time a Guardsman joined the Battalion he understood the importance of keeping up a good appearance. During training at the Depot recruits were allowed to walk out after three weeks and, as Jim Magee from Cardiff remembered from his days at Caterham in 1937, they had to pass a saluting test. Even so, 'we didn't get past the Sergeant of the Guard very often'.[18]

Inevitably perhaps, the slowness of promotion and the general torpor within the Army dominated memories of the inter-war years and characterized them as a time of boredom, frustration and playing for time. Although he was not a Guardsman, the actor David Niven, who was commissioned into the Highland Light Infantry in 1929, recalled his four years of military service, before he left for Hollywood, as a time of the four Ps – 'polo, piss-ups, parade and poking'.[19] For those prepared to look beyond the social scene, though, there were other opportunities. In September 1923 Sammy Stanier was appointed Adjutant of the 1st Battalion. It was, he recalled, 'a wonderful period in my career because the Adjutant, in practical terms, was almost more the Commander of the Battalion than the Commanding Officer'.[20] Stanier was determined that the Battalion's officers should excel at field sports, arguing that they would then receive invitations to ride with the best hunts and in so doing the Welsh Guards would enjoy enormous prestige as a sporting regiment. Being a good soldier as well as an excellent horseman, Stanier was also aware of the military benefits to be found on the hunting field:

> You got to know other people and you got to know what the country looked like. It's the best way: if you ride a horse you see the whole thing in front of you. In a motor car you're going so fast that it's impossible to understand what the land looks like. For example, the good horseman will know to jump a fence near a tree because it will have been weakened by the water dripping from it.[21]

Rugby Football

With its national background it is not surprising that the Regiment should hold one of the best records in the history of Army rugby. It has won the Army Cup (later the Army Premiership Cup) thirteen times and in April 1964, it became only the fourth regiment to perform a hat-trick of wins in the competition. In addition, the 1st Battalion has been runner-up on eleven occasions and the 2nd Battalion was runner-up in the 1946–7 season. This is an impressive record of twenty-five appearances in the final over the 107-year history of the competition.

In the 1920s and 1930s the Regiment produced a number of outstanding players who represented the Army as well as their country. The best known were T. E. Rees, who played for Wales in the 1926 season, and W. C. 'Wick' Powell, although the latter won his twenty-seven caps while playing for London Welsh at scrum-half after leaving the Army to train as an architect. Sergeant Tim Boast, who started his playing career as a soccer player, got a trial for England and Lieutenant (later Brigadier) W. D. C. Greenacre played in a Welsh trial in 1924 and might have proceeded further but for injury. During the Second World War there were further caps for Welsh Guards rugby players – Captain Peter Hastings (England), Lance Corporal G. Williams (Wales), Lance Corporal H. Pimblett (Wales) and Lance Corporal C. Jeffries (Wales). Perhaps the finest player of that period was the great Welsh and British Lions scrum-half Haydn Tanner who served as a PT Instructor with the Training Battalion at Sandown Park. In the 1940–1 season a team from that Battalion won the Middlesex Sevens tournament and also beat an Army XV 8–6 in a game played at the Richmond Athletic Ground on 29 November 1941. Another notable Welsh Guards international scrum-half was Maurice Turnbull who played twice for Wales in the 1933 season before he joined the Regiment.

In the period following the war, the Regiment enjoyed some sustained success, producing a number of notable players who represented the Army, including Lieutenant J. M. H. Roberts, the Army captain in 1950–1 and 1951–2. He was also selected for both a Welsh and an English trial, chose the former and only narrowly missed winning a cap. Others who represented the Army during this period were Sergeant M. Jones, Lance Corporal J.

Anderson, Lieutenant P. R. G. Williams, Captain J. B. B. Cockcroft, Lance Sergeant Don Hearne, Lieutenant Charles Guthrie, Lance Sergeant Wally Powell, Lance Corporal Dai Bowen and CSM A. Dando.

Mention should also be made of the British Army of the Rhine (BAOR) which came into being in the aftermath of the Second World War and which quickly spawned its own competitive cup competition. When the Army Cup competition resumed in the 1946–7 season, it was arranged that the final should be played between the winning UK team and the champions of BAOR, an arrangement which became the norm after 1952. In 1992, as a result of the end of the Cold War and troop reductions in Germany, BAOR was re-designated British Army (Germany). During the period of BAOR, 1st Battalion Welsh Guards won the BAOR cup in 1951, 1952, 1953, 1962, 1963 and 1971. They were runners-up in 1961, 1972, 1979 and 1985.

The twenty-two-year gap in final appearances between 1987 and 2009 was probably due to a lack of representation at Senior Army level. This was addressed during the period 2007–11 when WO2 Andy Price 80 was Army Head Coach, with WO2 Byron Cordy as Senior Army Team Manager and WO2 Conrad Price 13 as the team analyst. Army caps quickly followed in 2007. Guardsman Robert Sweeney was capped at outside half on the back of holding the Swansea record for most points scored in one game. Guardsman Melvin Lewis was also capped in 2007, going on to become the Army skipper, and he was joined by Lance Sergeant Scarf and Lance Sergeant Matthew Dwyer, with the latter being appointed as the Army skipper in 2014. WO1 Andy Campbell was appointed as the 2014 Head Coach of the Army Under-23 squad. This investment in players, coaching and management has seen a resurgence in Welsh Guardsmen being capped for the Army and also in success in the Army Premiership Trophy.

1st Battalion Welsh Guards – Army Rugby Cup Wins

Year	*Opponents*	*Result*
1923	2nd Bn, Welch Regiment	6–0
1932	2nd Bn, Leicestershire Regiment	11–3
1934	5th Bn, Royal Tank Corps	4–0
1952	Depot & Training Regt, RAMC	14–0
1962	1st Bn, Duke of Wellington's Regiment	9–3
1963	28 Coy, RAOC	9–6
1964	1st Bn, Somerset & Cornwall Lt. Inf.	25–3
1970	1st Bn, Royal Regiment of Wales	18–6
1971	1st Bn, Royal Regiment of Wales	6–3
1973	7 Signal Regiment, Royal Signals	22–9
1982	21st Engineer Regiment, RE	12–6
2011	2nd Bn, Royal Welsh	28–9
2013	39 Engineer Regiment, RE	11–3

The Battalion Rugby XV, Winners of the Army Cup, 1923. *Standing:* (*L to R*) Mr T. H. Vile (Welsh Rugby Union), Gdsm H. Griffiths, Gdsm H. Kavanagh, Gdsm G. H. Sheppard, Gdsm A. Comley, Gdsm W. H. Fisher, Dmr A. Gumbley, Gdsm W. Slimmons, Gdsm J. W. Davies, Capt G. C. H. Crawshay. *Seated:* Gdsm W. E. Morgan, Gdsm W. C. Powell, Lt W. D. C. Greenacre, Lt G. D. Young (Capt), Sgt F. A. Pates, Gdsm W. J. Murphy, Gdsm V. O. Griffiths. *(Welsh Guards Archives)*

Welsh Guards Uniforms Inter-war Period.
(From illustrations by Charles C. Stadden and Eric J. Collings commissioned by the Regiment)

1921, Lieutenant, Guard Order, Austerity Tunic. After the Great War, there was extended controversy about the reintroduction of full dress uniforms across the Army. In the end reissue was restricted to the Household Brigade. 1WG wore home service clothing for the first time at Warley Barracks in the autumn of 1920. Officers' bearskin caps are taller and more tapered than those of other ranks. They also have a 9 in. plume of cut feathers on the left side: 4 in. white, 2 in. green and 3 in. white. The tunic illustrated here is the 1920 austerity pattern, which is plainer than the pre-1914 pattern, and lacks the solid panels of gold lace and embroidery on the collar, cuffs and skirt. Trousers are barathea, shaped to fit over the boots. They have a 2 in. red stripe. The crimson silk waist sash has a bow with two tasselled heads on the left side.

1927, Lieutenant Quartermaster. The Quartermaster's cocked hat is made of black silk velour on a felt base. Its right or 'cock' side is decorated with a black pleated silk cockade, a double gold lace loop and button, and diagonal black bands of oak leaf pattern braid. The 5 in. plume with white and green swan feathers is worn on the right side. The tunic is the reintroduced pre-1914 pattern which features more gold lace than its comparatively austere predecessor.

1923, Field Officer, Mounted. The colonel is in guard of honour order. His austerity tunic is the same pattern as that of a junior officer, except for the badges of rank. The waist sash is made of gold and crimson silk. He is wearing butcher boots with brass spurs unique to officers of the Foot Guards. The saddle is the universal type with plain stirrups and blue web girths. The wallets have black bearskin flounces. Mounted field officers of the Foot Guards have their rank badges embroidered at each hind corner of the blue saddle cloth or shabraque, which is edged with two stripes of gold lace. Bridoon and bridle are of brown leather, with brow band and rosettes of blue silk. The brasses on the branch bit and the breastplate carry the regimental device: a leek surrounded by a motto circlet, and ensigned with the crown.

1926, Drummer, Guard Order, London. The Corps of Drums uniform has a character of its own. The drummer's bearskin cap is the same squarer pattern worn by other ranks, made with dyed fur and with a 6 in. white and green bristle plume. The distinctive white worsted Foot Guards pattern tunic survived unchanged despite inter-war austerity measures, since its lacing was not made of gold. It has the blue fleur de lys motif, which commemorates the lilies of France, which were carried on the Royal Arms until 1801, when all claims to sovereignty over France were laid aside. The front of the drummer's tunic is trimmed with horizontal button loops in two groups of five with lace covering every seam back and front. The collar, shoulder straps and wings are edged with lace. Since the collar and wings have blue and white fringes, leek badges are worn only on the epaulettes. The side drum with its rolled ticking cover is the old pattern, its head tensioned by the laborious operation of 'pulling up' the white buff lugs on the zig-zag ropes. Grouped on either side of the Royal Arms, painted on its brass shell, are ten of the twenty battle honours won by the Regiment in the Great War.

1928, Ensign, King's Birthday Parade. In 1928, the Regiment provided the Escort to the Colour for the first time. The King's Colour trooped the same year was one of the second pair presented to the Regiment in 1925. The antique system of Colours still preserved in the Foot Guards is the reverse of that of the line infantry regiments. The senior field officer's colour, known as the Sovereign's Colour, takes precedence over the 'Great Union' flag, which is thus the Second or Regimental Colour. To maintain their individuality, the Foot Guards have successfully appealed against all previous Clothing Warrants and King's/Queen's Regulations, except in 1868, when a reduction in the size of Colours was accepted. The King's Colour illustrated is of crimson silk bearing a golden Dragon of Wales passant. Beneath the dragon is a scroll inscribed with the regimental motto, *Cymru am Byth*. The Colour pike is 8 ft 7 in. long and has as its finial a gilt Royal Crest, from which laurel wreaths are hung on special occasions, or on the anniversaries of honours granted. The ensign is wearing guard of honour order with the gold and crimson sash created for state occasions.

Stanier was also convinced that those officers who hunted made better commanders in the field because they had a good feel for the lie of the land, being natural countrymen. The other great sport encouraged by the Regiment was rugby football and success soon became synonymous with the Welsh Guards. The game had been introduced to Wales in the late nineteenth century and following the foundation of the Welsh Rugby Union in 1881 it became a national pastime, attracting large numbers of players and supporters for club and international matches. From the outset close links were established between the Regiment and the London Welsh Rugby Football Club. Success came in the 1922–3 season when the Regimental Rugby Team, captained by Lieutenant Gavin Young, won the Army Cup, defeating 2nd Battalion Welch Regiment 6–0 in a close match at Aldershot. Lieutenant Young captained the Army team against the Royal Navy in the following year. Other successes in the Army Cup came in the 1931–2 season with an 11–3 win over 2nd Battalion, Leicestershire Regiment, and a narrow 4–0 win over 5th Royal Tank Corps in the 1933–4 season. Two Welsh Guardsmen represented their country during this period: T. E. Rees and W. C. 'Wick' Powell, although the latter won his twenty-seven caps after leaving the Army.

Success of a different kind came the Regiment's way when one of its wartime Guardsmen, Watcyn Watcyns, a former miner from Ton Pentre in the Rhondda valley, received widespread praise for his rich bass-baritone voice following an invitation to give a number of concerts in London between 1923 and 1926. Later he became a renowned recording artist and concert performer, especially of traditional English folk songs, his rendition of 'Drake's Drum' being particularly admired. In time he became a wealthy and well-known personality but he never forgot that his success had begun while serving in 1st Battalion Welsh Guards. As a former member of the Regimental Choir, Watcyns's talent had been first recognized by his company commander, Captain Geoffrey Crawshay, who had encouraged him to attend the Royal Academy of Music.[22]

In Watcyns's company commander the Welsh Guards had an officer who influenced the Regiment's excellence at both rugby football and at choral singing. Crawshay was the great-great-great-grandson of Richard Crawshay, the ironmaster who oversaw the first major expansion of the Cyfarthfa Ironworks and in time became one of Wales's great social benefactors.

The Battalion marches past HM King George V and HM Queen Mary, Windsor Castle, during the Presentation of New Colours, 1925. Lt Col R. E. K. Leatham, DSO (Commanding Officer); Capt Sir Alexander Stanier, Bt, DSO, MC (Adutant). *(Welsh Guards Archives)*

One of the first officers in the new Regiment, Geoffrey Crawshay not only founded the Regimental Choir but raised the first rugby XV. Having been badly wounded at Loos, he left the Army in 1924 and devoted the rest of his life to carrying out good works in Wales. His lasting monument is Crawshay's Welsh Rugby Football Club, an invitation side renowned for its emphasis on playing attacking rugby, which came into being in April 1922 when Crawshay took a Welsh Guards XV to Plymouth to play Devonport Services. His other great interest was the National Eisteddfod of which he held the office of Herald Bard, a key figure in the Gorsedd Beirdd Ynys Prydain, the 'Assembled of Bards of the Isle of Britain'. During his time in the Regiment and in retirement, Crawshay encouraged the Regimental Choir to participate at the National Eisteddfod, a tradition that had begun at Neath in 1918, the first time that the Welsh Guards had appeared at the event.

The role played by the Regimental Choir is noted elsewhere but it would be wrong not to acknowledge the sheer delight which generations of Welsh Guardsmen have taken in communal singing and the impact that this has had on morale, often in taxing situations. Crawshay remembered one such moment in September 1915 when the Regimental Choir gave a superb concert at the Army's Machine Gun School at Wisques near St-Omer in the presence of the Prince of Wales, but he also acknowledged that the real impact was always made from the heart of the Battalion before going into the attack or after battle. One well-remembered occasion occurred after the bloody engagement at Gouzeaucourt in November 1917 when the singing of 'In the sweet bye and bye, we shall meet on the beautiful shore' brought a hush over the survivors:

> ... when the shattered Battalion was withdrawn to a wood behind the village. The singers were hidden amongst the trees in the moonlight and the air was frosty and still. This was not a concert, but a message, a song of hope and faith. There were many similar dramatic moments.[23]

One of the few reminders of the changing times came in the spring of 1926 when the 1st Battalion was required to take part in giving aid to the civil power during the General Strike, a nine-day national stoppage which saw almost two million workers withdraw their labour in protest at conditions in the mining industry. On 8 May, Welsh and Coldstream Guardsmen wearing steel helmets and active service uniforms guarded a convoy of 158 motor vehicles, 'each of them heavily laden with bags of flour', as it made its way from Hyde Park to the docks where arrangements had been made by the Board of Trade for the flour to be distributed to London's bakeries by volunteer workers. This was a key moment in the strike as picket lines were broken during the operation and the show of strength was an attempt to underline the fact that the government was in control of the situation. Trouble had been anticipated but, as a Welsh Guards officer told a newspaper reporter, 'there was no sign of any hostility, no demonstration of any description, and the great majority of the crowd seemed to regard the armoured cars and the lorries merely as a Sunday morning diversion that did not often come their way'.[24] The occupation of the London docks by the Welsh Guards and Coldstream Guards was one of the factors that led to the collapse of the General Strike on

Captain Sir Geoffrey Cartland Hugh Crawshay

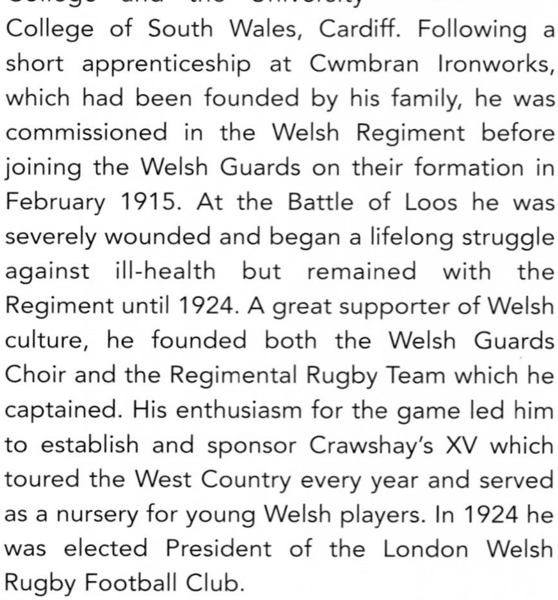

The son of Codrington Fraser Crawshay of Llanfair Grange, Abergavenny, he was born on 20 June 1892 and educated at Wellington College and the University College of South Wales, Cardiff. Following a short apprenticeship at Cwmbran Ironworks, which had been founded by his family, he was commissioned in the Welsh Regiment before joining the Welsh Guards on their formation in February 1915. At the Battle of Loos he was severely wounded and began a lifelong struggle against ill-health but remained with the Regiment until 1924. A great supporter of Welsh culture, he founded both the Welsh Guards Choir and the Regimental Rugby Team which he captained. His enthusiasm for the game led him to establish and sponsor Crawshay's XV which toured the West Country every year and served as a nursery for young Welsh players. In 1924 he was elected President of the London Welsh Rugby Football Club.

Crawshay was also an enthusiastic supporter of the National Eisteddfod and a member of the Gorsedd under the bardic title 'Sieffre o Gyfarthfa', and until 1947 he was an impressive mounted herald bard. He died on 8 November 1954.

The Corps of Drums at Wellington Barracks. In the centre of the front row is Capt F. A. V. Copland-Griffiths, MC, Adjutant, October 1924–October 1927. To the Adjutant's right is WO2 DMaj Osbourne, Drum Major, January 1926–December 1932. To the Adjutant's left is RSM Stevenson. To 'Stevo's' far left is 2731279 E. Burdett who replaced WO2 Osbourne as Drum Major. *(Welsh Guards Archives)*

13 May and, despite the fact that many Welsh Guardsmen came from mining communities in south Wales, there was no truth in rumours that the 1st Battalion was on the point of mutiny. These were written off by the authorities as propaganda by the Communist Party of Great Britain in an attempt to cause unrest in the police and armed forces.[25]

Otherwise, for the most part, ceremonial duties were still very much the order of the day for all Guards regiments based in London during the post-war period. On Monday, 4 June 1928, the Welsh Guards trooped the colour for the first time with the Regimental Lieutenant Colonel, T. R. C. Price, CMG, DSO, as the Field Officer in Brigade Waiting. Eleven years earlier HRH The Prince of Wales had become Colonel and he later recounted his memory of his first experience of being on parade:

Having accoutred myself with considerable care at York House, I mounted my horse and rode over to Clarence House to fetch my great-uncle [HRH The Duke of Connaught, Colonel Grenadier Guards] and accompany him to Buckingham Palace to join the King's procession to the Horse Guards. He was waiting for me on the steps. As I saluted, his eagle eye darted over my uniform to rest finally for interminable seconds at my waist. 'My dear boy,' he said coldly, 'don't you realise that you are improperly dressed? You are in "guard order" when you should be in "review order".' … I turned my horse around and trotted shamefacedly back to York House, where, without dismounting, the correct accoutrements were girt around me. The idea of having thus saved

Welsh Guards Uniforms Inter-war Period.
(From illustrations by Charles C. Stadden and Eric J. Collings commissioned by the Regiment)

1930, Regimental Sergeant Major, Drill Order. The 1920s was a politically controversial time for the Regiment, with Winston Churchill seeking its disbandment. A second attempt was made to disband the Regiment at the end of the 1920s. It had long been customary to send regiments overseas when they became controversial and so in 1929 1WG embarked on a two-year tour of Egypt. The battalion arrived at Kasr-el-nil Barracks in Cairo in May 1930. The helmet worn by the Regimental Sergeant Major is the late Wolseley pattern foreign-service helmet, of compressed cork covered in khaki drill. This helmet was taken into Army service at the time of the Indian Mutiny. Its folded cloth pagri was originally detachable and worn with one end hanging loose to protect the neck.

1930, Lieutenant, Review Order. The helmet is of the same pattern for other ranks, with a plume of similar nature to the one worn in a bearskin cap, but half the size. Field officers wore breeches and boots. At the end of the battalion's first year of service in Egypt, HRH The Prince of Wales, the Colonel of the Regiment, visited and inspected the battalion in Cairo.

1930, Guardsman, Residency Guard. The Battalion took its turn mounting guard at the residence of the British High Commissioner, overlooking the Nile in Cairo. Trousers were worn in place of shorts.

1930, Guardsman, Training Order. The Battalion excelled at musketry and won several competitions in Egypt.

1929, Lance Corporal, Regimental Police. The lance corporal is in drill order with buff waist belt and sidearm. He carries a regimental cane of jointed polished bamboo, and wears his rank badges on both sleeves with good conduct badges on the left cuff. Since the reign of Queen Victoria, Foot Guards lance corporals have worn two chevrons in place of a single one.

The Regimental Choir

The first recorded mention of a Welsh Guards Choir was in May 1915 when two organized choirs were formed, the first with the 1st Battalion at Esher and the second at the Guards Depot at Caterham. A number of public concerts were given there, as well as in the Park Hall in Cardiff on 24 July 1915. Later the Reserve Battalion also formed its own choir.

An early highlight was the Regimental Choir's first appearance at the National Eisteddfod, held in Neath in August 1918. Amidst great local excitement they were played to their quarters by the Band of the Welsh Guards under Lieutenant Andrew Harris, LRAM, and were subsequently invited to take part in the Ceremony of the Gorsedd, joining in the procession and forming a circle around the Maen Llog (Logan Stone). On 8 August the choir took part in the male voice competition, their test piece being 'Here's to Admiral Death', composed by Dr Vaughan Thomas. Although they came second to the winners, the incomparable Williamstown Choir, it was generally agreed that they had acquitted themselves superbly, given their short existence and lack of time for practice.

Operational service in the Second World War prevented the Choir from appearing in public on a regular basis and it was not until 1957 that it was revived with the arrival from the Regimental Band of Sergeant W. H. Carpenter. Under his direction a number of public engagements followed and in 1960 the Choir returned to the National Eisteddfod at Llangollen where they achieved third place in the male voice competition – a creditable performance given the international standard of the participating choirs. This was followed by the release of their first LP record and a performance on the BBC Light Programme's popular *Friday Night is Music Night*. This set the seal for the many public performances and broadcasts which followed, even when the 1st Battalion deployed to West Germany later in the decade.

The Choir remained vibrant in the 1970s and 1980s, with a large membership regularly singing in four parts, and successful rehearsals dependent on musical support and the free flow of beer to ease vocal chords and inhibitions. During the following decade the Choir performed twice at the Festival of Remembrance in the Royal Albert Hall in London, and continued to perform during the deployment at Ballykelly in Northern Ireland. In the period 2002–12 the pace of operations saw a falling off in the Battalion's ability to sustain a choir, but No. 3 Company (Major Llewellyn-Usher and CSM Williams 205) created a company choir during Operation Herrick 16, which continued to flourish well beyond the tour. It is hoped that this may be the basis for a successful Regimental Choir in the future.

The Battalion choir, 1922. *(Welsh Guards Archives)*

Welsh Guards Uniforms Inter-war Period. *(From illustrations by Charles C. Stadden and Eric J. Collings commissioned by the Regiment)*

Early 1936, King's Guard, Guard Order. The Battalion inherited the Slade Wallace equipment of 1888 at an intermediate stage of its evolution. In 1905 the black leather valise worn as a large pack had given way to a folded greatcoat worn above the shoulders. The badge from the old valise, now termed 'greatcoat ornament', was worn on the centre strap. Capes were carried to the back of the waist belt, rolled with coat straps and fastened by brace straps with the ends neatly rolled on top. Considerable effort was required to maintain this equipment and ensure that it looked smart on parade.

1934, Captain, Mess Dress. Despite austerity measures introduced after the end of the Great War, the War Office insisted that all officers in the Foot Guards should be in possession of mess dress by January 1922. A dark blue field service cap, piped in gold, was worn with it.

1937, Time-beater, Regimental Band. The proper designation for bass drummers of the Foot Guards. Other instrumentalists are called musicians, not bandsmen, as is the case in regiments of the line. The uniform illustrated here, which no longer survives, was worn by both bass and side drummers. A drum carriage of blue facing cloth, edged with gold, was worn over both shoulders, crossed at the back and fastened centrally above the waist belt by means of a snake hook. The drum is suspended from a brass hook at the collar, and a scarlet cloth is worn to protect the tunic front.

1933, Drum Major, State Dress. The drum major's state clothing has changed little since the mid-seventeenth century. It is shown here as worn at the King's Birthday Parade in 1933, when the Battalion found No. 6 Guard and lined the Mall. The jockey-style cork cap is covered with blue silk velvet, similar to that worn by musicians and trumpeters of the Household Cavalry. It is a Royal Household livery. The knee-length state coat is of crimson cloth, heavily braided with gold lace and gimp. Around the waist is a sash of crimson silk, edged with gold thread fringes and fastened with a gilt regimental brooch. Over the shoulder is a belt or 'baldric' in the blue facing colour of the regiment. It is lined with red leather, edged with gold lace and elaborately embroidered with the royal crest, military emblems, the regimental badge and battle honours. Two miniature drumsticks fit into gold loops. Underneath all this are knee-length blue trousers and white cloth gaiters which fit over the boots in the manner of spats. The drum major's staff is large with a heavily ornamented silver head. Foot Guards drum majors never twirl the staff or throw it in the air.

Late 1936, King's Guard, Guard Order. During his brief reign, King Edward VIII approved a significant change to guard order. The cape replaced the greatcoat, folded and secured in the same manner, with the brass cape ornament in the centre. With the pouches discarded, both arms drill and marching became more precise.

> me from so embarrassing a military error evidently gave him satisfaction. At every Birthday Parade thereafter he never failed to remind me of the incident. 'Do you remember that day when you started the parade in the wrong order? Wasn't it lucky I spotted it in time?'[26]

Despite that small sartorial mishap the new Colonel 'showed the keenest concern in the welfare of the Regiment and always displayed a lively interest in our activities'.[27] This extended from wandering into Wellington Barracks looking for company to playing golf and polo with the Regimental teams and attending field training days. Scrapbooks held by Regimental Headquarters also attest to the Colonel's prowess at steeplechases and race meetings but he always had to be aware of military protocol. As the Prince of Wales told Murray-Threipland, he was 'Colonel-in-Chief of several other regiments, and though naturally being a Guardsman, the Welsh Guards will always be my special regiment, I have to be careful not to make this too evident and cause jealousy'.[28] The connection continued after the Prince became King Edward VIII in 1936, but ended that same year when he abdicated following his decision to marry Mrs Wallis Simpson. Even so, as Duke of Windsor, he remained fond of his 'special regiment', not only by keeping in touch but by wearing the Regiment's service dress when he was serving in the rank of major general during the Second World War.

Another landmark was reached in the following year when 1st Battalion Welsh Guards deployed to Egypt, thereby becoming the first Foot Guards battalion to serve there in the period between the two world wars. The garrison, British Troops Egypt (BTE), was the largest outside India and it consisted of the Cavalry Brigade, Canal Brigade and Cairo Brigade, each commanded by a temporary brigadier (colonel) and the whole under a lieutenant general's command. The main British bases were at Abbassia on the outskirts of Cairo, Kasr-el-Nil (in central Cairo), Moascar (Ismailia) and at Alexandria. Britain had had a strategic interest in Egypt since 1869 when the Suez Canal was opened, a move that speeded up voyage times to India and the Far East but which had to be protected to safeguard British interests. Matters came to a head six years later when Britain gained a controlling interest in the canal after Prime Minister Benjamin Disraeli arranged the purchase of the shares owned by the Khedive, the Egyptian ruler under the Ottoman Empire. This relationship was cemented further in 1882 when a group of Egyptian Army officers led by Arabi Pasha attempted a coup against Khedive Tawfiq and this was repulsed with the help of a British expeditionary force under the command of Sir Garnet Wolseley. At one swoop Britain had become virtual ruler of Egypt, even if the Khedive remained in nominal control. Although Egypt achieved independence under King Fuad in 1922, to all intents and purposes it remained a British protectorate and Britain's influence in the country was paramount. Nevertheless, until the late 1930s when the Italians began to threaten Egypt from their own colony in neighbouring Libya, it would be true to say that the Egyptian garrison was a bit of a backwater.

By tradition Foot Guards regiments had not served abroad in peacetime before the First World War, giving rise to an Army saying to describe something unusual as being 'as rare as a Guardsman's sweat in India'. An exception had been made for campaigns such as the Crimean War and the Boer War in South Africa but in the inter-war period all five Foot Guards regiments served at various times in the Mediterranean area, either in Egypt or Gibraltar. On 19 April 1929, resplendent in pith helmets, 1st Battalion Welsh Guards marched out of Chelsea Barracks under the command of Lieutenant Colonel R. T. K. Auld for entrainment to Southampton where the troopship *Neuralia*, 'looking rather like a vulture waiting for its prey' (according to one Guardsman), was ready to take them to Egypt. The Regimental Headquarters Diary described the deployment as a 'short tour of foreign service'. On embarkation the Battalion consisted of 21 officers and 710 other ranks, plus their families.[29] Following an uneventful voyage through the Mediterranean, their first landfall was Alexandria, where special trains waited for the six-hour journey to Cairo. They arrived to find themselves in one of the most severe heatwaves in living memory. The Battalion's home was the sprawling redstone Kasr-el-Nil barracks which backed on to the Nile, and is currently the site of the modern Nile Hilton Hotel. Although in disrepair and possessing neither electricity nor modern sanitation – the barracks

Officers at a point-to-point: (*L to R*) Billy Fox-Pitt, Peter Ackroyd, Cyril Heber-Percy. *(Welsh Guards Archives)*

Recruits Harris, Lefort and Jeffreys, members of the same squad as LCpl Fairbanks, MM, No. 9 Coy, 'Tin Town', September 1919. *(Welsh Guards Archives)*

had been built in the middle of the nineteenth century – the facilities included an open-air swimming pool (of 'dubious quality' according to one Guardsman) and an open-air cinema with regular changes of programme.

The tour was marked by a succession of training exercises in the desert and by a steady round of sporting activities involving other regiments in Egypt. The Battalion won the BTE rugby and boxing cups while the Cricket Team were runners-up in the Inter-Regimental Cricket Cup. Having managed 'to defeat those common enemies, the bug, the shite-hawk and the flies', the Battalion got a lot out of the tour and for many of the Guardsmen it was their first experience of a deployment to an exotic location. Although it had been planned that the Battalion would remain in Egypt for up to two years, the tour came to an end in December 1930 when it returned to Britain to be replaced by 1st Battalion Grenadier Guards. Before leaving the country, a number of farewell parties were held as well as a final horse show at Abbassia where there was 'a little sorrow in the air' at the departure of a battalion which had made its mark in Cairene society.

HRH The Prince of Wales with (*L to R*) Drummer Davies and Sgt Jenkins 99 who accompanied him on his tour of Australia in the early 1920s. This photograph was taken at the Old Comrades Race, at the Battalion Sports Meeting, 22 July 1933. *(Welsh Guards Archives)*

On arriving at Southampton with 19 officers and 741 other ranks, the Battalion was met by the Regimental Adjutant and went into quarters at Warley Barracks, Brentford. For the rest of the decade, under an arrangement known as Annual Change of Quarters, the Battalion moved each year to a succession of new stations, visiting places such as the Tower of London, Albuhera Barracks at Aldershot, Victoria Barracks at Windsor, Wellington Barracks and Pirbright. For the Guardsmen this was part of a familiar routine, but for their wives and families it could be arduous and time-consuming as Regimental

Sergeant Major Stevenson's wife remembered:

> From then onwards we moved every year about October, packing up and laying out the utensils and furniture as per the inventory board for the Quartermaster to check, and wondering what our next quarters would reveal. Of course, no one left theirs as spick and span as you did, or so we imagined, but it was all great fun and I thoroughly enjoyed it all, particularly the Children's Christmas Tree and Party each year.[30]

In 1928 Stevenson had been commissioned as Lieutenant and Quartermaster and was replaced as Regimental Sergeant Major by L. Pownall. Nine years later 'Stevo' retired from the Army.

Other changes were also in the air. By the late 1930s it was obvious to most people that once again the world was drifting towards war. In 1934, following the death of President von Hindenburg, the Nazis had taken full control in Germany under the leadership of Adolf Hitler and their presumptuous territorial claims were soon trying the patience of the rest of Europe. A rash of sabre-rattling quickly followed. The Rhineland was reoccupied in March 1936, thereby breaking the terms of the Treaty of Versailles of 1919, and later that year Germany entered into treaties of friendship with Italy and Japan, the countries which would eventually become the wartime Axis powers. But it was in 1938 that Hitler's aggression increased with alarming intensity and rapidity. On 11 March, German forces moved into neighbouring Austria to complete the *Anschluss*, a political union which was also expressly forbidden by the Versailles Treaty. Emboldened by the lack of any opposition from the western powers, Hitler orchestrated another crisis in the summer by exploiting the demands of German-speaking Sudetens to cede from Czechoslovakia and to join the German Reich, as Germany had become.

Brigade of Guards contingent at King George VI's coronation in 1937. *(Welsh Guards Archives)*

Escort to the Colour, King's Birthday Parade, 1928. (*L to R*) Capt W. A. F. L. Fox-Pitt, MC (Captain of the Escort), 2Lt G. L. R. Jenkins (Ensign) and Lt A. W. A. Malcolm (Subaltern). The Field Officer in Brigade Waiting was Colonel T. R. C. Price, CMG, DSO and the Adjutant in Brigade Waiting was Lt W. D. C. Greenacre, MVO. *(Welsh Guards Archives)*

Recruiting poster,1925. *(National Army Museum)*

HRH Edward, Prince of Wales Colonel, Welsh Guards, 1919–36

From 3 June 1919 to 26 January 1936 HRH The Prince of Wales served as Colonel of the Regiment, a period later recalled by his near contemporary Brigadier J. C. Windsor-Lewis: 'He showed the keenest concern in the welfare of the Regiment and always displayed a lively interest in our activities. What pleasure it gave to see him playing polo for the Regiment in the Madrid Cup, riding with enthusiasm and guts at race-meetings and point-to-points and playing golf for the Regimental team. He was equally keen on other aspects of Regimental life and often came to watch the Battalion training and to sound out life in barracks – quite apart from his formal inspection on St David's Day. His charm of manner, his ability to remember names (and not all of them were Jones, Davies or Thomas) greatly endeared him to all of us. On ascending the throne in 1936 he became our Colonel-in-Chief, but alas we were destined to see very little more of him.' After his abdication in 1936 HRH The Duke of Windsor (as he became) retained links with the Regiment and visited the 1st Battalion for the last time in October 1951.

HRH The Prince of Wales inspects the Battalion, 1935. *(Welsh Guards Archives)*

Colonel W. Murray-Threipland, DSO, painted by Simon Elwes. *(Estate of Simon Elwes)*

At the time Britain had no obligation to defend Czechoslovakia. It was not part of its sphere of interest, there was no treaty with the country and Prime Minister Neville Chamberlain told parliament that there was no point in going to war against Germany 'unless we had a reasonable prospect of being able to beat her to her knees in a reasonable time and of that he could see no sign'. But the claims of *Realpolitik* also had to be addressed. France was in alliance with Czechoslovakia and would have to be supported if a wider European conflict broke out. As the situation deteriorated, Chamberlain decided to regain the initiative by flying to Germany to meet Hitler at his summer retreat at Berchtesgaden in the Bavarian Alps.

It was a parlous mission and a worrying moment for the people of Britain who still had vivid memories of the previous conflict and were desperately anxious to avoid going to war again. As it turned out, it took two further meetings at Bad Godesberg and Munich for Chamberlain to come to an agreement which appeared to force Hitler to back down, albeit without any intention of keeping his word. At a meeting on 29 September, attended by delegates from Britain, France, Italy and Germany (but not Czechoslovakia),

L to R: Maj H. Talbot Rice, Lt Col K. Auld (Commanding Officer), and Capt C. Richmond Brown en route for Egypt, 1929. Side caps were only authorized for wear on ships or in tented accommodation. *(Welsh Guards Archives)*

Mop fighting, Chelsea Gardens, 1920s. *(Welsh Guards Archives)*

the Sudeten Germans were given self-determination within agreed boundaries and for the moment the crisis was over. Chamberlain had bought much-needed breathing space at a time when Britain was not in any position to wage a continental war.

It was, though, the calm before the storm. On 15 March 1939, German forces occupied Bohemia and Moravia, and Hitler was driven in triumph through Prague as Czechoslovakia fell into Nazi hands. At the same time he made threatening noises about Poland, a move which was countered on 1 April by Britain and France, which jointly guaranteed Polish territorial integrity as a trip-wire to deter further German threats. War was now more or less inevitable and the mood of the country began to change. By then second-in-command of the Battalion, Sammy Stanier 'clung to a hope that the Prime Minister would be able to persuade Adolf Hitler to restrain himself', but 'by the spring of 1939, it was becoming clear that that was not going to happen'.[31] The euphoria of Munich became a memory, opposition to rearmament evaporated and appeasement became an unmentionable word. The Committee of Imperial Defence began to plan for war – the first steps were taken to create an expeditionary force to serve in France, and the Royal Navy moved to a war footing. Attempts were also made to woo the Soviet Union into an alliance that would be similar to the Triple Entente of the First World War, but these ended on 23 August when Josef Stalin entered into a non-aggression pact with Hitler, who had already decided that Poland should be his next target and did not want to trigger Soviet animosity. As the month neared its end there was a sense that war was not only imminent but necessary, both to save the country and to check Hitler's growing domination of Europe.

Having signed his peace deal with the Soviet Union, Hitler felt free to invade Poland at the

Guard mounting from Horse Guards in 1937 with the Escort to the Colour found by Prince of Wales Company. Captain of the Escort: Capt W. V. Makins; Subaltern of the Escort: Lt A. C. W. Noel; Ensign of the Escort: 2Lt C. A. la T. Leatham; RSM: WO1 Bray. *(Welsh Guards Archives)*

Drum Major C. Birch photographed for the *Sunday Dispatch*, 1932. His moustache was reputed to be over a foot in length from tip to tip, each tip supposedly rising and falling as he waved his drum major's stave, enabling the bandsmen behind to keep time. *(Welsh Guards Archives)*

beginning of September 1939. A border incident was fabricated and German divisions of Army Group North poured into the 'Polish Corridor', the disputed territory which provided Poland with access to the Baltic while cutting off Germany from East Prussia. Other German formations attacked further south. In response to the German aggression, Lord Halifax, the British Foreign Secretary, sent a curt message to Hitler, informing him that Britain would fulfil its obligations to Poland unless German forces withdrew, but the German leader was not in the mood to respond to firm words. There was a short delay as the diplomatic response had to be finalised with France but, at 09.00 hours on 3 September, Sir Nevile Henderson, Britain's ambassador to Germany, delivered an ultimatum stating that if hostilities did not stop by 11.00 hours a state of war would exist between Great Britain and Germany. Hitler did not respond and a quarter of an hour after the deadline Chamberlain went on BBC radio to announce to the British people that they were at war with Germany.

RIGHT: Lt Col R. Auld by his bivouac tent, in the desert near Cairo, 1930. *(Welsh Guards Archives)*

BELOW: The Officers' Mess in the desert near Cairo, 1930. *(Welsh Guards Archives)*

Field training camp by the Pyramids, Egypt, 1930. *(Welsh Guards Archives)*

CHAPTER FOUR

Phoney War, 1939–1940

1st Battalion, Arras, Dunkirk; 2nd Battalion, Hook of Holland, Boulogne

In September 1939 Chamberlain had had no option but to declare war against Hitler's Germany, but the country's armed forces were hardly in a fit condition to fight a modern conflict. The British Army could only put together four regular divisions as an expeditionary force for Europe, two divisions in the Middle East, four divisions and several cavalry brigades in India, two brigades in Malaya and a modest scattering of imperial garrisons elsewhere. Years of neglect and tolerance of old-fashioned equipment meant that the Army was ill-prepared to meet the modern German forces in battle and British industry was not yet geared up to make good those deficiencies. Once again in the nation's history it seemed that Britain was going to war with the equipment and mentality of previous conflicts.

As part of the hurried expansion of the armed forces which had been put in place after Munich, the Welsh Guards had also been enlarged. On 26 April 1939, King George VI signed a special Order of the Day authorizing the creation of a second regular battalion, which formed at Chelsea Barracks and was stationed initially at the Tower of London. (Other regiments which were strengthened by new or restored 2nd battalions were the Irish Guards, Royal Irish Fusiliers and Royal Inniskilling Fusiliers.) The race was then on to prepare the new battalion for operational service and it moved to camp at Theydon Bois on the edge of Epping Forest in Essex under the command of Lieutenant Colonel Sir Alexander Stanier, Bt, MC. Reservists (officers and men) arrived to rejoin the Regiment and a steady stream of recruits started arriving from the Guards Depot, some of whom would be sent as drafts to the 1st Battalion, which had deployed to Gibraltar on 22 April under the command of Lieutenant Colonel F. A. V. Copland-Griffiths. Many of those joining both battalions towards the end of the year were conscripts. On 2 September parliament passed the National Service (Armed Forces) Bill which made able-bodied men between the ages of 18 and 41 liable for military service 'for the duration of the hostilities'. By the end of December the strength of the Army was 1,128,000, of whom 726,000 were conscript soldiers.

For the new recruits there were other novelties. For the first time in the history of the British Army they were supplied with an active-service uniform, known as a 'battledress'. Including a waist-length blouse jacket with two patch pockets on the chest and voluminous trousers often braced up to the armpits, with side, hip and thigh pockets, it had come into being in 1938 after five years of trials. Initial patterns were made of denim but this was replaced by serge which was considered warmer for operational service

2nd Battalion at Buckingham Palace, 1940. *(Imperial War Museum)*

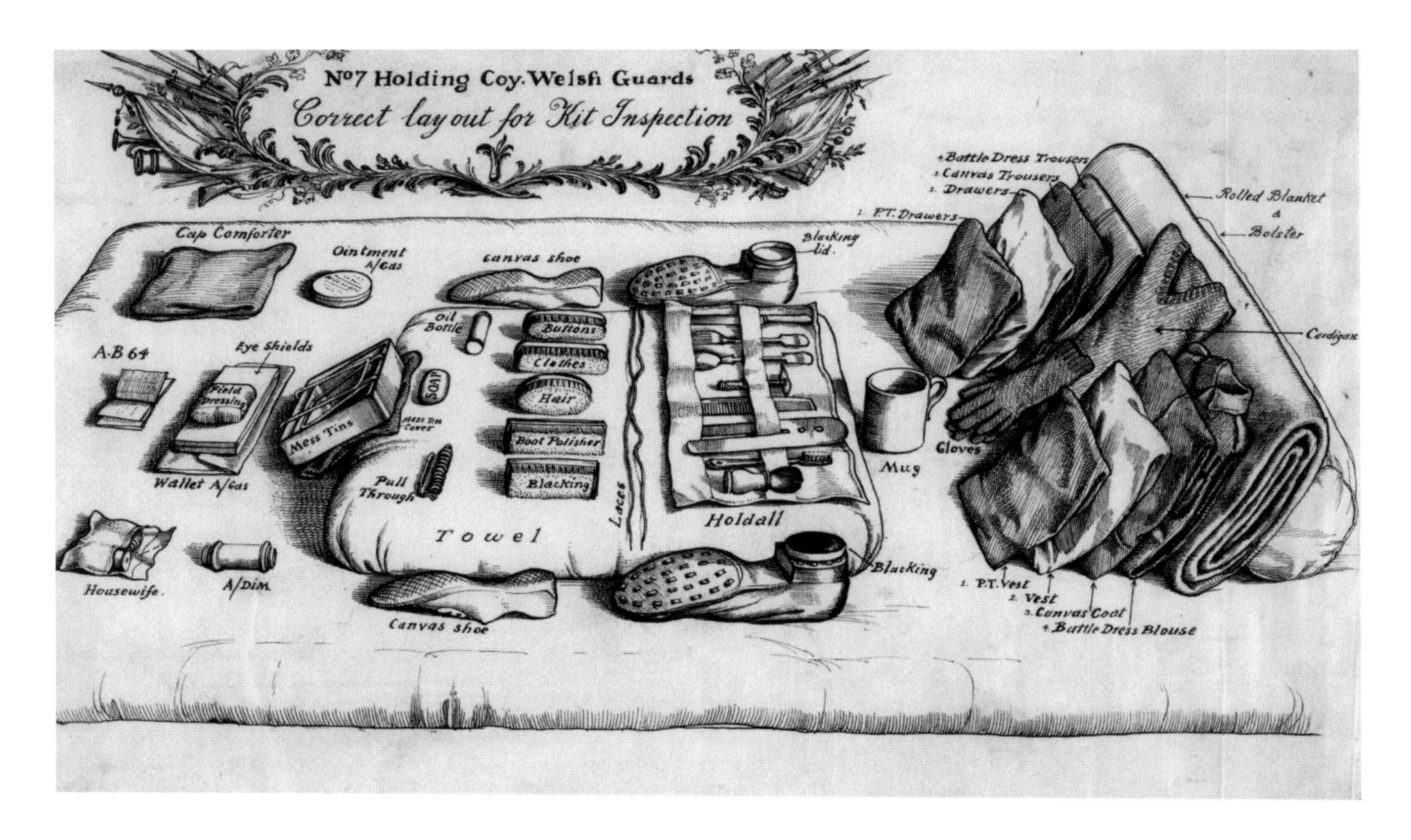

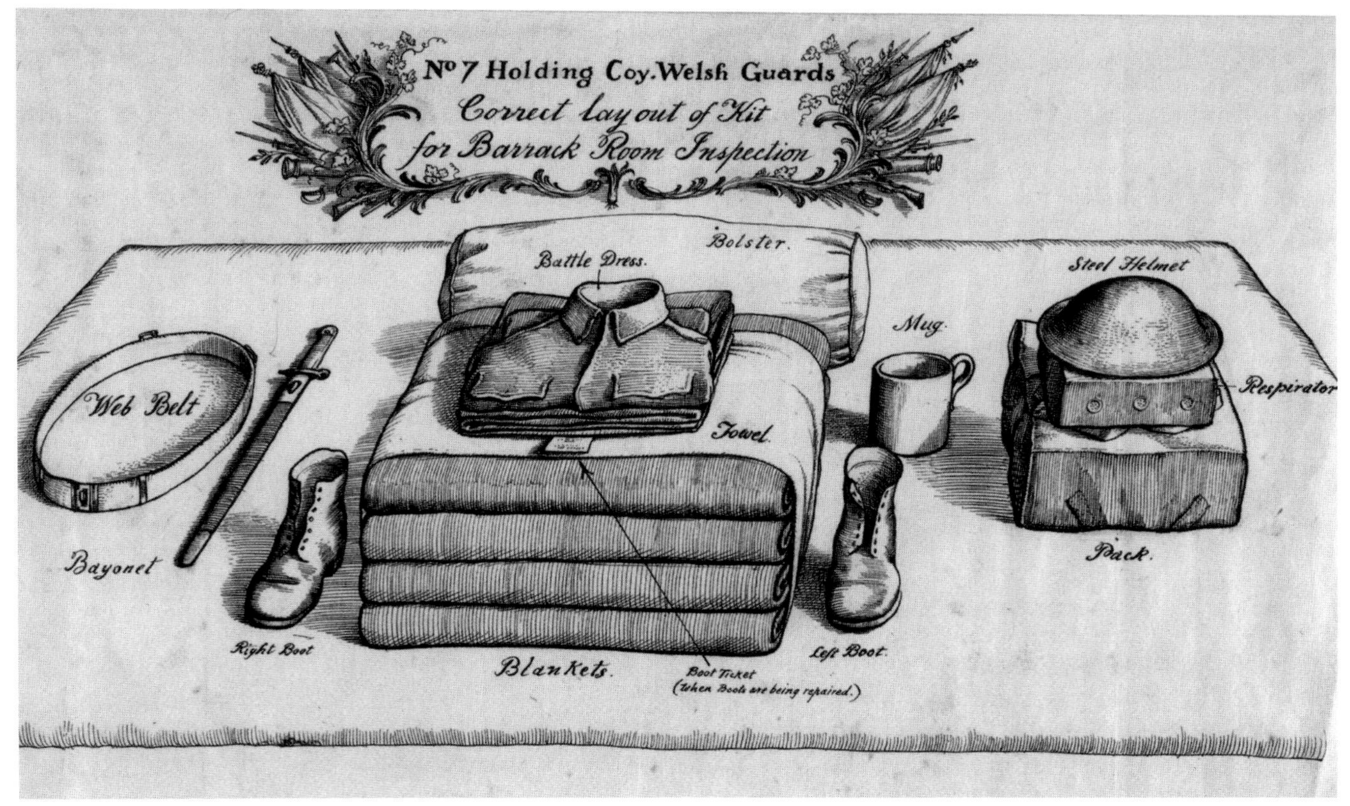

Correct layout of kit for barrack room inspection, 1940. Drawn by Rex Whistler. *(Estate of Rex Whistler)*

in Europe. Guardsmen's blouses were buttoned up to the neck, while officers wore theirs open to expose collar and tie; trousers could be buttoned to the blouse at the waist and were also buttoned at the ankle, further protection being provided first by short puttees and later by canvas gaiters. All Guardsmen were issued with service dress, cap and grey greatcoat, as well as other items such as shirts, underwear and toilet necessities (razor, safety and brush, shaving), but not pyjamas which were only introduced later (see Chapter 8).

Regimental Headquarters was also kept busy during this period interviewing potential officers and sifting through applications from retired officers who wanted to be re-employed and also from younger men seeking emergency commissions for the duration of hostilities. One of the more illustrious candidates was the novelist Evelyn Waugh, who noted in his diary on 18 October that he had been interviewed by 'two delightful officers of enormous age' who 'thought there might be something for me in six months'. Waugh was still a relatively young man at the time and, in common with many others of his age and

2 Lieutenant Gilbert Ryle painted by Rex Whistler. One of the distinguished men who joined the Regiment for the duration of the war,. Ryle's Army service interrupted his career as a philosopher at Oxford. His most famous work, *The Concept of Mind*, was published in 1949. *(Estate of Rex Whistler)*

background, was keen to be commissioned – his diaries speak wistfully of friends 'in uniform' – and had recently missed out on a job at the War Office where his fellow novelist Ian Hay, pen-name of John Hay Beith, had been appointed Director of Public Relations in the rank of major general. However, three days later there came a letter from 'the Welsh Guards unaccountably telling me that their list had been revised and that they had no room for me'.[1]

Waugh's description of the interview is difficult to verify but elements have a ring of truth. On 2 September 1939, with war imminent, Lieutenant Colonel R. E. K. 'Chicot' Leatham had returned for his second stint as Regimental Lieutenant Colonel, replacing Colonel W. A. F. L. Fox-Pitt, who was required for duties elsewhere. From his office in Wilton Crescent in Belgravia, Leatham assiduously scrutinized candidates for emergency commissions, accepting and rejecting according to his own judgement, and bypassing the War Office by not requiring the chosen young men to attend Officer Cadet Training Units. ('I know an officer when I see one and I don't need to be told by one of these new-fangled OCTUs whether he is any damned good.')[2] A great character who attracted a vast number of anecdotes to his name, Leatham is supposed to have rejected Waugh following a private telephone conversation with the Regimental Lieutenant Colonel of the Irish Guards who had interviewed Waugh earlier in the afternoon. In this version Leatham replaced the receiver and informed the hopeful novelist: 'Colonel Fitzgerald tells me you're a shit.' However, at the time the Regimental Lieutenant Colonel of the Irish Guards was not Colonel J. S. N. 'Black Fitz' Fitzgerald, as is recalled in the story, but Colonel the Hon. T. E. Vesey. Another reason for the rejection might have been that Waugh's fate was sealed by wearing suede shoes.[3] Incidentally, although described as being of 'enormous age', Leatham was fifty-six and the Regimental Adjutant, Major C. H. Dudley-Ward, was sixty-seven; at the time Waugh was thirty-six.

More fortunate in his dealings with Leatham was Rex Whistler, who had built up an outstanding reputation as a decorative artist and designer, and who applied for a commission in October 1939, at the same time as Waugh. This was an interesting coincidence as the novelist may have used the artist as the model for the character Charles Ryder in his novel *Brideshead Revisited*, which was published in 1945. For similar reasons, both men were desperately keen to join the Army. Earlier Whistler had applied unsuccessfully for

The Master Cook, Sgt J.W. Isaacs by Rex Whistler, Sandown Park, 1941. One of the Regiment's characters, Sgt Isaacs served with the 1st Battalion and Holding Battalion. *(Estate of Rex Whistler)*

Rex Whistler

One of the country's best-known artists of the inter-war era, Reginald John ('Rex') Whistler was born on 24 June 1905 at Eltham in Kent and educated at Haileybury where he showed a precocious talent for art and set design. He proceeded to the Royal Academy where he was considered 'unpromising' and transferred to the Slade. After he left, it soon became clear that his talents lay in imaginative decoration and he embarked on an extremely successful career as an interior designer, creating impressive wall decorations for the Marquess of Anglesey, at Plas Newydd, Anglesey, and for Lady Louis Mountbatten (later Countess Mountbatten of Burma) at Brook House, Park Lane, London. He was also interested in book illustration and set design, creating scenery and costume for ballets such as *The Rake's Progress* (1935) and *Le Spectre de la Rose* (1944), both for Sadler's Wells.

Whistler could have applied to become an official war artist, but at the outbreak of war he applied for a commission in the Welsh Guards and joined the 2nd Battalion. However, he still kept up a vigorous output as an artist until the invasion of France in June 1944. Many of his paintings are held by the Regiment. According to his entry in the *Oxford Dictionary of National Biography*, Whistler 'won quick popularity by his warmth, his wit, his enthusiastic and persuasive conversation, his charm of manner, and the modest distinction of his bearing. He was not married but children took to him instantly, and he loved them wholeheartedly, spending hours making drawings for their entertainment.' He was killed on 18 July 1944 near the village of Le Mesnil, during his first hours in action.

ABOVE RIGHT: *Self-portrait* by Rex Whistler, May 1940, painted on the terrace of a Regent's Park flat in London which he used as a temporary studio. *(Estate of Rex Whistler)*

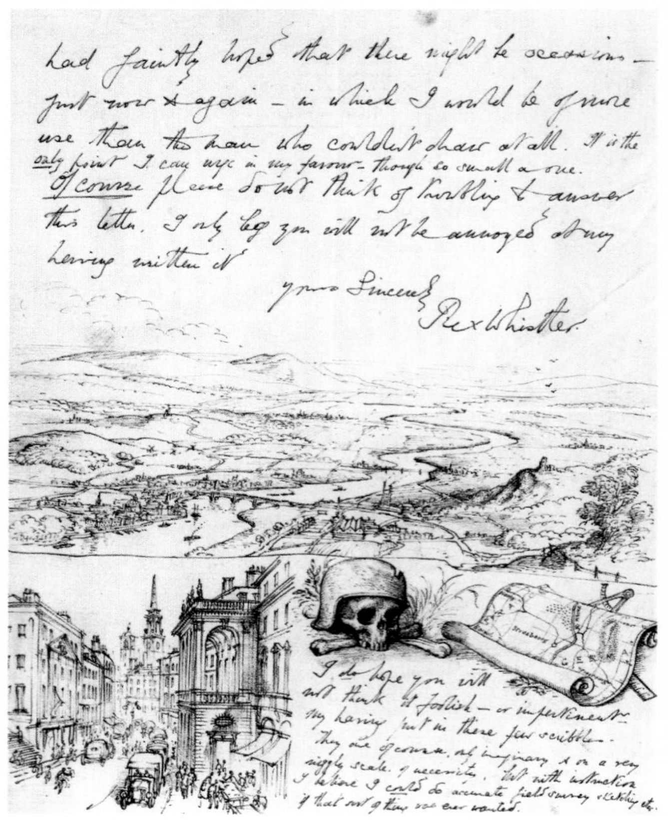

had faintly hoped that there might be occasions – just now & again – in which I would be of more use than the man who couldn't draw at all. It is the only point I can urge in my favour – though so small a one.
Of course please do not think of troubling to answer this letter. I only beg you will not be annoyed at my having written it

Yours Sincerely
Rex Whistler.

I do hope you will not think it foolish – or impertinent – my having put in these few scribbles. They are of course, all imaginary & on a very niggly scale. of necessity. but with instruction I believe I could do accurate field survey sketching etc. if that sort of thing was ever wanted.

RIGHT: An illustrated letter written by Rex Whistler to Col R. E. K. Leatham, dated 19 October 1939, applying for a commission in the Welsh Guards. The letter was written from Mottisfont Abbey, Romsey, Hampshire, the home of society hostess Maud Russel. *(Estate of Rex Whistler)*

Richard Llewellyn Lloyd

Richard Dafydd Vivian Llewellyn Lloyd was born in London in 1906, the son of Welsh parents. After working for a time in the catering industry, he fell out with his father and spent six years in the Army, serving in India and Hong Kong. On returning to London, he started writing under the pseudonym 'Richard Llewellyn', and his first stage play, *Poisoned Pen*, was also made into a film. In October 1938 he moved to Llangollen in Denbighshire, to assist Lord Howard de Walden (Evelyn Scott-Ellis) in his unsuccessful attempt to establish a Welsh national theatre. While in Wales, Lloyd completed his novel *How Green Was My Valley* (1939), which is set in a South Wales mining community, and became an instant bestseller. It was also made into an equally successful film by John Ford in 1940.

That same year Lloyd was commissioned in the Welsh Guards and was posted to the Training Battalion at Sandown Park. During this time with the Regiment he wrote his second novel, *None but the Lonely Heart* (1943), which is set in the East End of London; it was also a successful film, starring Cary Grant. Lloyd worked for the Ministry of Economic Warfare, but towards the end of the war served with the 3rd Battalion in Italy. After the war he returned to writing; he died in Dublin on 30 November 1983.

a commission in the Grenadier Guards but, as he told his friend the Marquess of Anglesey, 'the only remaining chance was the Welsh Gs'. His application to the Welsh Guards was probably also helped by two other factors: a former girlfriend, Penelope Dudley-Ward, was a niece of the Regimental Adjutant and on 18 October Whistler took the trouble to create an illustrated letter which he sent to Leatham outlining the military value of his artistic training. It included a fine sketch of a distant landscape with the skull and cross-bones of a helmeted German officer and it can have done him little harm.[4] During the interview, on discovering that Whistler was an artist, Leatham had said, 'I shall send you to the 2nd Battalion. There's a fellow there called Elwes. He's clever with his fingers too!' Having trained at the Slade School of Fine Art, Simon Elwes had emerged as a leading and much sought-after portrait painter and, after transferring from 2nd Battalion Welsh Guards to 10th Royal

Captain the Lord Mildmay of Flete

Anthony Bingham Mildmay, 2nd Baron Mildmay, was an amateur national hunt jockey who rode in the Grand National before and after the war. In 1939 he enlisted in the Royal Artillery and after a short period in the commandos he was commissioned in the Welsh Guards in 1941. With them he served as a captain in the 2nd Battalion from the invasion of Normandy until the end of the war. He was wounded twice, and mentioned in despatches.

'Nitty' Mildmay enjoyed great popularity as a persistent 'trier' and was often foiled by bad luck. In the 1936 Grand National, while riding the outsider Davy Jones, he was leading at the second-last fence when a buckle on the reins broke and the horse ran out. In 1947 he fell at Folkestone and injured his neck. The accident also gave rise to disabling attacks of cramp, one of which scuppered his chances at the 1948 Grand National when he finished third on his favourite horse Cromwell – named after the type of tank he had used. During his career, he rode thirty-two winners in one season and eight winners at Cheltenham Racecourse, including three at the Festival.

However, Mildmay's most notable legacy was in kindling Queen Elizabeth, the Queen Mother's passion for National Hunt racing. At a dinner in Windsor Castle in 1949, Mildmay sat next to the then Queen Elizabeth and persuaded her that he should buy her a horse, to share with her daughter, Princess Elizabeth. Mildmay's trainer, Peter Cazalet, selected Monaveen for them and he won his first race for them, at Fontwell Park, finished second in the Grand Sefton Chase at Aintree, and then took the prestigious Queen Elizabeth Chase at Hurst Park.

In 1950, Mildmay suffered an attack of cramp while swimming off the south Devon coast and drowned at the age of forty-one.

Major Maurice Turnbull, MC

Born in 1906 into a large sporting family – his father played hockey for Wales in the 1908 Olympics – Maurice Joseph Lawson Turnbull was educated at Downside and Trinity College, Cambridge. A talented all-round athlete, Turnbull excelled in several sports. He captained the Cambridge University cricket team and the Glamorgan County Cricket Club for ten seasons. In rugby union he represented Cardiff and London Welsh and gained two full international caps for Wales in 1933, playing at scrum half. Turnbull also represented Wales at hockey and was squash champion for South Wales. He is the only person to have played cricket for England and rugby for Wales, albeit before he joined the Regiment.

He was commissioned in the Welsh Guards in 1940 and served with the 1st Battalion in the rank of major. He was killed by a German sniper on 6 August 1944 near the village of Montchamp in Normandy.

Major the Revd Hugh Lister, MC

After education at Lancing and Trinity College, Cambridge, Hugh Evelyn Jackson Lister worked as an engineer before taking Holy Orders in 1929. On leaving college he became Curate of All Saint's Church, Poplar, before resigning his position in 1931 to become London Secretary of the Student Christian Movement. In 1934 he became Senior Curate at St Mary of Eton, the Eton College Mission at Hackney Wick, a post he held for three years. While there, he served as chairman of the Hackney branch of the Transport and General Workers' Union and organized a series of strikes in the East End in the late 1930s across a range of industries. He was outspoken in his opposition to the fascist movement which was growing in the East End and was a supporter of military conscription which was introduced in March 1939.

Lister was commissioned in the Welsh Guards in October 1939 after attending Sandhurst, and served with the 1st Battalion in Europe in 1944.

He was killed in action on 9 September 1944 during the Battle of Hechtel when he and Lieutenant John Arnold Alexander Henderson came forward to assist in the fighting to take possession of a house in the west of the village. Lister had wanted to make sure the way was clear for his men but he and Henderson were killed by German machine-gun fire.

Hussars, he became an official war artist, an option which Whistler had refused to consider.

In addition to selecting Whistler and Elwes, who were amongst the best-known artists of the pre-war generation, Leatham cast his net widely until early 1940 when the War Office put a stop to the practice of bypassing official officer-selection procedures. Other officers selected during that brief period of freedom included the controversial philosopher A. J. 'Freddie' Ayer, who had already published his iconoclastic study *Language, Truth and Logic*, which contained the first exposition of logical positivism in the English language; the amateur steeplechaser Anthony Bingham Mildmay, who had narrowly missed winning the Grand National in 1936; and Maurice Turnbull, a Fellow of All Souls, who had played Test cricket for England and rugby for Wales. Other notable officers granted emergency commissions at the beginning of the war included the philosopher Gilbert Ryle, the well-known film actor Anthony Bushell, who had been at Oxford with Waugh, and Richard Llewellyn Lloyd, author of the bestselling novel *How Green was My Valley* (1939). Perhaps the most outstanding (yet unexpected) personality amongst the intake was Hugh Lister, who had worked as a priest in the East End of London and had been an outspoken opponent of the growth of the fascist movement in Britain. He rose to the rank of major; on his death in France in September 1944, the 1st Battalion War Diary said of him: 'All who served with him will always remember him as a man who was truly great.'[5] Shortly before Lister was killed in action at Hechtel, someone asked him if he wanted to be killed since he continually exposed himself to hostile fire. Lister thought for a second before

Newly recruited Welsh Guards officers in the Royal Box, Sandown Park. Painted by Rex Whistler in 1940. *(Estate of Rex Whistler)*

responding with the thought: 'Not really, but I am curious to know what happens on the other side.'[6]

In making its selection of officers, the Regiment also played an unlooked-for role in the creation of the Special Air Service (SAS) Regiment which came into being in the Middle East during the summer of 1941 as a special force trained to operate behind enemy lines. One of these officers was John Steel Lewes, better known as 'Jock', an Oxford rowing blue who had been born in Calcutta and had served as a TA officer before joining the Welsh Guards in 1940. The other was David Carol Mather, who joined the Regiment on the outbreak of war and was trained at Sandhurst before volunteering for service with the newly raised Commando forces at their base at Lochailort in the Highlands of Scotland. Both men ended up serving in L Detachment which formed the basis for the later SAS Regiment. Although the credit for the raising of the SAS Regiment is usually given to David Stirling of Keir, an unconventional Scots Guards officer who argued that small groups of specially trained soldiers could exact greater damage on the enemy's ability to fight than an entire battalion, Stirling also conceded that Lewes should be counted with him as a co-founder.[7] On 10 October 1941 Admiral Sir Walter Cowan, RN, who had defied his age to work with the special forces as a small boats adviser – he was seventy at the time – wrote to his neighbour Lord Glanusk about his pleasure at Lewes's and Mather's activities: 'These two of yours, Jock Lewes and Carol Mather, were outstanding and offered themselves for any and every enterprise and, in particular, the "forlorn hope" sort entailing one or two officers and a handful of men to be pitched on shore

Welsh Guards Uniforms 1939–40. *(From illustrations by Charles C. Stadden and Eric J. Collings commissioned by the Regiment)*

1939, Gibraltar, Review Order. the 1st Battalion was kept busy in Gibraltar on the ranges and building defences. The Wolseley helmet, which had not been used since 1929, was worn for ceremonial parades. It had a brass regimental cap badge with a red cloth backing.

1940, France, Marching Order. From Gibraltar the 1st Battalion went to Marseilles and northern France, marching through Paris en route. From March until May 1940 their role was that of GHQ troops deployed against an airborne threat to Arras. Their new standard uniform was the 1937 battledress and web equipment, the result of lengthy pre-war development. The 1937 web equipment replaced the 1908 pattern. The old cartridge carriers gave way to larger pouches suitable for Bren gun magazines, and cotton bandoliers.

1940, France, Fighting Order. Following the German invasion of Belgium, the Regiment experienced three exhausting weeks of combat. The 1st Battalion defended the BEF's GHQ at Arras before withdrawing to Dunkirk. Meanwhile the 2nd Battalion engaged in a rearguard action in Boulogne that finished on the quays of the port. Officers wore their own brown boots and carried the standard sidearm of the time, the Enfield .38 Revolver, No. 2 Mark I.

1940, France, Fighting Order. From 1937, the Bren .303 Light Machine Gun largely replaced the heavier Lewis Gun. The Bren gun was issued to all platoons at the rate of one per section. Accurate and reliable, it dramatically increased the firepower of infantry battalions. A highly innovative design when adopted, it remained in British Army service until the 1990s. With an integral bipod, it could also be attached to a tripod or pintle mount for use in vehicles, such as the Universal Carrier, seen behind, or for static anti-aircraft fire.

1940, London, Guard Order. The 2nd Battalion made history by mounting King's Guard from the Tower of London. The soldier here is wearing the older pre-war uniform with puttees, but 1937 pattern web equipment. Field training was carried out in Richmond Park. In April 1940, the 2nd Battalion was warned for service in Norway. May 1940 brought further changes with the successful evacuation of the Dutch royal family and British embassy from the Hook of Holland. Shortly afterwards the 2nd Battalion found itself fighting for its life in Boulogne.

Lieutenant John Steel 'Jock' Lewes

Born in Calcutta to an Australian mother and British father, he grew up in Sydney, where he attended the King's School, Parramatta. One of the founders of the Special Air Service Regiment, Jock Lewes was commissioned in the Welsh Guards in the autumn of 1939, having served previously in the Territorial Army with 1st Tower Hamlets Rifles. A product of Christ Church, Oxford, he was an accomplished oarsman and had been president of the Oxford University Boat Club. In 1940 Lewes transferred to the newly created Commando forces and was deployed to North Africa. Disillusioned by the poor training facilities, he joined a small group of volunteers led by Lieutenant David Stirling, Scots Guards, which became known as L Detachment of the Special Air Service Brigade. The main objective of this group was to conduct raids against the lines of communication of Axis forces in the Western Desert. Lewes was killed in action by a German aircraft in December 1941 while returning from a raid in Libya. David Stirling wrote to Jock's father on 20 November 1942 paying posthumous tribute to him: 'Jock could far more genuinely claim to be the founder of the SAS than I ... There is no doubt that any success the unit had achieved up to the time of Jock's death and after it was, and is, almost wholly due to Jock's work. Our training programmes and methods are, and always will be, based on the syllabuses he produced for us.'

In November 2008 a memorial to Jock Lewes was unveiled at Hereford by HRH Prince William of Wales.

Jock Lewes painted by Rex Whistler at Sandown Park in 1940. *(Estate of Rex Whistler)*

and left to shift for themselves in the hope of wrecking an aerodrome or destroying an ammunition dump.'[8]

In the spring of 1941 the *Daily Sketch* was sufficiently impressed by the number of interesting cultural, sporting and academic figures in the Welsh Guards to write a feature article on the phenomenon. Leatham himself had already told one hopeful young candidate: 'So you want to join our Regiment. Well, we've got all sorts of people joining: jockeys, brewers. Polo players, artists; I think we can find room for you later on.'

1st Battalion

Following its deployment to Gibraltar in April 1939, the 1st Battalion settled into the routine of providing a garrison force, strengthening defences and helping to deal with the succession of prize ships brought in by the Royal Navy. At the time, Gibraltar Command consisted of three infantry battalions (1st Battalion Welsh Guards, 2nd Battalion King's Regiment, and 2nd Battalion Somerset Light Infantry), a heavy artillery brigade and supporting troops. For everyone living in Britain and its dependencies during the winter of 1939–40 there was a period which came to be known as the 'phoney war' or 'bore war'. It lasted from the outbreak to the early summer of 1940 when Hitler unleashed his forces to complete the conquest of France, following which the first serious enemy air raids were mounted against the main cities of the United Kingdom. It was a pleasant, even bucolic

Lieutenant Colonel Sir Carol Mather, MC

David Carol MacDonnell Mather spent twenty-two years in the Welsh Guards and was one of the founding members of the Special Air Service. He was educated at Harrow School and Trinity College, Cambridge, after which he went to Royal Military Academy, Sandhurst. Soon after he was commissioned in 1940, he transferred to the newly formed Commando Force. He saw action in North Africa where he joined Colonel David Stirling's fledgling L Detachment of the Special Air Service Brigade.

In October 1942 he was transferred to General Montgomery's staff as a liaison officer, but, unable to resist the chance of further action, he took part in one last raid behind enemy lines in Tripolitania. He was taken prisoner and sent to Italy by submarine. His war seemed to be over, but he managed to escape in September 1943 following the Italian surrender and made his way back to the Allied lines. He returned to Montgomery's staff in North-West Europe and was almost killed when his Auster light aircraft was shot down. After the war he returned to the Welsh Guards and served with the 1st Battalion in Palestine. He retired from the Army in 1962 having reached the rank of lieutenant colonel.

Mather then joined the Conservative Party and was elected Member of Parliament for Esher in 1970. He quickly became known and respected for his trenchant views on membership of the Common Market (which he opposed) and the reintroduction of capital punishment (which he supported). In 1987 he received a knighthood and retired from politics that same year. He died in July 2006.

Capt Carol Mather is in the passenger seat of this Willis jeep, equipped with twin Vickers K machine guns. Capt Gordon Adston, RA, is in the driver's seat. This photograph was taken before Mather's last raid into Tripolitania in December 1942, before he was captured in March 1943. He escaped forty-five days later. Mather was later awarded the MC for his role in the capture of the Nijmegen Bridge. *(Imperial War Museum)*

The 2nd Battalion cheering HM King George VI after being presented with Colours at the Tower of London, 1940. *(Welsh Guards Archives)*

existence on 'The Rock', as Guardsman Eric Coles, 2734458, told his father, even if it did not last long.

> It was on the roof of South Barracks I first heard that war was declared. It was obvious that our stagnant days in Gibraltar were over; no more sunbathing at the only two small beaches; no more visits to Tangiers; no more free fights with any other regiments who wanted to make trouble. Our evenings in the Trocadero and Alameda Gardens would soon only be a memory.[9]

On 7 November 1939, 1st Battalion Welsh Guards moved by HMT *Devonshire* to Marseilles and then on by train to Paris to join the British Expeditionary Force (BEF) whose move to France had been planned as early as February 1939 when the Cabinet had taken the decision to commit its armed forces to a continental role by forming an expeditionary force in support of the French Army. The BEF consisted of four Regular infantry divisions, a mobile division and, later, four Territorial divisions, but it was very much a race against time: by summer 1939 even the Regular divisions had only half of the required numbers of anti-tank and anti-aircraft weapons, and scales of ammunition were completely inadequate to fight a protracted campaign. Nevertheless, plans were pushed ahead for the BEF to take its place on the left of the French Army in north-eastern France with the Territorial divisions joining it as they became 'ready'. In fact the initial deployment was completed fairly quickly and within five weeks of the outbreak of hostilities the four Regular divisions were in position under the overall command of Field Marshal Lord Gort, VC.

Having arrived in Paris, 1st Battalion Welsh Guards became the first British troops to march through the French capital following the outbreak of war, before moving north-west to Arras to become the defence battalion of General Headquarters, serving alongside 9th Battalion West Yorkshire Regiment and 14th Battalion Royal Fusiliers. With the Battalion based at the commune of Izel-les-Hameau, the rifle companies were scattered across the rural canton of Aubigny-en-Artois. In common with the rest of the BEF, the Welsh Guards settled down to a winter of watching and waiting. New drafts arrived throughout the period from the Depot at Caterham. Boredom was the biggest enemy for the Guardsmen, with guards of honour for visiting dignitaries, including HM King George VI and

Members of the 1st Battalion at Arras, 1940. They have been issued with 1937 pattern web equipment, but not the new battledress uniform. *(Imperial War Museum)*

Welsh Guards recruiting meeting, Cardiff, November, 1939. The Regimental Lieutenant Colonel, Col R. E. K. Leatham is in the trench coat, with Lord Portal on his right, and Major C. H. Dudley-Ward, DSO, MC, on his left. *(Welsh Guards Archives)*

The Colonel-in-Chief, HM King George VI visits the Training Battalion at Sandown Park, Surrey, 1941. *(Welsh Guards Archives)*

HRH Duke of Windsor, being the only break from a succession of exercises and fatigues. There were compensations, however. For all the Guardsmen it was a novel experience to be deployed in France where wine was served with food in all local *estaminets* and the good rate of exchange meant that the Guardsmen enjoyed a better standard of living than their families did in England and Wales. Other excitements included the local brothels which left recruits like Guardsman Bill Williams 2735036 'wide-eyed and dumbfounded. There was nothing like them in Pontypool!'[10] Eric Coles had also ventured into a brothel with his mates and had been 'amazed when we entered that place to find everything so plush we felt out of place in our drab army uniforms'. By his own admission, he left a 'virgin soldier'.[11] Another recruit, Harold 'Bert' Harrison, was from Gloucestershire; he had joined the Welsh Guards by chance and never forgot his arrival with the Battalion at Arras where he was introduced to his 'favourite' Sergeant Major, 'Daddy' Larcombe. 'He had a dark complexion, black hair, hatchet face and eyes that looked right through a man,' remembered Harrison later. 'You look like a party of Girl Guides, I'll sweat that wine out of you,' were Larcombe's far from encouraging words of welcome.[12]

War came to the BEF with a vengeance on 10 May 1940 when the Germans subjected France and the Low Countries to the devastating tactics of *Blitzkrieg*, using armour and air power to back a rapid ground assault into Belgium, Luxembourg and the Netherlands. The surprise was total and the resistance was negligible. Early in the morning, German airborne units of Army Group B began landing in the Netherlands to capture The Hague and the vital crossings of the Meuse. Two days later, all Dutch resistance near the border was at an end and, as their forces fell back towards Rotterdam and Amsterdam, they left the Belgian left flank unprotected. At the same time seven German panzer divisions of Army Group A pushed through the Ardennes and began an unexpected move west towards the Channel ports. While this was happening the BEF began its pre-arranged move into Belgium towards a defensive position known as the Dyle Line, passing such well-known battlefields of earlier wars as Waterloo, Ypres and Mons. Even at that stage Gort was confident that the Germans could be held, issuing an order of the day on 13 May, telling his troops that 'the struggle will be hard and long, but we can be confident of final victory'.[13]

A newly manufactured Universal Carrier developed by Vickers Armstrong. These were issued on a scale of 10 per battalion. It mounted the .303 Bren light machine gun and became known as the Bren Gun Carrier. The Universal Carrier's ability to carry personnel was limited, but it was effective as a mobile machine gun platform or scouting vehicle. Later it was armed with a flame-thrower and rechristened the 'Wasp'. *(Welsh Guards Archives)*

The Commanding Officer addresses the 1st Battalion at Cherry Tree Camp, 27 May 1940, shortly after the evacuation from Dunkirk. *(Welsh Guards Archives)*

On 17 May 1st Battalion Welsh Guards (less No. 3 Company at Gort's headquarters) moved into Arras where the defence of the area was under the command of Major General R. L. Petre. On arrival, Copland-Griffiths took over command of the garrison and his second-in-command, Major Sir William Greenacre, took over temporary command of the Battalion which was deployed as follows: Prince of Wales Company (Captain Sir W. V. Makins), defence of the Doullens Road; No. 2 Company (Captain J. E. Gurney), defence of roads from Cambrai and Bapaume; No. 4 Company, defence of river bridges at St-Catherine and St-Nicolas; Headquarter Company (Captain H. C. L. Dimsdale), defence of the St-Pol road. Two days later Arras was subjected to enemy aerial bombardment and the station was destroyed while two trains were loading refugees, a distressing experience for all concerned. Orders came for the Battalion on 19 May to move to Aire, some thirty miles to the north-west, but these were quickly countermanded. On the following day German armoured cars and infantry attacked the Welsh Guards' positions but these assaults were repulsed, causing the Battalion's first casualties – thirteen killed in No. 4 Company, when their shelter received a direct hit, and six were killed and nine wounded in Headquarter Company.

The next day a British counter-attack was made south-west of Arras and, although it enjoyed some initial success thanks to the well-armoured Matilda tanks, it faltered after French V Corps failed to join the attack southward from Douai. Gort concluded that the attack had 'imposed a valuable delay on a greatly superior force', but the British position in Arras was becoming untenable and on 23 May Copland-Griffiths gave the order to withdraw. The Germans had crossed the Somme and cut the railway near Noyelles; the British defensive line along the River Escaut had had to be abandoned. Initially a thick ground mist helped to conceal the British retreat from Arras towards Douai during the night of 23/24 May but three miles out of the town the Battalion's transport under the command of Quartermaster Lieutenant J. C. Buckland found that the road was blocked, forcing the column to turn.

Welsh Guardsmen arrive in England from Dunkirk. Unlike so many other units, they returned in good order, with most if not all of their personal equipment. *(Welsh Guards Archives)*

To give the forty vehicles a better chance, the

Battalion's Carrier Platoon and a troop of light tanks under the command of Lieutenant the Hon. Christopher 'Dickie' Furness attacked the German positions near a wooded area on the rising high ground above the road. They must have known that the odds were hopelessly stacked against them but Furness and his small force pressed home their attack at close quarters, thereby engaging the enemy's attention and allowing the transport to make good their escape. The light tanks were knocked out early in the engagement but, attacking in a 'V' formation, the carriers circled the German positions until Furness's carrier was put out of action and his gunner and driver were killed. Despite having been wounded in the leg the previous day, Furness jumped down and engaged the Germans in hand-to-hand combat. The action was later described by Sergeant G. E. Griffin, the commander of the second carrier, in the Carrier Platoon's War Diary:

> By now Driver Griffiths could steer only with his left arm and we were going out of control, firing whenever a target appeared. Those kaleidoscopic images are impressed on my mind: Lieutenant Furness standing up in his carrier, revolver drawn, shooting at a German officer he was holding by the throat. We kept circling – two Germans appeared from behind the haystacks and made towards the now silent carrier where the German officer was squirming on the ground in his death throes. I shot the two just before we crashed into the back of Lieutenant Furness's carrier which showed no signs of life. There was now a hail of crossfire from both flanks.[14]

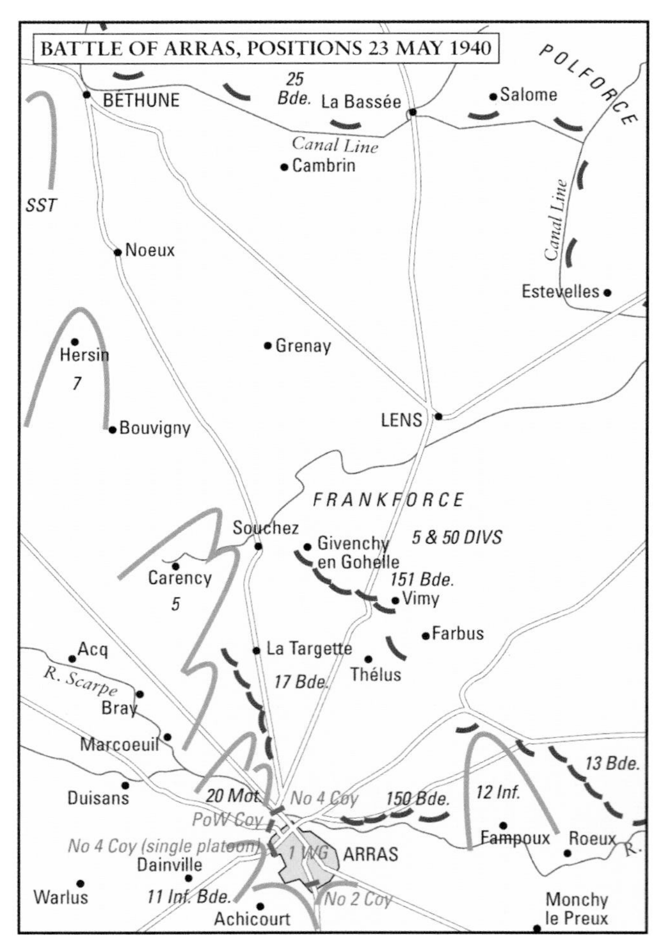

The Battle of Arras, showing positions on 23 May 1940 at the time of the British withdrawal.

Of the men in the carriers, four were killed (Lieutenant Furness, Guardsman J. W. Berry, Guardsman J. P. Daley and Guardsman D. Williams), four were wounded (Guardsman C. D. Griffiths, Guardsman G. Roberts, Guardsman I. L. Thomas and Guardsman T. Griffiths) and one was taken prisoner (Lance Sergeant A. E. Hall) but the battalion transport was saved as were many other lives. For his supreme courage, Furness was posthumously awarded the Victoria Cross in 1946.

HM King George VI visits the 1st Battalion after the evacuation from Dunkirk. *(Welsh Guards Archives)*

Welsh Guardsmen at Arras, 1940. *(Imperial War Museum)*

For the next week, the Battalion operated over a twenty-mile stretch of land between the small hill town of Cassel, now the site of GHQ, and the port of Dunkirk as the BEF made its last retreat towards the Channel coast and safety. During this period strong-points were created to allow the retreating formations to move along a protected corridor towards Dunkirk, the theory being that a secure defence could check the German advance for a sufficient length of time to enable an orderly withdrawal. Under the temporary command of Brigadier Charles Norman (1st Army Tank Brigade), 1st Battalion Welsh Guards was ordered to hold a four-mile front running from Vyfweg to West-Cappel in order to delay any German forces hoping to use the roads that passed through the villages. The Battalion was supported by the light tanks of 1st Fife and Forfar Yeomanry and by 6th Green Howards, a labour battalion consisting of half-trained soldiers armed only with rifles and deployed in the pioneer role. The position occupied by the Battalion formed a triangle bounded by the villages of Vyfweg, West-Cappel and Ratte Ko. It was in this arena that the Battalion fought a rearguard action which Major Ellis described in his history as 'a sobering reminder of the price which must be paid by a thin screen of infantry left to protect retiring forces from a thrusting enemy's advance'.[15]

After a quiet night, the first German attack was directed against No. 2 Company's positions in the morning of 29 May, with the brunt being born by 5 Platoon in the centre under the command of Lieutenant W. H. R. Llewellyn. As the position was in danger of being overrun, Captain Gurney gave the order to his depleted company to make a fighting retreat into the grounds of the local château which became his company headquarters. German armour had also attacked the positions held by Prince of Wales Company and No. 3 Company in Vyfweg and at one point Brigadier Norman's headquarters was almost overrun. At the same time the Battalion's overall position had been weakened by the withdrawal of 48th Division on the left of Bergues and by the growing intensity of the German attacks. At the end of the day, with the light tanks of the Fife and Forfar Yeomanry providing cover on the flanks, Prince of Wales Company and the bulk of Headquarter Company were moved to the beaches in lorries while No. 3 Company were also able to withdraw without much difficulty. However, not everyone was able to get away and small parties of Welsh Guardsmen were surrounded and taken prisoner while attempting to break out. The majority were destined to spend the rest of the war in German prisoner-of-war camps.

A similar fate faced the men of the depleted No. 2

Lieutenant the Hon. Christopher 'Dickie' Furness, VC

'Dickie' Furness was the son of 1st Viscount Furness, chairman of the shipbuilding firm of Furness Withy. His stepmother, Thelma Viscountess Furness (née Morgan), had been the mistress of the Prince of Wales and had introduced him to Mrs Wallis Simpson, whom the former King would marry in 1937. Furness joined the Welsh Guards in 1932 but was obliged to resign three years later. He rejoined the regiment on the outbreak of war. When the BEF crossed over to France, Furness took with him two greyhounds with labels attached to their collars, reading 'By order of Lord Gort'. At the time he was engaged to Princess Natasha Bagration, a cousin of Marina Duchess of Kent. He was killed in action in May 1940 and awarded the VC posthumously, after the war, once all the facts of that event had become known (see citation opposite).

Lieutenant the Hon. Christopher 'Dickie' Furness, VC, painted by Oswald Birley. *(Estate of Mark Birley)*

Extract from the *London Gazette*, 7 February 1946

Lieutenant the Honourable C. Furness was in command of the Carrier Platoon, Welsh Guards, during the period 17th–24th May 1940, when his Battalion formed part of the garrison of Arras.

During this time his Platoon was constantly patrolling in advance of or between the widely dispersed parts of the perimeter, and fought many local actions with the enemy. Lieutenant Furness displayed the highest qualities of leadership and dash on all these occasions and imbued his command with a magnificent offensive spirit.

During the evening of 23rd May 1940, Lieutenant Furness was wounded when on patrol but he refused to be evacuated. By this time the enemy, considerably reinforced, had encircled the town on three sides and withdrawal to Douai was ordered during the night of 23rd–24th May. Lieutenant Furness's Platoon, together with a small force of light tanks, were ordered to cover the withdrawal of the transport consisting of over 40 vehicles.

About 0230 hours on 24th May, the enemy attacked on both sides of the town. At one point the enemy advanced to the road along which the transport columns were withdrawing, bringing them under very heavy small arms and anti-tank gunfire. Thus the whole column was blocked and placed in serious jeopardy. Immediately Lieutenant Furness, appreciating the seriousness of the situation, and in spite of his wounds, decided to attack the enemy, who were located in a strongly entrenched position behind wire.

Lieutenant Furness advanced with three carriers, supported by the light tanks. At once the enemy opened up with very heavy fire from small arms and anti-tank guns. The light tanks were put out of action, but Lieutenant Furness continued to advance. He reached the enemy position and circled it several times at close range, inflicting heavy losses. All three carriers were hit and most of their crews killed or wounded. His own carrier was disabled and the driver and Bren Gunner killed.

He then engaged the enemy in personal hand-to-hand combat until he was killed. His magnificent act of self-sacrifice against hopeless odds, and when already wounded, made the enemy withdraw for the time being and enabled the large column of vehicles to get clear unmolested and covered the evacuation of some of the wounded of his own Carrier Platoon and the light tanks.

Company who were pinned down at the château but refused to let the seriousness of the situation affect them. As the building started to collapse through the weight of the German bombardment, Guardsman Bill Williams of 5 Section remembered Lance Corporal Warwick shouting to his mate and (alleged) fellow womanizer: 'Cummings, Cummings, there's a tart asking for you at the front gate.'[16] Everyone burst out laughing and tensions were relieved, albeit temporarily; but the reality was no laughing matter. Unable to continue fighting, Captain Gurney decided to break out across the fields under the cover of darkness and eventually a small handful of men reached the safety of the beaches and evacuation back to England. Guardsman Williams provided a vivid account of No. 2 Company's withdrawal towards the coast:

> We were now approaching masses of abandoned British and Allied tanks, guns, vehicles and equipment, all abandoned with breeches removed from guns and engines on vehicles smashed beyond repair. It was not the most heartening sight but by now nothing surprised us. We had lived a lifetime in a traumatic week.[17]

The last entry in the Company's War Diary gives an indication of the ferocity of the fighting that day: 'Coy had 3 officers (2 wounded), 21 ORs left.'[18]

Dunkirk was described as a 'miracle' in that 338,226 Allied soldiers escaped from the beaches but, as Churchill cautioned at the time, 'wars are not won by evacuations'. However, as the following story shows, self-respect and discipline remained intact within 1st Battalion Welsh Guards. As it was considered essential for morale to remove the remains of the BEF from the Channel ports as quickly as possible, special trains took the survivors to inland towns and camps where they were housed, fed and given fresh clothes. One such train pulled into the station at Leamington Spa where the RTO noticed that the end carriage remained closed while dishevelled groups of soldiers stumbled out onto the platform from the others. On enquiry, the carriage was found to contain a subaltern and an ensign from the 1st Battalion with some fifty men who now formed up in two ranks complete with rifles and helmets 'as if on parade'. On being told that they were to proceed into Leamington Spa where the officers would be billeted, while the men were bussed to a camp on the far side of town, the subaltern explained that there could be no question of the group splitting up. They also refused the offer of transport and, led by the two officers, the Guardsmen marched through the centre of Leamington Spa to the camp. That was not all. The camp commandant, a colonel from the previous world war, had been told of the incident, and as the Welsh Guards marched in he turned out the guard and stood at the salute with tears running down his face.[19]

Holding Force, Dovercourt, commanded by 2Lt J. D. Gibson-Watt, MC**. This platoon later formed the nucleus of the 3rd Battalion Welsh Guards. 2Lt Gibson-Watt served as adjutant of the 3rd Battalion during the North African and Italian campaigns. *(Welsh Guards Archives)*

King George VI, Queen Elizabeth, Princess Elizabeth and Princess Margaret with officers of Prince of Wales Company, at Sandringham, 1940s. *(Welsh Guards Archives)*

2nd Battalion

Having settled into their training camp at Theydon Bois, the 2nd Battalion began training in earnest. The War Diary shows the intensity of effort: a refresher course for returning signallers was held on 19 September, a Battalion route march was held the following day, a platoon commanders' course was held on 2 October, and four days later the Corps of Drums paraded for the first time. Towards the end of the month the Battalion sent a draft of ninety-seven Guardsmen to join the 1st Battalion and a new training battalion came into being under the command of Major H. Talbot Rice, who took them to more permanent quarters at Roman Way Camp in Colchester. He then handed over command to Lieutenant Colonel Lord Glanusk before moving to command the Welsh Guards Company at the Depot. These changes allowed the 2nd Battalion to return to the Tower of London where it took over duties from 11th Battalion, Royal Fusiliers. Training was limited by the confines of the physical environment and men had to be sent on training courses further afield, but on 16 October the Battalion found King's Guard for the first time, relieving 1st Battalion Irish Guards, making the Welsh Guards the first regiment to mount guard from the Tower of London. Captain of the King's Guard was Major G. W. Browning and the entire party travelled to Buckingham Palace by underground train, another novelty. In addition to finding King's Guards and other public duties, this period included escorting German prisoners-of-war to camps in the north of England. A highlight of the deployment to the Tower of London was the presentation of Colours by the Colonel-in-Chief, King George VI, on 14 February 1940. The ensigns were 2nd Lieutenant J. D. A. Syrett and 2nd Lieutenant P. J. McCall.

In the middle of April the Battalion received 'verbal news of an imminent move to a Theatre of War'. In preparation for mobilization the Battalion handed over public duties on 17 April to the Canadian Royal 22e Regiment, and the War Diary noted that the occasion was 'without precedent in that it was the first time that Colonial Troops of Non-British descent had mounted

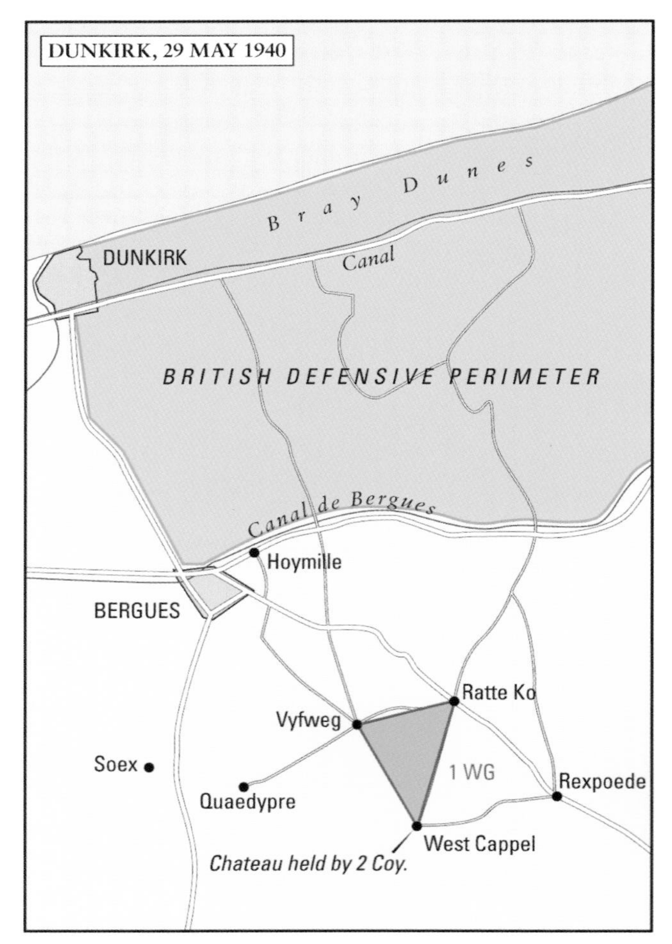

Withdrawal to Dunkirk, the rearguard action fought by 1st Battalion at Vyfweg and West Cappel on 29 May 1940.

Welsh Guards reservists reporting for duty, 1940. *(Welsh Guards Archives)*

Dawn at Dunkirk. Drawn by Sgt C. Murrell while he and other Guardsmen were waiting on the beaches to be evacuated. *(Courtesy of the Murrell family)*

the King's Guard'.[20] (Although Commonwealth regiments had mounted King's Guard in the past, the Royal 22e Regiment was French-speaking, considered itself French and had its headquarters in Quebec.) The Captain of the King's Guard that day was Major H. M. C. Jones-Mortimer, the Subaltern was Lieutenant J. A. Meade and the Ensign was 2nd Lieutenant Baron de Rutzen.

Mobilization of the Battalion had been completed by 21 April 1940 and the following day 2nd Battalion Welsh Guards, 'at War Strength and fully mobilized', moved to Old Dean Common Camp at Camberley where an intensive programme of platoon training with live firing was put into effect. There they were brigaded with 2nd Irish Guards to form an under-strength 20th Guards Brigade, commanded by Brigadier Sir Oliver Leese, Bt, DSO. On 11 May, both battalions were due to go on Whitsun leave but the German attack into the Netherlands and Belgium put a stop to those plans. At 10.00 hours orders arrived from the War Office cancelling leave and ordering the brigade to prepare for embarkation to The Hague to assist the Dutch government evacuating the country. In the confusion it proved impossible to halt all the special leave trains and it quickly became clear that neither battalion would be at full strength. As a result, a composite Irish/Welsh Guards Battalion was formed under the command of Lieutenant Colonel J. C. Haydon, OBE, Irish Guards, with Major G. St V. J. Vigor, 2nd Battalion Welsh Guards, second-in-command. The Welsh Guards' contribution to the 651-strong battalion was a company of 201 under the command of Captain C. H. R. Heber-Percy. One of their number, 'Dai' Tilley, remembered that the first inkling of their destination came when Sergeant Obie Walker asked him if he liked tulips. On being told that they were not Tilley's favourite flower, Sergeant Walker replied: 'Well, you'd better bloody well learn to like them and real quick, as we're leaving for Holland tonight.'[21] The following day, 13 May, the composite battalion, designated 'Harpoon Force', landed at the Hook of Holland with orders to assist the Dutch royal family and to help personnel at the British embassy to withdraw from the Netherlands.

On landing, the small force started digging in at Walcheren where they came under attack by German Ju 87 Stuka dive-bombers. The men were lightly armed with some Bren guns and a few 3-inch mortars and Boys anti-tank rifles. There was no transport apart from two signals trucks and it soon became clear that they would be unable to withstand a German armoured attack. Shortly after midday, a party of the Dutch royal family led by Queen Wilhelmina arrived at the docks and embarked on the destroyer HMS *Hereward*. They were followed later in the day by members of the Dutch government, and the next day the decision was taken to withdraw Harpoon Force on board the destroyers HMS *Malcolm*, HMS *Vesper* and HMS *Whitshed*. Forty-eight hours after their departure, the small force was back in Camberley; their casualties were seven killed in action and twenty-three wounded.

This was only the beginning of the 2nd Battalion's war. A week later, on 21 May, while they were on exercise, a signal arrived ordering them to be ready at two hours' notice for embarkation at Dover as part of a move to France by 20th Guards Brigade, now under the command of Brigadier W. A. F. L. 'Billy' Fox-Pitt. Later in life he recalled how the instruction from the War Office arrived: 'I've got rather a hot one for you. You've got to go out and hold on to Boulogne as long as you can. There's very little information as to what's happening.'[22] Behind those terse words lay a worrying situation for the BEF. Following the German breakthrough at Sedan on 14 May, the way to the Channel ports was now open and Boulogne would be the first target. Getting the brigade there was another matter. Having received their orders, the two battalions proceeded immediately to Dover, arriving at midnight to begin embarking on two cross-

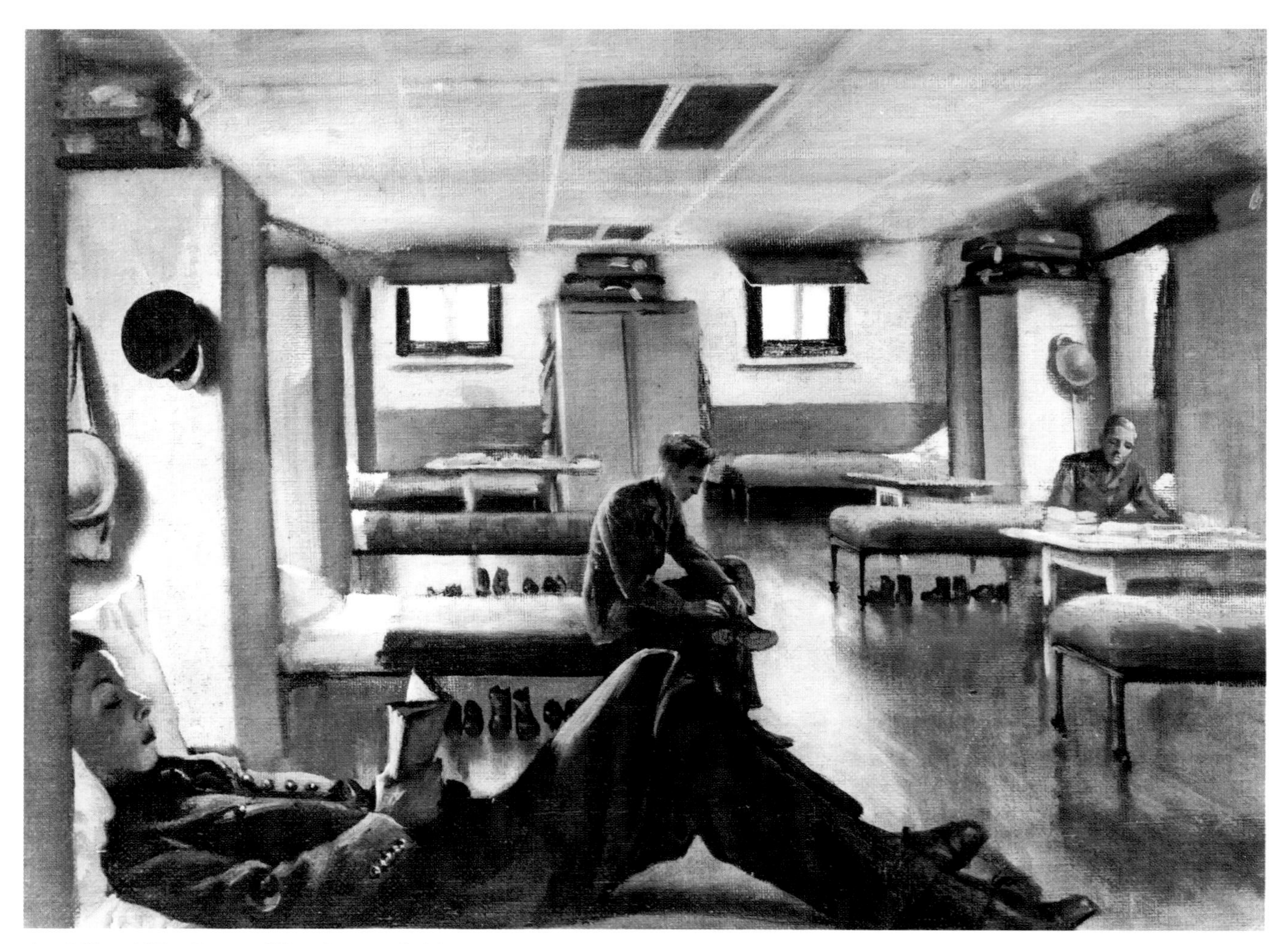

An Officers' Hut, Roman Way Camp, Colchester, 1940. This picture was painted by Rex Whistler and donated to RHQ in 1944 by the artist's family. It shows 2Lt Adrian Pryce-Jones reading in the foreground, while the artist himself is sitting on the bed in the middle. The third figure is unidentified. Pryce-Jones and Whistler shared a taxi from the Colchester station to Roman Camp. Upon arrival Whistler recorded that their 'spirits were at zero'. *(Estate of Rex Whistler)*

Channel steamers, SS *Biarritz* and SS *Mona's Queen*. The confusion awaiting them on the quay was described in the Battalion's 'Report on Operations, 21–24 May 1940':

> From the outset little or no Staff arrangements had been made on the Quay for loading personnel, weapons or stores. Eventually most of the personnel of the Bn were embarked in S.S. 'BIARRITZ', but the M.T. column carrying the weapons, ammunition and stores was still well back in DOVER and when it finally came forward, was allowed to double bank, causing a complete breakdown in loading. Thereafter, with time getting short, weapons, ammunition and digging tools at the expense of all other stores were manhandled forward along the Quay and loaded. The Master of the 'BIARRITZ' then stated that he could load no more. He had previously announced that he had 60 tons of ammunition and stores already on board, which he had not unloaded on his last trip to BOULOGNE, as the stevedores there had ceased to work the cranes and the position there was reported to be very uncertain.[23]

Escored by the destroyers HMS *Whitshed* and HMS *Vimiera*, the convoy arrived at Boulogne early the following day and began disembarking the two battalions. Having reconnoitred the front with the only maps available (one large-scale and two small-scale), the Commanding Officer, Lieutenant Colonel 'Sammy' Stanier, made his disposition to cover a front which was 6,000 yards wide. The Battalion's line ran eastwards from the River Liane, with No. 2 Company (Major H. M. C. Jones-Mortimer) on the right, followed by Headquarter Company (Captain R. B. Hodgkinson) at Ostrohove, No. 3 Company (Major J. C. Windsor-Lewis) at La Madeleine to Mont Lambert and No. 4 Company (Captain J. H. V. Higgon) on the left at St-Martin Boulogne. When No. 1 Company (Captain C. H. R. Heber-Percy) arrived in mid-afternoon, it took position on the left at the crossroads at St-Martin Boulogne. One of Heber-Percy's first actions was to use an anti-tank gun to destroy a church tower which was being used by a French fifth columnist to signal to the Germans. As Stanier recalled, 'at first he [Heber-Percy] couldn't get the elevation but then he propped the gun on a ramp and knocked the top of church tower off and down came the man with it'.[24]

During the day (22 May) the German 2nd Panzer

Major General William Augustus Fitzgerald Lane Fox-Pitt, CVO, DSO, MC, DL

'Billy' Fox-Pitt was born on 28 January 1896 and educated at Charterhouse. After the outbreak of the First World War he was commissioned into the Cheshire Regiment and saw action in France while serving with 2nd Battalion, Sherwood Foresters. He was one of the founding officers of the Welsh Guards in February 1915. Wounded during the fighting on the Hohenzollern Redoubt in October 1915, he was invalided home but returned to the Western Front as Adjutant of the 1st Battalion in May 1917. He was wounded again at Flesquières in September 1918. After the war he served in London and Egypt and as Commanding Officer of the 1st Battalion between 1934 and 1938, after which he was appointed Regimental Lieutenant Colonel, a post he held until September 1939. On the outbreak of the Second World War, he returned to operational duties in command of 20th Guards Brigade. Other wartime commands included 5th Armoured Brigade and East Kent District. In September 1945 he was appointed Aide-de-Camp to HM King George VI, retiring from the Army two years later in the rank of major general. He died in 1988. His sons Mervyn Fox-Pitt and Oliver Fox-Pitt, both noted horsemen, also served in the Regiment.

Captain Dabell with Brigadier W. A. F. L. Fox-Pitt at a garden party in Wimbledon in 1940, shortly after Dunkirk. *(Welsh Guards Archives)*

Division had reached the southern outskirts of Boulogne and the following morning initiated a determined attack at 07.30 hours. The brunt of the initial assault was mounted against the line held by No. 2 Company and No. 3 Company, which managed to hold off the enemy for two hours during which a number of tanks and armoured fighting vehicles were destroyed or disabled. At 11.30 Stanier sent a despatch rider with a situation report to brigade headquarters, informing them of the position facing his battalion: 'Strong enemy pressure on my front is being resisted. Have made slight readjustment to 2 Coy's position. Ammunition running short.'[25] Before the adjustment could be made, a liaison officer arrived from brigade headquarters ordering Stanier to withdraw his Battalion into the town to defend and block the approaches to the harbour facing north-east from the sea as far as the main railway bridge over the River Liane. By then, with German forces breaking into the town, the only option was withdrawal by sea. At around 18.00 the Battalion was ordered to withdraw to the quay to await the arrival of the destroyers which would carry out the evacuation. HMS *Whitshed* and HMS *Vimiera* went into the harbour first, followed by HMS *Wild Swan*, HMS *Venomous* and HMS *Venetia*. In the words of the Official History, 'All three ships engaged in a most unusual naval action, firing over open sights at enemy tanks, guns and machine guns only a few hundred yards away while they took the troops on board.'[26] At the same time the ships were coming under fire from heavy guns on the hills above the town which the retreating French had neglected to destroy: *Venetia* was so badly damaged that she had to back out of the harbour with her guns still firing.

At 23.00 HMS *Windsor* sailed in to complete the evacuation, but unfortunately in the confusion No. 2 Company and No. 4 Company had become isolated in a large customs shed where they received erroneous information that the last ships had departed and that they were on their own. Also left behind in Boulogne was a disparate force of Welsh Guardsmen from No. 3 Company, unarmed pioneers and French stragglers, all under the command of Major Windsor-Lewis who ordered them to create a defensive position for a last stand against the rapidly approaching German forces. It was a courageous decision but, as Windsor-Lewis wrote in his after-action report, the reality was that the position was hopeless.

> At noon the enemy, now strongly reinforced, opened up an intense fire upon my position and I was compelled to withdraw from my front line of breastworks into the Station itself, protected only by glass overhead and by a train on the left flank.
>
> There was a little food and ammunition left and no more water, and after another hour of

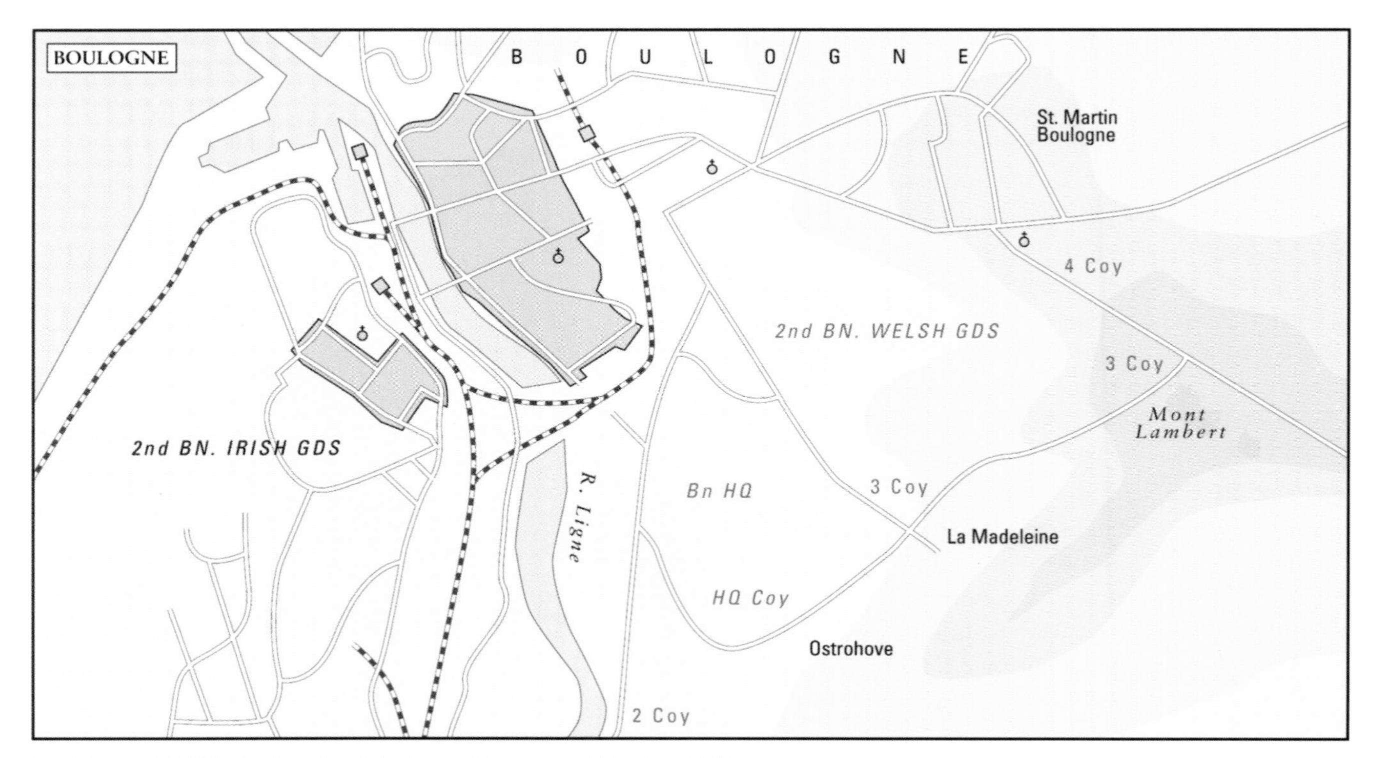

Positions of 2WG during the fighting at Boulogne, 24 May 1940.

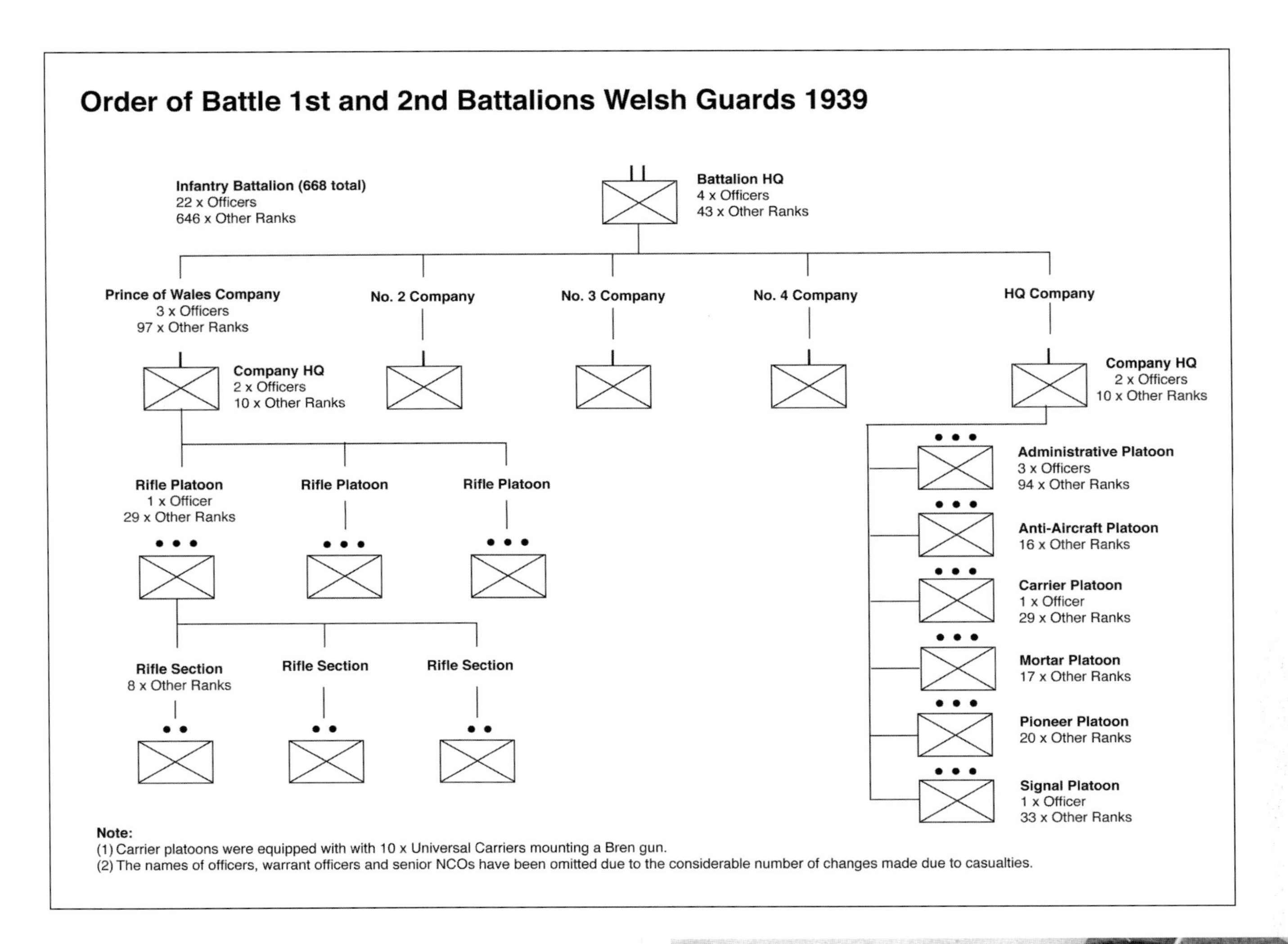

> the greatest discomfort I decided that my position was now quite hopeless and that a massacre would ensue if I did not capitulate. Having an eye to the number of refugees under my care and the big percentage of unarmed men, I decided to surrender.[27]

By the end of the day on 24 May all resistance in Boulogne had come to an end and, although No. 2 Company and No. 4 Company attempted a breakout, they too went into enemy captivity. However, a small party led by 2nd Lieutenant R. C. Twining and 2nd Lieutenant E. G. F. Bedingfield had managed to reach the quay and were evacuated on board HMS *Windsor*. They then returned to Dover and were transported by train to Fleet for a brief respite at Tweseldown Racecourse before moving on to Cherry Tree Camp at Colchester, which was to be their home until 12 June. On 29 May, Brigadier Fox-Pitt wrote to Vice Admiral Sir Bertram Ramsay, commanding the evacuation, expressing his thanks for the skill and courage of the destroyer crews who had rescued the Irish and Welsh Guardsmen:

The château at West-Cappel. A German tank became stuck on the bank of the moat, in the foreground of the picture, but was still able to train its gun on the defending troops. *(Welsh Guards Archives)*

Brigadier James Charles Windsor-Lewis, DSO, MC

The son of a Welsh Guards officer killed in action in June 1916, Jim Windsor-Lewis was born on 18 March 1907 and educated at Eton. He followed his father into the Army, being commissioned in the Welsh Guards, and served as Adjutant from October 1932 to October 1935. In the Battalion in those days there was 'much enthusiasm for hunting and racing' and Lewis emerged as a skilled rider who competed as an amateur in the Grand National.

In 1938 he was appointed ADC to Lord Gowrie, who was serving as Governor-General of Australia, but returned to England on the outbreak of war, travelling by ship and flying boat. While commanding No. 3 Company during the attempted breakout from Boulogne, he was forced to surrender to the Germans on the dockside, having noted in his later report that his 'little force had fought most splendidly in the face of heavy odds. Exhausted and without proper nourishment, they never lost heart.' Although he was taken prisoner, he managed to escape from a hospital in Liège and was subsequently helped by Mary Lindell, the English-born Comtesse de Milleville, and her Marie-Claire escape organization based at Ruffec, who supplied him with an Irish passport. Windsor-Lewis made it to Lisbon and returned to England the following year. Between December 1943 and May 1946 he was Commanding Officer of the 2nd Battalion and led them in the liberation of Brussels in September 1944.

Windsor-Lewis ended his Army career in the rank of brigadier and served as Regimental Lieutenant Colonel between May 1951 and June 1954. He died in October 1964 and is buried in the graveyard of All Saints Church at Crondall in Hampshire. His grandson, Lieutenant Colonel David Bevan, joined the Regiment in 1997.

The Regimental Band entertaining crowds outside St Paul's Cathedral during WW2. *(Welsh Guards Archives)*

I am writing to you on behalf of every officer and man in this Brigade to express our admiration and gratitude for the part played by H.M. Destroyers in the evacuation from Boulogne.

But for their skill and courage in action I am certain that the evacuation would have resulted in far higher casualties, and might well have proved impossible. The debt which my Brigade owes to the British Navy will not be forgotten, and I can assure you that the way in which they performed their duties will be remembered by us all.[28]

The experiences in France had been an expensive lesson for both battalions. It would take another four years before they went into action again on the European mainland. During the fighting in France the Regiment's casualties were 72 officers and men killed and 88 wounded, while 453 went into enemy captivity as prisoners-of-war. The Battle for France was over and Britain was now on its own, supported only by the forces of the Dominions and Commonwealth of Nations and the armies which had managed to escape from Nazi-dominated Europe.

CHAPTER FIVE

The Tide Turns, 1940–1943

1st and 2nd Battalions, England
3rd Battalion, North Africa, Fondouk and Hammam Lif

FOR THE NEXT FOUR years the 1st and 2nd Battalions spent their time in various locations in the south of England training and re-equipping in anticipation of the day when Britain's armed forces could take the war back to Nazi-occupied Europe. On its return to England, the 1st Battalion was stationed at Wimbledon from 5 July 1940 and the rifle companies were spread over a wide radius including Roehampton and Putney. For No. 3 Company there was the pleasure of being billeted in the Church Road area of Wimbledon, which included the fabled All England Tennis Club. 'Dare I say it,' remembered Meiron Ellis who had joined up the previous year, 'a game of rugger was played on the hallowed ground for a short period until we were shooed away'.[1] During the frequent air raids in the late summer of 1940 both battalions gave aid to the civil authorities when bombs hit and destroyed a number of local public houses, including the Arab Boy in Upper Richmond Road and the Cricketers in Wandsworth Road. Training, too continued apace, although during a visit by King George VI Guardsman Bill Williams remembered that a demonstration of the use of Molotov Cocktails was 'pure *Dad's Army*, although I think even the scriptwrit-

Officers' Mess Tent, 1942, painted by Rex Whistler during a lull in training in the long, hot summer of 1942. The 2nd Battalion was on exercise on Salisbury Plain. In the foreground, the officer reading in a deck chair is Lt Lord Lloyd. *(Estate of Rex Whistler)*

Welsh Guards Uniforms, 1942–5. *(From illustrations by Charles C. Stadden and Eric J. Collings, commissioned by the Regiment)*

1943, North Africa, Fighting Order. This sergeant is a member of the 3rd Battalion, which was sent to North Africa (and then Italy) after heavy casualties were suffered by 1st Guards Brigade during the Tunisian Campaign. The Battalion left England in February 1943 and became an infantry battalion in the 6th Armoured Division. While the primary personal weapon remained the .303 SMLE rife, NCOs often carried sub-machine guns. Seen here is the .45 Thompson sub-machine gun. Often called the 'Tommy gun', it was purchased from the US and issued on a limited scale until the newly developed and less expensive 9mm Sten gun arrived. It had a 20-round box magazine.

1944, Holland, Fighting Order. After the Battle of Normandy was over, both the 1st and 2nd Battalions advanced in Holland and Germany with the Guards Armoured Division. This LCpl wears the new 'Invasion Pattern' steel helmet, which had a deeper profile to offer increased protection. He also wears a leather jerkin for increased warmth in the winter months. By this time, the new .303 No. 4 rifle had entered service, but here a Sten Mk II sub-machine gun is carried.

1942, England, Officer, Training Order. Soldiers of the Guards Armoured Division adopted the same uniforms as those of the Royal Armoured Corps. This officer of the 2nd Battalion wears a black one-piece 1942 pattern tank crewman's oversuit. Other tank regiments wore a tan version of the same suit, but the black beret shown here was universally worn by all armoured regiments. He wears a cut-down version of the 1937 webbing equipment with a pistol holster, braces and ammunition pouches. Members of the Guards Armoured Division wore the 'ever open eye' divisional badge on their battledress blouses. This was adapted from a WWI design by Rex Whistler. Initially, the 2nd Battalion trained with the obsolete Covenanter tank, then the Crusader and Centaur, before finally receiving the 30-ton Cromwell tank, with which they deployed to Normandy.

1940, France, Despatch Rider. Despatch riders wore a unique uniform including a Mk 2 helmet, leather jerkin, gauntlets, motorcycle breeches and leather boots. The Guardsman illustrated here carries a map case. A side arm, such as a .38 Enfield No. 2 revolver or 9mm Sten gun, was carried. The most commonly used motorcycles were Triumphs and BSAs. On the narrow, crowded roads of Normandy, motorcycles often proved much more suitable means of delivering messages than jeeps.

1944, Belgium, Fighting Order. By the time the 2nd Battalion deployed to Normandy in 1944, a new one-piece oversuit had been issued. As well as being of a more practical design with improved zips and pockets, it was made from a three-layer material comprised of an outer layer of reinforced cotton denim, an inner lining of wool flannel and a water-resistant oilskin interwoven between them. The suit's hood was made of the same material and was attached to the stand-up collar with press studs. The 2nd Battalion entered Brussels ahead of the Allied Armies on the fifth anniversary of the war. The firepower of their Cromwell tanks was adequate rather than exceptional. In early 1945, the 2nd Battalion received a limited number of Challenger tanks armed with the same 17-pounder gun fitted to the Sherman Firefly. This weapon was the first British tank gun that could engage German Tigers and Panthers on equal terms at longer ranges. Both Cromwell and Challenger were superseded by the Comet, the first truly excellent British tank, but the war ended before it could be issued to the 2nd Battalion. In the background is the Cromwell. A low and compact design, its 600-hp Rolls-Royce engine gave it a road speed in excess of 40 mph.

ers of that series would have thought it unbelievable'. When the Guardsmen were ordered to throw their petrol-filled bottles at a nearby stone wall the exercise descended into farce:

> Two or more bombs missed the wall altogether and hit and then set fire to a dozen decorative red blossomed apple trees in the next garden. In seconds flames engulfed the whole orchard, they spat and crackled like Chinese fireworks, spreading from one to another like wildfire. CSM 'Orace yelled, 'Don't stand there laughing you silly bastards, get the stirrup pumps and buckets of water.'[2]

While this was going on the 2nd Battalion moved to West Byfleet in Surrey. Both battalions were gradually brought up to strength by drafts from the Training Battalion and both received new commanding officers, respectively Lieutenant Colonel Julian Jefferson and Lieutenant Colonel G. St V. J. Vigor. Behind the urgent need to regroup and refit there was a very real fear that the Germans would follow-up their capture of France and the Low Countries by launching a seaborne

Captain the Hon. Robert Pomeroy *(centre with moustache)* and other officers. *(Welsh Guards Archives)*

invasion of England before the summer came to an end. On 16 July Hitler issued Directive No. 16, 'Preparations for the Invasion of England', which gave the go-ahead for the operation to commence in the first half of September when conditions would be favourable – a dark passage and a rising tide on arrival. The operation was codenamed Sealion. In retrospect, and with the benefit of hindsight, the Germans' ability to mount a cross-Channel invasion was exaggerated, not least because they lacked specialist amphibious equipment, but at the time the threat was taken very seriously indeed. It also suited Churchill's purpose to unite the nation behind him to act in common cause, both through his own rhetoric and by the rapid expansion of measures to stem any enemy assault on the British mainland.

After the rapid fall of Norway in April, followed by the collapse of France, there was a need to rebuild national confidence and morale and that necessity accounts for the enthusiasm and determination which suffused the nation's mood in the summer of 1940. Hitler's ambition to mount an invasion was not entirely delusional. With Poland and western Europe in his hands it seemed inconceivable that the German leader would not turn his attention to Britain and attempt an invasion. His commanders were cock-a-hoop following their easy victories in the Low Countries and northern France: the litter of military equipment on the Dunkirk beaches told them all they needed to know about Britain's plight. Such was the disarray in the home forces that the 1st (London) Division, responsible for defending the Channel coast from Sheppey to Rye, had only eleven modern 25-pounder field guns as well as four obsolete 18-pounders and eight 4.5-inch howitzers which had been used in the previous conflict. In the whole of the United Kingdom there were only eighty heavy tanks, all incapable of engaging modern German panzers with any hope of success, and 180 light tanks useful largely for reconnaissance purposes. There were fifteen divisions available for combat after the fall of France, but as well as lacking weapons they had insufficient transport and contingency plans had to be put in place to use civilian buses.

It was not until the end of the summer that the threat receded when the RAF's success in the Battle of Britain and the German failure to win air superiority put paid to any immediate thought of a cross-Channel invasion. By then, too, Hitler's strategic goals had also changed: on 18 December he issued a new Directive, No. 21, in which he ordered his forces to turn their attention eastwards and be prepared to 'crush Soviet Russia in a rapid campaign'.

The 2nd Battalion drives past the Regimental Lieutenant Colonel, Salisbury Plain, 1943. *(Welsh Guards Archives)*

(From L) the Regimental Lieutenant Colonel, Col R. E. K. Leatham, with the Commanding Officer and Adjutant of the 3rd Battalion, Lt Col D. E. P. Hodgson and Capt G. D. Rhys-Williams. *(Welsh Guards Archives)*

Colonel A. M. Bankier, DSO, OBE, MC

Albert Methuen Bankier (always known as Bertie) was originally from Edinburgh. He was born in 1894 and educated at Winchester. After training at Sandhurst he was commissioned into the Argyll and Sutherland Highlanders in September 1914.

He had an immensely gallant record in the First World War. He was Adjutant of the 2nd Battalion Argyll and Sutherland Highlanders, for almost two years, from August 1915 to June 1917 and was awarded the Military Cross in June 1916, the French Croix de Guerre in November 1918 and the Distinguished Service Order in June 1919.

In July 1919 he transferred to the Welsh Guards and joined the 1st Battalion at Purfleet, before moving to Wellington Barracks. He was appointed Adjutant of the Battalion in May 1922. After a year with the Palestine Gendarmerie and two years at Headquarters, London District, he retired from the Army in December 1927.

He was recalled in June 1939, aged forty-five, and joined the 2nd Battalion. After a variety of staff appointments, for which he was awarded the OBE, he was promoted lieutenant colonel and in September 1941 took over command of the Holding Battalion before becoming Regimental Lieutenant Colonel in February 1942.

He was a greatly respected officer who, with his distinguished record in the First World War, was admired by all ranks. He worked tirelessly for the Regiment before finally retiring in August 1945.

Bertie Bankier had two half-brothers who both served in the Regiment. Pip, serving in the 3rd Battalion, was killed, aged twenty-three, at the battle of Monte Piccolo on 27 May 1944. Michael, who was also in the 3rd Battalion, was severely wounded in the same battle. He was killed in a car crash in the USA in 1966.

Lt Col F. A. V. Copland-Griffiths (*left*) and Lt Col A. M. Bankier, DSO, MC, at Pirbright, early 1930s. *(Welsh Guards Archives)*

During the summer and autumn of 1940 and continuing into 1941 both battalions were engaged in intensive training. Their war diaries provide vivid testimony to the rapid technological advances taking place in the Army, with officers and NCOs being sent on courses to cover training on subjects as various as aerial photography (Farnborough), anti-aircraft gunnery (Northolt), gas warfare (Winterbourne) and camouflage (Old Sarum). Recruits arrived in both battalions relatively well trained following four months at the Guards Depot at Caterham and up to four more months with the Training Battalion at Sandown Park, Esher, where, according to Major L. F. Ellis, the fledgling Guardsmen 'could take part in schemes and large-scale exercises and realize in the field the significance of earlier lessons and their application to the new ways of war'.[3] In April 1941 a Holding Battalion was formed at Hounslow, under the command of Lieutenant Colonel W. D. C. Greenacre, MVO, to provide a home for men who had completed their preliminary course. Six months later, on 24 October, it was reconstituted as the Regiment's 3rd Battalion, under the command of Lieutenant Colonel A. M. Bankier, DSO, OBE, MC.

By then an even greater change had taken place with the formation on 17 June 1941 of the Guards Armoured Division (Major General Sir Oliver Leese, Bt, CBE, DSO) in which two Welsh Guards battalions served, the 1st Battalion in the infantry role and the 2nd Battalion in the armoured role. This innovation followed a call for the creation of additional armoured formations to meet the threat from German panzer divisions and eventually ten new British armoured divisions were formed. When it was suggested that the Foot Guards regiments should form one of the new armoured divisions, there was concern that the average height of the Guardsmen would hinder their employment in tanks, but these fears proved to be

Covenanter tank with anti-mine roller crossing a scissors bridge, Salisbury Plain, 1942. LSgt Roberts is on the right. *(Welsh Guards Archives)*

groundless. It helped that the Chief of the Imperial General Staff, General Sir Alan Brooke, was in favour of the proposal:

> The Regiments of Foot Guards had the officers and men of the right type available, they could be converted more quickly than other units, and a large number of officers in the Brigade, holding the view that armour rather than infantry was becoming the predominant arm, were keen to embark on the venture.[4]

For the 2nd Battalion this transformation involved a steep learning curve as all ranks had to be instructed not just in the mysteries of armoured warfare but also had to learn to drive and maintain tanks and to shoot their guns. In September 1941 the 2nd Battalion moved to Codford St Mary, some seven miles from Warminster on the road to Salisbury. Suddenly the Battalion War Diary showed officers and NCOs attending courses on gunnery, wireless telegraphy, tracked vehicle driving, and driving and maintenance – 'that unfathomable mystery known to tank lovers as D and M'.[5] Throughout 1941 it was estimated that at any one time at least 80 per cent of the 2nd Battalion were attending courses on armoured warfare. The Division also required a symbol for its badge and Leese invited all his senior officers to suggest something suitable. One option was to use the 'ever open eye' symbol used by the Guards Division in the previous conflict; another was to choose something completely new and different. Eventually Rex Whistler

A Guardsman from the 3rd Battalion in a captured German position at the summit of a hill near Fondouk. Note the mailed fist badge of 6th Armoured Division. This position had been occupied by a German machine-gun team well supplied with hand grenades. *(Welsh Guards Archives)*

Covenanter tank of the 2nd Battalion crosses the River Avon, Codford, September 1943. *(Welsh Guards Archives)*

The Commanding Officer with the Regimental Lieutenant Colonel, inspecting Centaur tanks of the 2nd Battalion on Salisbury Plain, September 1943. *(Welsh Guards Archives)*

Training on Salisbury Plain, 1943. *(Welsh Guards Archives)*

was ordered to produce a fresh design; he came up with a revised version of the 'ever open eye' which was approved as the division's badge.[6]

Other changes involved the 2nd Battalion adopting the terminology and uniform of an armoured regiment: black berets were worn with tank overalls, companies became squadrons and platoons became

Lt Col D. E. P. Hodgson with Maj Gen C. F. Keightley, GOC 6 Armoured Division, before the Battle of Fondouk, 1943 *(Welsh Guards Archives)*

troops. Initially the equipment was the Cruiser Tank Mark V, also known as the Covenanter I, but in March 1942 these were replaced by the upgraded Covenanter III. Constructed under the aegis of the London Midland and Scottish Railway from 1939 onwards, the cruiser tanks were designed to be of welded construction, with many aluminium components, a special low-profile engine and innovative Merritt-Brown controlled differential transmissions for the steering. By the time the tank entered production, riveting had replaced welding, armour thickness had been increased and aluminium could no longer be used as by then it was a priority material for the RAF. This all conspired to increase the weight of the Covenanters. To make matters worse, the transmission systems were prone to failure and the tanks suffered cooling problems through having the engine at the rear and radiators at the front. For all that it was a striking design with a low silhouette and well-shaped turret, the Covenanter was regarded as unbattleworthy in modern warfare and was declared obsolete in 1943.

It was replaced by the Crusader, armed with a superior 6-pounder gun which the 2nd Battalion began to receive in December 1942. But it was not until the following year that the Battalion was given its first modern and reliable tank in the Centaur, a variant of the Cromwell (Cruiser Tank Mk VIII), which equipped 2nd Battalion Welsh Guards from 1943 onwards. This was a huge improvement on anything that had been used before and in its different variants the Cromwell would prove the most successful British cruiser tank of the Second World War. It was equipped with a dual-purpose 75 mm gun and had a top speed of around 40 miles per hour, and while it was reliable, its speed and serviceability was bought at the cost of lighter armour which was only 4 in. at its thickest point. Nevertheless it was liked by its five-man crews for whom it was home for lengthy periods and, as Major Ellis revealed in the Regimental war history, 'many became greatly attached to their tanks by the end of the war'.[7]

At the same time as the 2nd Battalion started learning its new role, the 1st Battalion moved to Midsomer Norton in Somerset, travelling in two special trains from Wimbledon on 15 September 1941. Although the 1st Battalion was not required to come to terms with the intricacies of operating tanks, new techniques had to be learned for operating in the field as one of the infantry components of an armoured division and there was much experimentation to discover the most effective balance of arms and equipment. Even battle-hardened veterans such as Guardsman Bill Williams 36 of No. 2 Company had to master the new approach and soon found that the training made sense for soldiers who had faced modern weapons of war at Arras and Dunkirk and knew only too well their capabilities:

> Training with the ironclad monsters all over Somerset was really new to us infantry types. Instead of 3-tonners and jeeps we now roared into mock battle on APCs [armoured personnel carriers] and half-tracks accompanied by scout cars of the Household Cavalry. It was a long way from Buck Guard and Horse Guards but everyone was impressed, especially those of us who had fought with rifles and Brens against German panzers in 1940.[8]

For everyone this new regime meant a life which involved 'harder and harder training, fitter and fitter, tougher and tougher, our standards continually improved, runs and walks in fighting order, virtually daily, exercises for days on end, there were few breaks, weapon training never ending'.[9] Not that it was all work and no play. Many Guardsmen recall with fondness Saturday nights in the Redan public house at Chilcompton, 'boozing, womanizing, singing, carousing and larking about … we were young, fit and full of mischief and we loved every sacred minute of nights like that'.[10] Other visitors to the pub were members of the Auxiliary Territorial Force (ATS) and the Land Army, some of whom were Free French. On one occasion when the Regimental Police were trying to clear the pub at closing time, a French girl who was sitting on a Guardsman's lap brought the house down when she shouted at a Company Sergeant Major in loud but broken English: 'Kapitan, kapitan, vous **** off!' The 1st Battalion's War Diary opted for a more austere tone:

1941 October 7
- Companies carry out three days Section and Platoon Training.
- Commanding Officer, Adjutant and all Company Commanders attend Special Group weekly lectures, 'Infantry in an Armoured Division', followed by Demonstration at MARSTON BIGOT of 'A Motor Company', carried out by 4th Battalion COLDSTREAM GUARDS.

1941 October 13
- Anti-Aircraft Platoon visit COLERNE Aerodrome, BATH.
- 4th Drill, Rifle, L.M.G. and W/T, 3rd Map Reading and Section Leading Courses begin.

1941 October 24
- Commanding Officer and Company Commanders attend Special Group weekly lectures 'Preparation of an Exercise'.
- Captain G.G. FOWKE takes part in Demonstration by 153rd R.H.A. REGIMENT.
- No. 2 Company carry out an Exercise in Co-operation with Troop 21st ANTI-TANK REGIMENT.

1941 November 6
- Assault Boat Exercise by No. 2 Company, one Troop 21st ANTI-TANK REGIMENT co-operating.

1941 November 26
- Drill Course for young N.C.O.s.
- Visit of Divisional Commander.
- Prince of Wales Company fire rifles.
- Assault Boat Exercise by No. 4 Company.
- No. 4 Company zero rifles.

1941 December 23
- H.Q. Company fire rifle at LANGPORT.
- No. 2 Company carry out field firing Exercise at VELVET BOTTOM Range.
- Prince of Wales Company carry out Exercise in co-operation with one Troop 21st ANTI-TANK REGIMENT.[11]

Through all the training during and beyond the unusually hot summer of 1942 the 1st Battalion gelled into a cohesive fighting formation, quickly assimilating even the most recently arrived recruits such as 2nd Lieutenant Peter Leuchars.

Major General Peter Leuchars, CBE

Peter Raymond Leuchars was born in London on 29 October 1921 and was educated at Bradfield College where he was head of school and got his colours for cricket and football. He joined the Welsh Guards in 1941 and saw action with the 1st Battalion in Normandy and North-West Europe where he had some lucky escapes before being wounded by 'friendly fire' during the fighting for Hechtel in September 1944. After recovering he was posted to the 3rd Battalion in Italy.

After the war he accepted the offer of a regular commission and was Adjutant of the 1st Battalion during its deployment to Palestine. Staff College followed and he was in due course appointed Brigade Major of the 4th Guards Brigade. In 1954, he re-joined the 1st Battalion in Egypt before returning to Staff College as an instructor. Leuchars took command of the 1st Battalion in 1963 before it was due to deploy to Aden in 1965 and his considerable operational experience ensured that the pre-deployment training was of a high order.

After an appointment as chief of staff to the director of operations in Borneo during the 'confrontation' with Indonesia, he commanded 11th Armoured Brigade in BAOR. A spell as deputy commandant of the Staff College was followed by promotion to major general upon his appointment as GOC Wales.

He retired from the Army in 1976 and from 1980 to 1989 was Chief Commander, St John Ambulance. During his retirement he wrote a number of personal memoirs about his life and career as a soldier in the Welsh Guards. He died on 17 July 2009.

Portrait painted by Alex Talbot Rice. *(Alex Talbot Rice)*

> There was a very close relationship indeed with the men in my platoon because we did a great deal of training and therefore got to know each other extremely well. I suppose I knew them better than my own family. You lived so close to them and did so many odd things and adventurous things and underwent a certain amount of hardship as well.[12]

During this period of hard training there were frequent changes in personnel in the two battalions, which saw Colonels Jefferson and Vigor exchange commands. At the same time, in February 1942 Colonel Leatham retired as Regimental Lieutenant Colonel and was replaced at Regimental Headquarters by Colonel Bankier. In September the 1st Battalion moved to Heytesbury near Salisbury while the 2nd Battalion moved to Fonthill Gifford, also in Wiltshire. Since becoming part of the Guards Armoured Division the 1st and 2nd Battalions had gone through many changes as they tried to find the correct balance for engaging in armoured warfare. Of necessity, given its armoured role, these innovations had mainly affected the 2nd Battalion which had been transformed in October 1942 into the 2nd (Armoured Reconnaissance) Battalion – the eyes and ears of the Division – following the departure of the 2nd Household Cavalry Regiment. In the summer of 1943 the Guards Armoured Division moved to Snarehill near Thetford in Norfolk to take part in Exercise Spartan as part of II Canadian Corps in British Second Army. Later in the year the division moved north to Yorkshire as part of VIII Corps. Scarborough became the focus for off-duty activities but, while the tempo of training was maintained, many Guardsmen found time hanging heavily on their hands as they waited in the wintry seaside town which Bill Williams recalled as 'the most frozen, coldest, wettest area in Great Britain. The icy winds whipped across the North Sea, tore the blood from the veins and filled the eyes with tears.'[13] By way

of contrast, Meirion Ellis of No. 3 Squadron, 2nd Battalion Welsh Guards, remembered the prevailing feeling of togetherness within the two battalions: 'We had a camaraderie par excellence, we were more like brothers and we were itching to go to war, we never doubted our cause, we knew we could beat the Germans and beat them in Europe at that.'[14]

A highlight of the deployment in Yorkshire was Christmas 1943 when Lieutenant Rex Whistler suggested that, as the 2nd Battalion had been denied leave, they should put on a party for the local children in the town of Pickering. The suggestion was accepted by the new Commanding Officer, Lieutenant Colonel J. C. Windsor-Lewis, DSO, MC, and as a result 294 children were entertained in the town's Memorial Hall, which was specially decorated by Whistler. (After being captured during the operations at Boulogne, Windsor-Lewis had managed to escape in November 1940, while being held in Belgium, and made his way to Marseilles before returning to Britain through Madrid and Lisbon.[15] On his return he found that he had been posted 'missing, presumed dead'.)

3rd Battalion, Tunisia

For the 3rd Battalion there was a move to Hampstead under Lieutenant Colonel D. E. P. Hodgson and with it came a change of role. Instead of being responsible for providing reinforcements for the 1st and 2nd Battalions, the 3rd Battalion was given an operational role and ordered to mobilize for service overseas in North Africa as part of 1st Guards Brigade, at that time under the command of a Welsh Guardsman, Brigadier F. A. V. Copland-Griffiths, DSO, MC. The brigade was part of 6th Armoured Division, under the command of Major General Charles Keightley, an experienced cavalryman. Having been constituted as an operational formation, the 3rd Battalion moved immediately into the familiar routine of training for war. Officers were sent on courses which ranged from Street Fighting (Chelsea Barracks) to Messing (London District School of Cookery), and in June 1942 the Battalion went into camp at Ilfracombe before moving first to Hillingdon Park, Uxbridge, and then to Hampstead, which was its home until the new year. From the very outset of its wartime service the

Brigade Commander's inspection, Scarborough, 1943. *(Welsh Guards Archives)*

3rd Battalion was anxious to maintain the high standards of the Welsh Guards and in this respect it was helped by the quality of its non-commissioned officers. When Lance Sergeant Thomas 63 and Lance Sergeant Davies 08 joined the Battalion in Hampstead where the Sergeants' Mess had been housed in a luxurious mansion, they were hailed by Regimental Sergeant Major A. Barter with the salutation: 'When you two joined up, you didn't have field marshals' batons in your knapsacks, you had bloody orchestra leader's batons, get your hair cut.'[16]

Another boost to morale was the presence of older pre-war Guardsmen such as Lance Sergeant Manchip 273, who had served in the 1st Battalion and was universally known as 'Pop'. At forty-five he was thought to be the oldest Welsh Guardsman on active service and had a wealth of experience and good advice which he gladly passed on to the younger recruits, although when necessary he had a sharp tongue and was not afraid to use it. A newcomer to Pop's section could not stop talking about the pleasures of food and the potential delights of the local female talent, but was silenced with the warning: 'For St David's sake, belt up, that's all you're on about, your bloody stomach and what hangs from it.' Another great character was Company Sergeant Major 'Woolly Bear' Roberts of No. 1 Company who was known throughout the Battalion for his malapropisms and his willingness to take a joke. On one occasion, during training in an area which was dotted with pylons carrying high-voltage electricity cables, he shouted a stentorian warning: 'Sergeant Price, keep clear of them pythons, you could get electrocuted!'[17]

All this was important for creating *esprit de corps* at a time when the new Battalion was working up for its active service role. At the beginning of February 1943, under the command of Lieutenant Colonel D. E. P. Hodgson, with Captain G. D. Rhys-Williams as Adjutant, the 3rd Battalion was taken north by train, travelling from King's Cross and St Pancras railway stations to Glasgow where they boarded HMT *Nea*

RSM Horace Cyril 'Phil' Phillips, MBE, MVO

The first Welsh Guardsman to hold the coveted post of Academy Sergeant Major at Sandhurst, Phil Phillips was born at Chepstow in Gwent on 27 March 1916 and joined the Welsh Guards in 1934. Taken prisoner near Arras during the retreat to Dunkirk in May 1940, he spent the rest of the war as a prisoner in Poland, in Stalag 383. In 1945 the Germans started moving their prisoners west to escape from the advancing Russians and Phillips was one of a group who wrested control from the SS shortly before they were liberated by the Americans. After service in Palestine, he acted as CSM in Prince of Wales Company, which was Escort to the Colour in the 1949 Sovereign's Birthday Parade. After postings in West Germany and Berlin he took part in the 1953 Coronation and then went on his first tour of duty at Sandhurst, as Regimental Sergeant Major of Old College, one of the three constituent parts of the Academy. He was then seconded to the King's African Rifles in East Africa, in the 1960s, before returning to Sandhurst as Academy Sergeant Major in succession to the legendary John Lord. Colleagues warned him that Lord would be a difficult act to follow, especially as Phillips was the first Welsh Guardsman to do the job, after a long line of Grenadiers. But, as *The Times* said in its obituary, 'When Phillips retired in December 1970, marching up Old College steps after the Sovereign's Parade, while the band played "Auld Lang Syne" he had carved out his own place in Sandhurst history.' After leaving the Army he accepted the post of senior Messenger Sergeant Major of the Queen's Bodyguard, twinned with that of Superintendent of St James's Palace. Phil Phillips was a notable rugby player in his youth, turning out as flanker for Newport, London Welsh and the Army and for the Welsh Guards when they won the Army cup after the war. He died in south Wales on Christmas Day 1992.

Major (QM) J. C. Buckland, MBE

The Quartermaster at the outbreak of the Second World War, John Clement Buckland served with the 1st Battalion at Arras and Dunkirk and remained in that post until September 1941. A native of Bangor, he joined the Welsh Guards in 1915 and ended the First World War in the rank of lance sergeant. In August 1937 he was commissioned and appointed Quartermaster. Being in charge of the Battalion transport, Buckland was one of the last Welsh Guards officers to see Lieutenant Furness alive. During the evacuation from the beaches at Dunkirk, Buckland went 'scrounging' at Lord Gort's headquarters and returned with two dixies of hot tea which were passed to grateful Welsh Guardsmen standing waist-deep in the sea. In 1941 he was appointed Quartermaster of the 2nd Battalion and served with it for the rest of the war. For later generations he contributed in 1965 his memories of life at the Guards Depot when he joined the Regiment: 'The instructors, the drill, the discipline, and above all the spirit and determination of the individual, not to be defeated by any power, spiritual or temporal. The training, extended order, Coulsdon Common, kit inspection, pay parade, saluting for something you hoped you would get, saluting for a meagre 1s. 1d. a day, the fear of losing one's name, the Short Arm Inspection in the shivering huts, the baths, "Tin Town", the wet canteen, the fatigues.' Buckland retired from the Army in July 1948 and died on 7 December 1965.

Major General Merton Beckwith-Smith, DSO, MC

Born in 1890, Merton Beckwith-Smith was educated at Eton and Christ Church, Oxford, before being commissioned in 1910 in the Coldstream Guards, serving with them throughout the First World War. He transferred to the Welsh Guards in 1930, becoming Commanding Officer of the 1st Battalion in October 1932. A first-class shot and keen horseman, he was Regimental Lieutenant Colonel from October 1934 to January 1938. In 1940 he was given command of 1st Guards Brigade, part of the British Expeditionary Force in France. After the evacuation from Dunkirk, Beckwith-Smith took over command of the Territorial 18th (East Anglian) Infantry Division, which he trained in preparation for duty overseas. Later he took his division to Singapore to be part of the garrison, but it saw little action before the surrender to the Japanese in February 1942. Together with the men of his division, Beckwith-Smith went into captivity. He died of diphtheria in a prison camp in Formosa (Taiwan). Many years later his grave was identified by the POW rights campaigner Jack Edwards on the request of Diana, Princess of Wales, one of whose ladies-in-waiting was Beckwith-Smith's grand-daughter Anne. Beckwith-Smith's sons Peter and John later served with the Regiment.

Major (QM) W. L. Bray, MBE, DCM, MM

The first Quartermaster of the 2nd Battalion, Walter Luke Bray was born on the Isle of Sheppey in 1896 and joined the Army in 1914, serving in the Grenadier Guards. Always known as 'Twinkle', he was awarded the Distinguished Conduct Medal and the Military Medal while serving on the Western Front. In 1934 he transferred to the Welsh Guards and was appointed Regimental Sergeant Major, a post he held until his retirement in 1938. On the outbreak of the Second World War he returned to the Regiment, serving as Quartermaster of the 2nd Battalion, where it was said of him that 'his experience and enthusiasm did much to help mould the newly formed battalion'. In 1941 he became Quartermaster of the 1st Battalion and served with it until the end of the war.

In January 1946 he was appointed Camp Commandant of HQ Eastern Command until his second retirement in August 1951. He died in December 1990.

Hellas, part of a convoy bound for the Mediterranean through the Bay of Biscay.[18] Built in Glasgow in 1922 as the *Tuscania* for the Anchor Line's transatlantic services, *Nea Hellas* ('New Greece') was later sold to the General Steam Navigation Company of Greece and requisitioned by the Allies for war service as a troopship. She was known throughout the Army as the 'Nellie Wallace'.

At the beginning of 1943 the war against the German and Italian armies in North Africa was reaching a climax. In the previous October, following months of setbacks, the British Eighth Army under General B. L. Montgomery had defeated the German Afrika Korps at the Battle of El Alamein and had started pushing the enemy back towards Tunisia. Next, the British First Army (Lieutenant General Kenneth Anderson) and US II Corps (Major General Lloyd Fredendall) had landed in Morocco and Algiers at the beginning of November 1942, as part of Operation Torch and had started moving eastwards towards Tunis and Bizerta. Although the joint British–US enterprise had enjoyed a promising start, by February 1943 the impetus had stalled following ignominious defeats at Sidi-Bou-Zid and the Kasserine Pass. Anderson's army consisted of three army corps, two British (V and IX) and one Free French (XIX); after the opening rounds they needed reinforcement. Amongst those which arrived early in 1943 was 3rd Battalion Welsh Guards, which had been despatched to North Africa to strengthen 1st Guards Brigade. This brigade had originally landed in November 1942, when it consisted of 3rd Battalion Grenadier Guards, 2nd Battalion Coldstream Guards, and 2nd Battalion Hampshire Regiment but the Hampshires had been badly mauled in the opening rounds and had had to be replaced. Casualties in the brigade had been heavy from the outset of the campaign and some idea of the ferocity of the fighting can be seen in the fact that the Hampshires emerged from their first battle at Tebourba on 26 November 1942 with only three officers and 170 men unscathed. Amongst those killed or wounded were six Welsh Guards officers who had been attached.

The 3rd Battalion landed at Algiers in the late afternoon of 16 February 1943 and immediately faced a fourteen-mile march to their first camp at Oued Zarga – much to the dismay of the Guardsmen who were wearing 'change of station order' (carrying large packs, blankets, rifles, ammunition, small packs and full water bottles). Almost immediately they had to come to terms with the enervating conditions which Sergeant (later Drill Sergeant) W. B. Davies 08 of 8 Platoon, No. 3 Company, described as 'body draining heat and irritating flies, and that blasted insidious North African dust that settled everywhere, and covered static objects with its greyish brown film of discomfort and annoyance'.[19] Despite the heat and despite being relatively unfit after the ten-day voyage, no one fell out during the lengthy march. A week later they were transported to the front by trains in trucks marked in the French fashion with the sign '*Hommes 30 – Chevaux 12*'; at Guardimaou they were reunited with their transport and carriers and arrived at El Arousa, south-east of Medjez el Bab, where they eventually joined up with their new brigade, suitably on St David's Day. By later in March the strategic situation had improved. To the south, Montgomery had attacked the Mareth Line, a defensive position between the towns of Medenine and Gabès in southern Tunisia, while to the north US II Corps (now commanded by Lieutenant General George S. Patton) had captured Gafsa to secure the right flank of First Army. In these new dispositions overall command of the Allied armies was coordinated by General Sir Harold Alexander, 18th Army Group. It was at this stage, as the German and Italian forces were being squeezed between the two advances, that 3rd Battalion Welsh Guards was first committed to battle at the end of the first week of April 1943.

The place was the Fondouk Pass, a strategically important position halfway along the road between El Guettar and Tunis where the River Maguellil threaded its way through a narrow defile less than a thousand yards wide. To the north lay high ground formed by the rocky pinnacles Djebel Houfia and Djebel Rhorab, while on the southern side was an equally precipitous

Commanders prepare for the Battle of Fondouk. (*L to R*) Brig F. A. V. Copland-Griffiths (Commander 1st Guards Brigade), Maj Gen C. F. Keightley (GOC 6th Armoured Division), Lt Col D. E. P. Hodgson, (Commanding Officer 1st Battalion) and Brig T. Lyon Smith (CRA 6th Armoured Division).

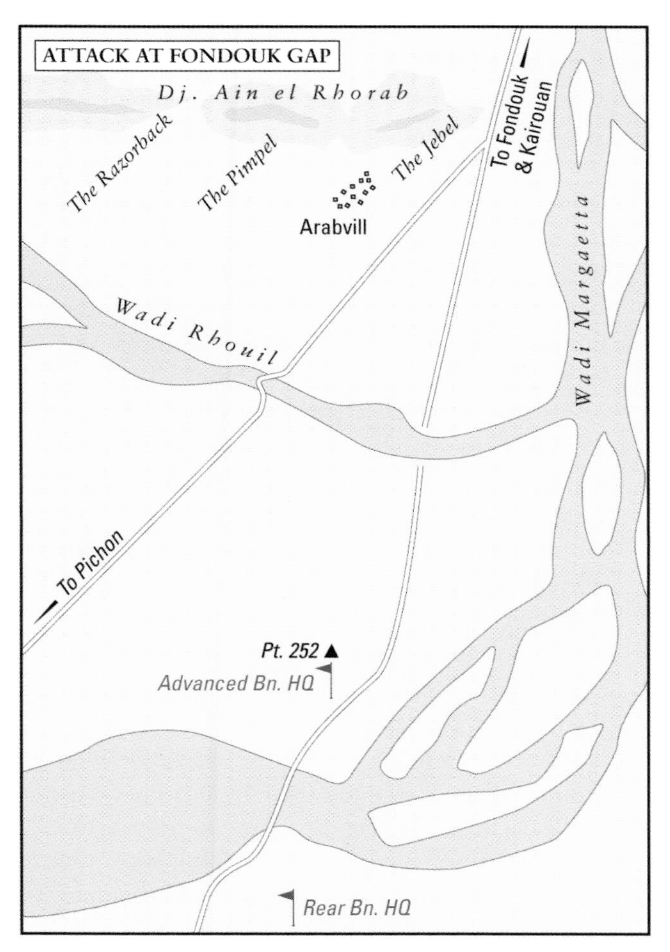

Attack by 3rd Battalion on Fondouk Gap, 10 April 1943. The main objective was a feature known as the Djebel Ain el Rhorab.

escarpment called the Djebel Haouareb, all of which were home to well-dug-in German artillery and mortar positions. In March the US Army had made a half-hearted effort to breach the pass but it quickly became clear that a larger and more powerful force would be required. The task was given to US 34th Division and British IX Corps under the command of Lieutenant General John Crocker, who had earned a reputation as a forceful brigade commander during the retreat to Dunkirk. Direct in his approach and normally competent in his preparations, he under-estimated the strength of the German positions at Fondouk and, as a result, his plan contained flaws which worked against the formations under his command. It did not help matters that he had a poor opinion of US troops in general, in consequence of their earlier indifferent performances in the campaign. Crocker's plan called for US forces to attack to the south of the pass, while British 128th Infantry Brigade assaulted the positions to the north. Once the pass had been secured, the tanks of 6th Armoured Division would sweep through along Highway 3 and move towards Kairouan and the coastal plain to link up with Montgomery's Eighth Army as it attacked north. According to one recently arrived Welsh Guards officer, Lieutenant W. K. (Keemis) Buckley, the planning for the attack on the Fondouk pass should have been 'simplicity itself' but it did not turn out that way.[20]

The attack began at 03.00 on 8 April and very quickly the US 34th Division came under intense German artillery fire and 'found themselves in the beaten zone almost immediately'.[21] To the north, the British brigade did better but failed to secure the vital summit of the Djebel Ain el Rhorab position which was divided into two separate features – the *djebel* (or *jebel*, Arabic 'hill') itself and another peak known as the 'Razorback'. This necessity forced Crocker to commit more infantry to the attack and the task was given to 1st Guards Brigade for the night of 8/9 April, with 3rd Battalion Welsh Guards leading the assault from the west. During the reconnaissance, Captain R. C. Twining, commanding No. 1 Company, was killed – Sergeant Davies was the last friendly face to see him alive as he 'blended with the darkness'.[22] He was replaced by Lieutenant A. G. Stewart, a large imposing man who had difficulty finding suitable cover in the wastes of the *wadi* (watercourse). At first light on 9 April, 3rd Battalion Welsh Guards went into the attack, No. 2 Company on the right with No. 4 Company in support, and No. 3 Company on the left with No. 1 Company in support. Although the initial goal, the Rhouil *wadi*, was taken without much difficulty, the intensity of the German machine-gun fire increased during the ascent of the *djebel*, leaving the rifle companies pinned down in the lower reaches. Within a very short time, every officer in No. 2 Company and No. 3 Company had either been killed or wounded, leaving the two companies severely incapacitated. Many of those wounded had received horrendous injuries from the seemingly unremitting mortar fire, but amidst the carnage, there were moments when the Guardsmen showed their pride and sense of humour. Having been badly wounded in the thighs by machine-gun fire and with blood streaming down his legs, Guardsman Davies 38, a newly married man, had only one concern: 'If I go home without them,' he shouted, 'Rena will be bloody raving.'[23]

Not far from the scene of this action, the officer commanding No. 3 Company, Captain C. A. la T. Leatham, had also been cut down by enemy fire and was unable to move. As he lay stricken, an NCO approached him across the bullet-swept ground. However, it was not to apply a field dressing but to administer the Last Rites. Tony Leatham was a nephew of Chicot Leatham, and like him did not suffer fools gladly. His response was swift and to the point: 'Stop that ridiculous nonsense and get back to your platoon at once!'[24] He survived and, thanks to the ministrations of the Battalion's highly revered Medical Officer, Captain David Morris, who served with the 3rd Battalion Welsh Guards throughout the war, was eventually evacuated home, recovered and later commanded the 1st Battalion.

At this stage in the battle some luck and good leadership came to the aid of the 3rd Battalion Welsh Guards. Realizing that his men in the forward positions needed artillery support but were without any workable radio contact, Lieutenant Colonel Hodgson made his way back to rear Battalion Headquarters where he met Brigadier Copland-Griffiths who promptly arranged for the deployment of tanks of 2nd Lothians and Border Horse (26th Armoured Brigade) to support the Battalion. Meanwhile, the Adjutant, Captain G. D. Rhys-Williams, went forward by carrier to inform the four rifle companies of the plans for a renewed assault which would see the tanks manoeuvre around their left flank and attack the reverse slopes of the Razorback position. Showing great presence of mind and no little courage, Rhys-Williams succeeded in making contact with all the companies and also managed to get through on the radio net to direct artillery fire onto the objective. All this was done while he was under enemy fire. His actions while leading the attack were observed by Lance Sergeant K. G. Summers, in command of the mortar platoon of No. 1 Company:

> I could see Captain Rhys-Williams dashing from one section to another encouraging them on to their objective. He never seemed to tire although he was covering twice as much ground as anyone else by dashing about from one to another. On reaching the bottom of the hill they paused for a few moments and then went on to the final assault. The whole way up the hill I could see Captain Rhys-Williams in the lead.[25]

The ferocity of the attack and the introduction of armoured and artillery support forced the Germans to withdraw from the *djebel* position; by midday it was in the hands of the Battalion. During the final assault Rhys-Williams was killed by enemy sniper fire, but it

Captain Glyn Rhys-Williams

Glyn Rhys-Williams came from a distinguished Welsh family of Miskin Manor, near Llantrisant in the Vale of Glamorgan. The family had a wonderful record of service to the Regiment. His father, Sir Rhys Rhys-Williams, was one of the original officers, having transferred from the Grenadier Guards in 1915, and his brother, Sir Brandon Rhys-Williams, served in the Regiment before becoming MP for Kensington in 1968 and a Member of the European Parliament.

Glyn Rhys-Williams was educated at Eton and joined the Regiment in April 1941. In November 1942 he was posted to the 3rd Battalion, later to sail for North Africa as part of 1st Guards Brigade in the 6th Armoured Division. His potential was soon recognized and in 1943 he was appointed Adjutant, aged just twenty-one. The advance eastwards had been held up and at the Battle of Fondouk on 8 and 9 April 1943 the 3rd Battalion was committed to taking the high ground which dominated the Fondouk Pass. The battle is described in the text, but when the attack was in danger of faltering, with the leading platoons pinned down by accurate enemy fire and losses mounting, it was the inspirational leadership and selfless bravery of Rhys-Williams which turned the course of the battle.

Rhys-Williams was killed by a sniper as he led the final assault on the hill. After the battle he was nominated by the Commanding Officer for the award of a Victoria Cross, but sadly, as was so often the case, this was downgraded by the Army Headquarters to a Mention in Despatches. He is buried in Enfidaville War Cemetery in Tunisia.

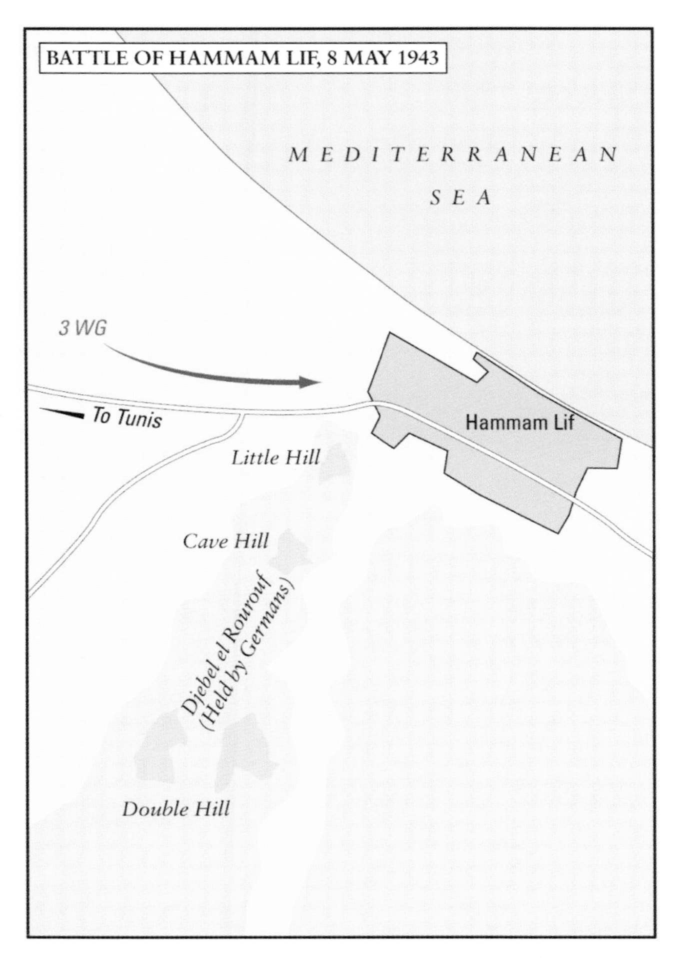

Attack by 3rd Battalion to secure Hammam Lif, 8 May 1943. To do this they had to attack German positions on the Djenel el Rorouf.

was due to his inspired leadership that the Djebel Rhorab was taken. (Rhys-Williams's undoubted courage prompted questions within the Regiment about why it was not recognized by a gallantry award. It later emerged that he had been nominated for a Victoria Cross but as he had already been mentioned in despatches and the War Office had 'an inflexible rule that no one may be awarded more than one decoration for the same action'.)[26] The action at Fondouk was typical of many fought in the campaign and in the subsequent fighting in Sicily and Italy. A well-defended enemy position had to be secured by infantrymen fighting over difficult ground and under constant fire. Although armour and artillery played their part, the enemy would not have been defeated but for the steadfastness and determination of those on the ground. Watching the attack from a neighbouring position was the Australian war correspondent Alan Moorehead, who never forgot what he had seen and recorded the incident in his subsequent history of the campaign:

> It was the Welsh Guards I especially remember that day, though there were others in the fighting as well. In a steady, unflinching line the Welshmen went up the last bare slopes on foot, and they faced a withering machine-gun fire all the way up. When a man fell, someone was always there to step in and the line went on until it reached the top.[27]

Losses in the Battalion were 114 officers, NCOs and Guardsmen killed. Their sacrifice encouraged Crocker to order that the pass should be stormed by Sherman tanks of the 17th/21st Lancers. Although the attack failed with the loss of thirty-two tanks, most of them having hit mines, a follow-up assault by 16th/5th Lancers eventually succeeded in breaching the pass by late afternoon with a fury which one tank commander remembered as 'engines revving, Browning rattling, guns crashing, being flung backwards and forwards and from side to side'.[28] As a result of the Axis withdrawal from Fondouk, 6th Armoured Division was able to move on towards Kairouan on 10 April but an opportunity had been lost – the delay in securing the pass had allowed the Germans and the Italians to head north and regroup for the defence of Tunis. As happens so often in war, a bad plan had been redeemed by the courage and tenacity of the men on the ground, in this case 3rd Battalion Welsh Guards and the tank crews of 17th/21st and 16th/5th Lancers. Even before the attack had gone in, Sergeant W. B. Davies overheard an honest tribute from a fellow soldier who was digging trenches as the Battalion moved up to Fondouk: 'Those are the Guards, the gen is they are putting in a big attack. They'll win, the Guards ALWAYS win, no bastard can beat them.'[29]

Ahead lay the approach to Tunis across the Goubellat Plain and, as Lieutenant W. K. Buckley recalled the details of the route, it was along 'a flat dead straight road [where] we were repeatedly machine gunned by enemy aircraft'.[30] On 14 April Brigadier Copland-Griffiths was replaced as commander of 1st Guards Brigade by Brigadier S. A. Foster. By then Alexander had switched the main axis of the attack to Anderson's First Army which was reinforced with elements of Montgomery's Eighth Army. This allowed him to change tactics from squeezing the perimeter around Tunis to making an all-out assault from the direction of Medjez el Bab and along the valley of the River Medjerda. With the Free French XIX Corps and British Eighth Army providing a stop line at Enfidaville, the attack began in the early hours of 6

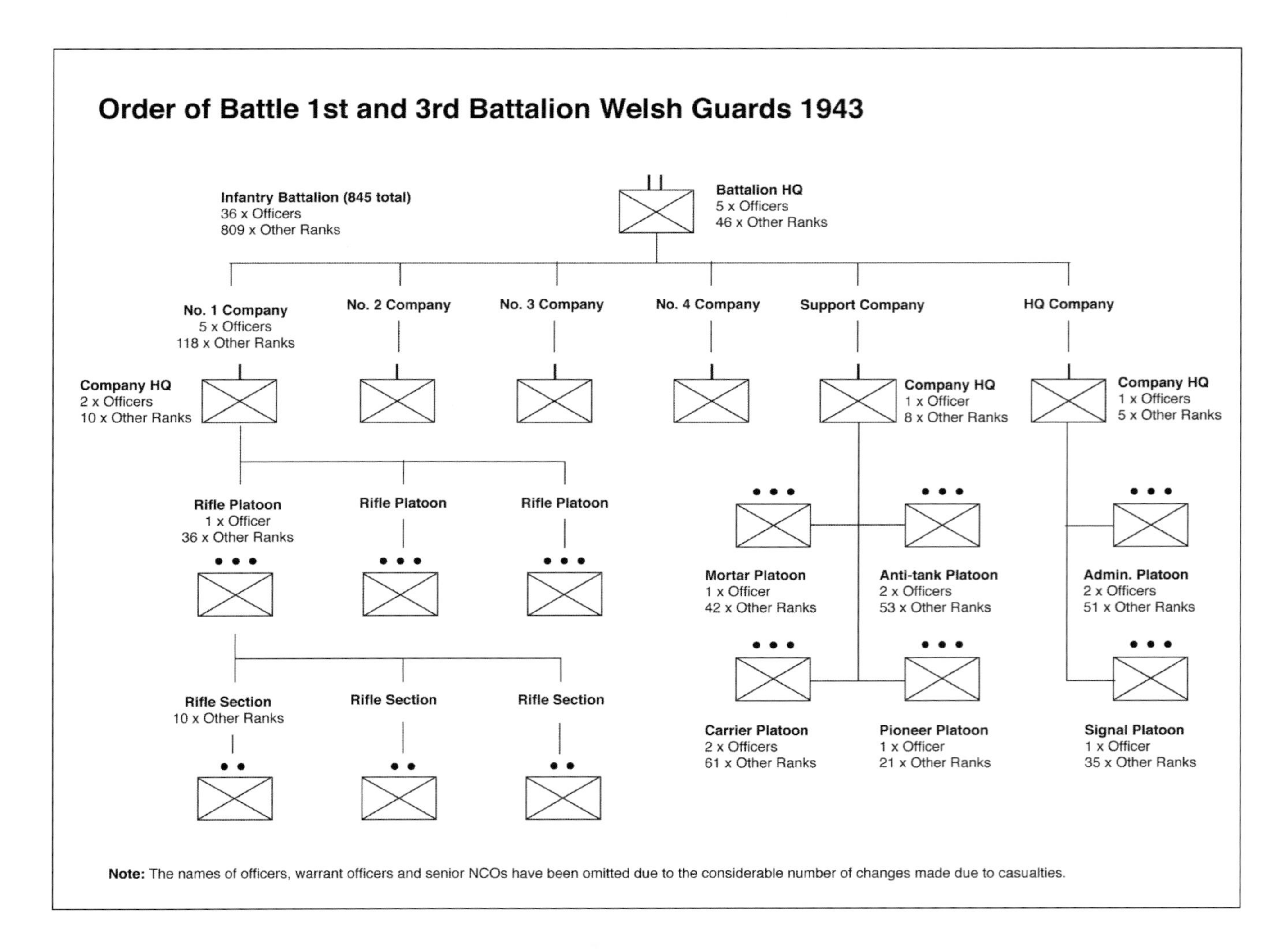

May. By the end of the day the first elements of 6th and 7th Armoured Divisions had entered Tunis, while the remainder swung north towards Bizerta and south-east towards the seaside town of Hammam Lif which guarded the approaches to the Cape Bon peninsula. Because of fears that this might become a German redoubt, it was decided that the town should be taken, but before any attack could be made, the German positions on the Djebel el Rorouf, a dominating ridge above the town, had to be captured. The task was given to 3rd Battalion Welsh Guards with 2nd Coldstream Guards in support. The ridge was a formidable objective: although less than a thousand feet high, the *djebel* was steep-sided and its summit consisted of three separate hills – named Double Hill, Cave Hill and Little Hill – all of which provided substantial challenges.

The attack was timed to begin at 13.30 on 8 May but was delayed by ninety minutes as the road below was choked with traffic. Following an intensive artillery barrage of 'undiscriminating generosity', the assault began with No. 3 Company (Captain J. R. Martin-Smith) on the right heading towards Double Hill, with No. 2 Company (Captain W. N. R. Llewellyn, MC) on the left heading towards Cave Hill.[31] Once the barrage had died down, both companies came under heavy machine-gun and mortar fire and, although the lead platoon of No. 3 Company succeeded in capturing its objective, both companies were soon taking casualties as they crossed the open slopes. The topography also caused problems: as the Guardsmen fought their way to the top of the ridge, they discovered that the main

RIGHT: The 2nd Battalion Welsh Guards is inspected by the Regimental Lieutenant Colonel in 1943. The tanks at the front of the picture are Covenanters from No. 1 Sqn. Beyond these can be seen newly issued Centaur tanks, which were in turn replaced by Cromwells, The main difference between the Centaur and Cromwell was that the latter was fitted with the Rolls-Royce 12-cylinder Meteor engine. This was a development of the Spitfire aircraft's Merlin engine. It gave the Cromwell 540 bhp and endowed the tank with a road speed in excess of 40 mph. *(Welsh Guards Archives)*

Major J. D. Gibson-Watt, MC**

James David Gibson-Watt, always known as David, was born on 11 September 1918, a descendant of the inventor James Watt and the eldest of five children of Major James Miller Gibson-Watt. He was brought up on his family's estate in Wales, where he quickly became an enthusiastic shot and fisherman; he was educated at Eton, where he played cricket, joined the OTC and became captain of his house, before going up to Trinity, Cambridge. At the outbreak of the Second World War, Gibson-Watt, who had joined the TA at university, was commissioned in the Welsh Guards. After a year with the Training Battalion, he joined the 2nd Battalion in October and later saw action with the 3rd Battalion in North Africa and Italy.

In May 1943 he was awarded the first of three Military Crosses for his 'outstanding gallantry and leadership' during the Battle of Hammam Lif. A first Bar was added during the Battle of Monte Cerasola in February 1944 and a second Bar followed in April during the fighting at Castel Gugliemo. In the words of the citation: 'Throughout this memorable day, Major Gibson-Watt continuously inspired and cheered all under his command, and his obliviousness to his own safety was complete.' His brother Andrew also served in the 3rd Battalion.

After retiring from the Army in 1946, Gibson-Watt entered politics and, following one or two false starts, was elected Conservative Member for Hereford in February 1956. Fourteen years later he was appointed Minister for State at the Welsh Office in Edward Heath's administration of 1970–74. He was created a life peer in 1979 and died on 7 February 2002. In its obituary the *Daily Telegraph* described Gibson-Watt as 'in many ways, the epitome of the Tory gentleman farmer of a vanished age, fitting in careers as a soldier, Member of Parliament and JP alongside continuous, less formal, services to the countryside and his local community'.

This painting by Rex Whistler is in fact a landscape target which was used for training soldiers to give fire control orders. *(Estate of Rex Whistler)*

features on the summit were separated by ridges and crests of rock which provided ample protection for the German positions, while the open ground was devoid of cover. When the attack stalled at 17.00 hours, the Commanding Officer ordered the reserve companies to reinforce those on the summit, supported by tanks of 2nd Lothians and Border Horse. On the left, No. 1 Company, under the command of Major H. C. L. Dimsdale, succeeded in taking Cave Hill, while on Double Hill No. 4 company, led by Captain J. D. Gibson-Watt, managed to reach the summit. As night fell the rifle companies consolidated their positions, allowing the 2nd Battalion Coldstream Guards to move through their lines to clear the rest of the ridge. By dawn the following day, 9 May, the enemy no longer occupied the heights of Djebel Rorouf and this freed the advancing tanks of 26th Armoured Brigade to complete the encirclement and capture of Hammam Lif. During the fighting, the 3rd Battalion Welsh Guards lost twenty-four Guardsmen who were killed or died from their wounds and fifty were injured. Both Fondouk and Hammam Lif were awarded to the Regiment as battle honours. A month after the fighting ended, the War Office recognized the role played by the 3rd Battalion Welsh Guards in an official communiqué:

> They had an important part to play in the final phase of our operations. The high ground south of Hamman Lif, south-east of Tunis, fell to their attack on 9th May, and their support did much to enable our armour to break through to the east and cut off the Cape Bon Peninsula from the enemy forces encircled behind Enfidaville.[32]

The capture of Hammam Lif signalled the end of the campaign in Tunisia and the Battalion War Diary

recorded the 'great welcome given by local inhabitants' as the Guardsmen entered the town. It also noted that amongst the many Italian prisoners-of-war who surrendered were three colonels.[33] Three days later, 12 May, saw the surrender of the German commander, Colonel General Hans-Jürgen von Arnim. Described as 'one of the last knights of the Old School' – one of his finest actions had been to send a message in clear to Alexander, asking him to call off an RAF bombing raid on the Italian ship *Belluno* in Tunis harbour as 700 wounded British prisoners-of-war were on board.[34] Although von Arnim had his faults as a battlefield commander, he was clearly a gentleman.

With the fighting stage of the operation at an end, the 3rd Battalion Welsh Guards occupied the nearby town of Nabeul where their main task was dealing with Italian prisoners-of-war – 3,000 passed through the Battalion's hands on 11 May. On the following day they moved into a rest area at Bou Ficha where the streams of captives continued, but the Guardsmen were also able to take advantage of the nearby beach and offers of wine and food, 'a terrific change after weeks in the bare hills, short of water, unwashed, unshaven and covered in dust'.[35] As Kemmis Buckley recorded in his memoir of his service with the 3rd Battalion, it was impossible to say if the end of the operation was climax or anti-climax or a mixture of both: 'Everywhere there were parties of enemy giving themselves up and we captured a group of Italian generals at an immaculately prepared breakfast table laid with a fine linen table cloth.'[36] The next stage of the Allied war effort was the invasion of Sicily as a precursor to the invasion of mainland Italy in September 1943, but 1st Guards Brigade was not required until early in 1944. For the rest of the year the 3rd Battalion remained in North Africa training and refitting, first near Sousse in Tunisia and then near Constantine in Algeria. On 28 July, command of the Battalion passed to Lieutenant Colonel Sir William Vivian Makins, Bt, who had been seconded to the Sudan Defence Force earlier in his career between 1928 and 1930.

CHAPTER SIX

Invasion of Europe, 1944

1st and 2nd Battalions, Normandy, Belgium
3rd Battalion, Italy, Cerasola, Cassino, Piccola and Arce

IN DECEMBER 1941, FOLLOWING Japan's unprovoked attack on the US Pacific Fleet at Pearl Harbor in Hawaii and Hitler's declaration of war, the USA joined the Allied war effort. From the very outset the American military planners made a strong case for an early attack on the European mainland and the decision to press ahead with the invasion of North-West Europe was confirmed as early as May 1943 at an Allied conference (codenamed Trident) in Washington. Planning for it began under joint US–British direction immediately after the summit had ended. The main desiderata for the cross-Channel amphibious attack were quickly established: a landing area with shallow beaches and without obstacles which was within range of Allied fighter aircraft, the neutralization of local defences to allow a build-up which would equal the strength of the German defenders; and the presence of a large port for reinforcement and re-supply. Deception also formed part of the plan: the idea was to persuade the Germans that the assault would be made across the narrowest part of the English Channel to land in the Pas de Calais region, where the beaches were shallow and led into the hinterland without the obstacles of cliffs and high ground. It also offered the opportunity to make a quick strike into the Low Countries and from there

The 2nd Battalion lands at Arromanches and moves forward into Normandy, June 1944. *(Welsh Guards Archives)*

Welsh Guards cap-badge painted by Rex Whistler, Brighton, Summer 1944. *(Estate of Rex Whistler)*

Men of No. 2 Company, 3rd Battalion Welsh Guards riding on a Sherman tank of 2nd Lothians and Border Horse. Arce, May 1944. *(Welsh Guards Archives)*

Lance Corporal Keogh, MM, by Augustus John, OM. *(Estate of Augustus John)*

into Germany. All those reasons made the Pas de Calais the ideal place for invasion, but because it was the obvious location, it was quickly discounted as the Allied planners realized that their German counterparts would deploy their strongest defensive forces there. At the end of the summer the plan was shown to the Allied leadership at the Quadrant conference in Quebec, which amongst other matters discussed the tactics to be used in the invasion of Europe. The chosen landing ground was the Baie de la Seine in Normandy between Le Havre and the Cotentin peninsula, an area which met all the criteria, including having a deep-water port nearby at Cherbourg.

It was eventually agreed that the initial seaborne assault should be made by the equivalent of five divisions, two US, two British and one Canadian, with one British and two US airborne divisions operating on the flanks. The D-Day invasion began on 6 June 1944 with the airborne forces securing the flanks overnight while the main assault went in at dawn, preceded by a mighty bombardment from warships in the Channel. By the end of the day the assault divisions were ashore and the five landing areas – Utah, Omaha, Gold, Sword and Juno – had been secured with the loss of under 10,000 casualties (killed, wounded or missing), fewer than expected.

In his appreciation of the post-invasion operations, the Allied land forces commander, General Montgomery, estimated that the battle for Normandy would require three phases lasting up to eighty days in total. The first would run for twenty days and would see the US First Army (Lieutenant General Omar N.

Bradley) capture its objectives in the Cotentin peninsula, while the British Second Army (Lieutenant General Sir Miles Dempsey) assaulted west of the River Orne, pivoting on Caen, to shield the US offensive. The second phase would be the beginning of the break-out, with the British forces pushing south through Falaise towards Argentan, while the Americans moved towards the Loire and Quiberon Bay. Phase Three would take the Allies to the Seine, with the US First Army heading towards Paris, while the British and Canadians would operate to the north between Rouen and the Channel. At the same time, Lieutenant General George S. Patton's US Third Army would move through the US First Army's front to clear Brittany and would then operate on the southern flank. Following the creation of a bridgehead on the Normandy beaches, the follow-up forces designed to achieve these goals began arriving in France almost immediately. Amongst them was the Guards Armoured Division, which crossed over between 18 and 29 June.

1st and 2nd Battalions, Normandy

For both Welsh Guards battalions the waiting came to an end on 1 May 1944 when they moved south from Yorkshire to the division's assembly points near Brighton and Eastbourne, with stops at Doncaster and Stevenage. All sixty-one of the 2nd Battalion's Cromwell tanks travelled under their own traction, a good test of their mechanical readiness, and once at the assembly point they were waterproofed for the landing. Before leaving England, the division was addressed by the Regimental Lieutenant Colonel, The Earl of Gowrie, VC, and also on 25 May by the Allied Supreme Commander, US General Dwight D. Eisenhower, who, according to the diary of Sergeant Charles Murrell, 'said just the right things, in just the right way. He spoke to us not as if to children and inferiors, but as man-to-man.' Murrell served in the intelligence section of the 1st Battalion and his diary plus his accompanying illustrations provide a graphic record of the progress of the war as the two battalions made their way into North-West Europe.[1]

Three weeks later, the division began the task of moving across to France, travelling in a variety of vessels and heading across the Channel to the huge Mulberry artificial harbour at Arromanches. Looking at the conditions on board the former Irish mail packet *Princess Maud* – the holds containing six tiers of bunk beds – Meirion Ellis simply noted in his memoirs, 'We expected no comforts and we got none.'[2] Poor weather

Sergeant Charles Murrell

Charles Stuart Murrell was born in Cardiff in 1912 of English parents and had a peripatetic childhood. After leaving school he found that he had a talent for art and worked for a time for a firm of commercial artists. Always restless as a young man, he felt that he 'needed a really hard kick in the pants' and joined the Army in 1933, signing on for four years' service in the Welsh Guards. After leaving in 1937 to work for the Ordnance Survey, he was called up as a reservist in June 1939 and after two months' training was posted to the Signals Section, 1st Battalion Welsh Guards. He served in the Battalion throughout the war, seeing operational service in 1940 in France and again in France and North-West Europe 1944–5. He left the Army after the war in the rank of sergeant and returned to work with the Ordnance Survey. Throughout the war he kept a personal diary and drew hundreds of sketches illustrating his and the Battalion's experiences. Many of these were used to illustrate Major L. F. Ellis's war history, *Welsh Guards at War*, and his own diaries were published in 2011 under the title *Dunkirk to the Rhineland: Diaries and Sketches of Sergeant C. S. Murrell, Welsh Guards*. Murrell died in 1987.

A jeep passes the Divisional Advanced Dressing Station near Kevelaer. Drawing by Sgt C. Murrell. *(Courtesy of the Murrell family)*

Guardsmen of the 3rd Battalion cleaning their .303 Vickers machine guns. These were carried by two sections of the Carrier Platoon. Standing (*L to R*) are Sgt D. Thomas, in charge of the machine guns, LCpl W. J. Rasbridge, Sgt J. Tumelty and LSgt T. Anzani. *(Welsh Guards Archives)*

Guardsmen of the 3rd Battalion at the Battle of Arezzo, July 1944. *(Welsh Guards Archives)*

Guardsmen of the 3rd Battalion learn of the Allied invasion of Normandy, June 1944. *(Welsh Guards Archives)*

Welsh Guardsmen resting during the battle for Arezzo. The NCOs shown include Sgt G. A. James, LSgt R. Davis and LCpl H. Lewis, July 1944. *(Welsh Guards Archives)*

conditions and a rough sea made for a difficult crossing but many of the hardier Guardsmen travelling on board LSTs (landing ships, tank) of the US Navy were astonished by the quantity and quality of the food that was available in the mess decks and made full use of what was on offer. The crossings were completed by 29 June.

After landing in France, the 1st Battalion proceeded to St-Martin-les-Entrées, to the east of Bayeux, while the tanks of the 2nd Battalion assembled in the fields nearby. With the battle to take Caen raging this was supposed to be a quiet area but the 1st Battalion received a foretaste of the unforeseen nature of war when a mortar salvo injured a number of officers and Guardsmen, including the Commanding Officer, Lieutenant Colonel G. W. Browning, who all had to be evacuated to England. The misfortune was compounded the following day when another mortar attack killed Browning's replacement, Major J. E. Fass; as a result Major C. H. R. Heber-Percy, MC, took over command of the Battalion. The opportunity to redress the balance came on 18 July when both battalions were involved in Operation Goodwood, Montgomery's attempt to end the stalemate at Caen by attacking the German positions to the east with three armoured divisions, including the Guards Armoured Division which was on the left flank of the attacking force.

Following a massive aerial bombardment, the division went into the attack at the beginning of what

German prisoners, Italy, 1944.
(Welsh Guards Archives)

Liberation of Brussels, September 1944.
(Welsh Guards Archives)

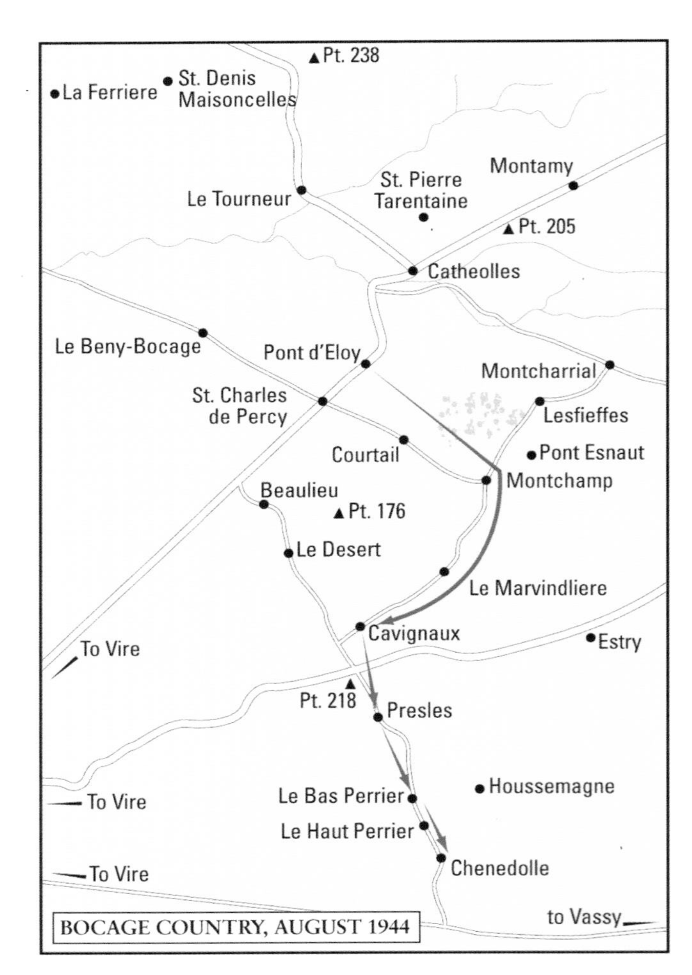

Guards Division advance through the *bocage* country of Normandy towards Montchamps and Estry, August 1944.

was to prove a day of confused fighting. The 1st Battalion reached the village of Cagny and then pushed on to Le Poirier to the south-west, while the 2nd Battalion supported a company of 1st Battalion Grenadier Guards on the left flank. During this phase of the attack Lieutenant Rex Whistler was killed by mortar fire after dismounting from his tank to clear wire from its sprocket. His last work had been completed before he crossed over to France – a fanciful depiction of King George IV as Prince of Pleasure which graced a room in the house at 39 Preston Park Avenue in Brighton where the artist had been billeted with other officers of 3 Squadron. After the war the mural was purchased by Brighton Council and is now in the Royal Pavilion Art Gallery.[3] Bad weather in the shape of heavy rain hampered the rest of the operation which was wound down two days later without achieving its objective of completing the break-out from Caen.

The next stage was an advance towards Cagny and Colombelles, both places of ill-repute, where the Germans put up fierce resistance. The weather, too, was bad and in his diary for 21 July Sergeant Charles Murrell did not attempt to disguise the deteriorating conditions and the effect they had on the men:

> Well, here we are, another day – my bones and my back aching – my sodden clothing sticking clammily to my tired body. To add to the miseries of mud and waterlogged trenches,

Three Wartime Chaplains

The Regiment was fortunate to have outstanding chaplains who served with all three battalions in England, France, North-West Europe and Italy. According to those who knew them, 'they were men of strong character who had already had some experience of the world. They were quick to adapt to the way in which the Foot Guards go about their business at home and on active service. They had a close affinity with Wales and related easily with Welshmen.'

The Revd. Percy Forde Payne, universally known as 'Padre Payne', served with the Regiment throughout the war, first with the 2nd Battalion at Boulogne and then with the 1st Battalion in the UK and North-West Europe. The 1st Battalion's War Diary recorded his departure on promotion in September 1944: 'He will be missed by every officer and man in the Battalion.' Payne was awarded the Military Cross in March 1945.

The Revd. Cecil Howard Dunstan Cullingford also served with the Regiment throughout the war, first with the 1st Battalion in France and then with the 2nd Battalion in England, France and North-West Europe. Described as 'a born thief', he earned this reputation during the retreat from Arras in May 1940 for the ingenuity of his raids on the NAAFI to provide additional rations for the hard-pressed Quartermaster. After the war he became Headmaster of Monmouth School and was an authority on caving. He died on 7 July 1990.

The Revd. T. M. H. (Malcolm) Richards joined the Regiment in the summer of 1944 and served with the 3rd Battalion in Italy and Austria. He enjoyed a good rapport with all ranks and was usually referred to by the officers as 'The Turbulent Priest'. During the evacuation of the Cossacks in Carpathia he spent time comforting the older men and women and giving chocolate to the children. After the war he was vicar of St John the Baptist in Bedwardine in Worcester.

constant rain, and all the other miseries of this place, whole squadrons of viciously buzzing mosquitoes have appeared, biting and tormenting us. But here we are still. A rainy, misty, chilly dawn … and always the dreadful, bumpy, shallow and waterlogged trench.[4]

Welsh Guardsmen provide aid to injured Germans after their truck was destroyed during the Allied advance. *(Welsh Guards Archives)*

Portrait of Lt R. G. Whiskard, painted by Rex Whistler. Like Whistler, Whiskard was killed in action during the Normandy Campaign. He is wearing the distinctive black overalls worn by 2nd Battalion officers during training. *(Estate of Rex Whistler)*

August brought both battalions into Normandy's *bocage* country. With its small hedge-bound fields and deep lanes running beside high banks it is an attractive enough landscape – the stone-built villages and farms added to the charm – but the topography favoured the defending Germans who used it to their advantage by positioning tanks and artillery in its hidden places. It

A 2nd Battalion Cromwell tank enters Trie-Chateau, near Gisors, France, 31 August 1944. *(Imperial War Museum)*

was certainly not a country made for rapid armoured warfare and the 1st Battalion's attack on the village of Montchamps on 5 August proved to be the 'grimmest' fight so far undertaken, with over a hundred Guardsmen killed, wounded or missing. With the 2nd Battalion operating in its reconnaissance role ahead of the division, 1st Battalion continued the move south, attacking the village of Le Haut Perrier on 11 August before moving on to Houssemagne. This phase of the fighting showed what could be done by armour and infantry acting in tandem. Even when tanks were hit and disabled, their crews dismounted and joined in the ground fighting, which was heavy and sustained. On 14 August the 1st Battalion's War Diary recorded the scene in the closing phase of the fighting:

Cromwell tanks of the 2nd Battalion are greeted by cheering crowds as they enter Eindhoven, 19 September 1944. *(Imperial War Museum)*

> The battalion area is certainly far from pleasant as its chief amenities consist of dead cows, dead Germans and burnt-out tanks. Thanks to the untiring labours of Padre [P. F.] Payne, all our dead have now had a proper and reverent burial and a small and carefully tended burial ground has been formed just by Battalion Headquarters.[5]

The mention of Padre Payne is a reminder of the important role played by the padres of the Welsh Guards during the war, especially in bringing succour to the wounded and dying. Another way of aiding the wounded was demonstrated by Lance Corporal F. W. Dyke, the company clerk of Prince of Wales Company, who was known as 'the company CIGS' (Chief of the Imperial General Staff, the Army's highest ranking soldier) on account of his military knowledge. Realizing that his close friend Guardsman John Rogers from Cardiff was missing – having been left behind in the buildings which the company had been attacking – Dyke ran back and bundled the popular storeman into a wheelbarrow before pushing him under heavy fire to safety.

Following the frequently confused fighting as the

Cromwell tanks of No. 3 Squadron, advancing through the *bocage* country, Normandy, July 1944. *(Welsh Guards Archives)*

No. 4 Company, moving up to the front line near Caen, 1944. The leading officer next to the mortar carrier is Lt A. J. Bland. Behind him is LSgt H. C. W. Brawn. *(Welsh Guards Archives)*

Welsh Guardsmen, Normandy, July 1944. *(Welsh Guards Archives)*

Cromwells in action, Normandy, July 1944. *(Welsh Guards Archives)*

Allies made their way south, both battalions spent the rest of August recovering from their efforts. By 22 August all German forces in the Falaise Pocket had been eliminated and the remainder of Hitler's Army Group B were retreating eastwards with the Allies in pursuit. Freed from the physical restrictions of the *bocage* country, the Allied forces moved gratefully into the open landscape, with the 2nd Battalion leading the charge ahead of the Guards Armoured Division and the lorried soldiers of the 1st Battalion following in hot pursuit. For Peter Leuchars, in temporary command of No. 3 Company, it was more like 'a triumphal drive' than a military operation:

> Once we passed Falaise where the whole area was littered with smashed German vehicles, we entered unspoiled countryside. Every town and village was crammed with people shouting, waving flags, offering us fruit and wine as we drove steadily eastwards, crossing the Seine by Bailey Bridge near Vernon. Here, on 1st September, Guards Armoured Division took over the advance and, led by 2nd Battalion Welsh Guards, the Armoured Reconnaissance Regiment, we followed close behind with Vimy Ridge as our objective.[6]

The 1st Battalion moves forward, Normandy, July 1944. *(Imperial War Museum)*

No. 2 Squadron during the liberation of Brussels on 3 September 1944 shows (*L to R*) Sgt T. E. Williams, LCpl R. Gibson, a Belgian girl, Gdsm R. T. Pedgeon, LCpl H. Thomas and Sgt G. H. Greenstock. *(Welsh Guards Archives)*

By the afternoon the division had reached the outskirts of Arras where the advance was halted and the decision was taken to allow Prince of Wales Company, the last to leave Arras in May 1940, to be the first to enter it four years later. Needless to say, it proved to be a memorable occasion, but more followed the next day when Major General Allan Adair, commanding the Guards Armoured Division, summoned an Orders Group to which he announced the ambitious objective: 'Guards Armoured Division will advance and capture Brussels – and a very good intention too.' This surprising announcement, delivered with what Colonel Windsor-Lewis described as Adair's 'most mischievous smile', was met with acclaim and 'roars of laughter'.[7] By then the two Welsh Guards battalions had formed a Welsh Guards Battle Group which would advance to Brussels along the southern route, while the Grenadier Guards Battle Group would take the old route through Tournai. The start line was at Douai. At first light No. 1 Squadron and Prince of Wales Company started engines and the race for the Belgian capital got under way with a squadron of 2nd Household Cavalry armoured cars in the van as the division set off for Brussels, ninety-two miles away. Both battalions entered the city at 20.00 on the evening of 3 September and, according to the War Diary of the 1st Battalion, normally a most sober document, the welcome exceeded anything the Guardsmen could have expected:

> The whole way along the route, the column had been greeted by cheering crowds, throwing fruit and covering with flowers every passing vehicle.
>
> It may incidentally be noted that it is not easy to operate a wireless set satisfactorily when unripe fruit continually hurtles through the window.
>
> Once the two Battalions had entered Brussels, however, the welcome exceeded anything the Battalion has ever seen before or is ever likely to see again.
>
> It was practically impossible to get through the streets, so great was the press of the crowd, and presents of every sort were showered on the troops.
>
> Scores of bottles of brandy, wines of all descriptions were pressed through the windows and nearly every vehicle had a blonde in it, but under the circumstances it was extremely difficult to eject them, even if the wish had been there.
>
> Drill Serjeant [*sic*] Blackmore had two on the back of his motor cycle.[8]

No. 4 Company, during the attack on Cagny, July 1944. The Company Commander, Maj J. D. A. Syrett, is seen indicating a mortar target to Sgt Vessey. Gdsm Kitchen is in the foreground and Gdsm Fenwick is the Bren gunner. Major Syrett was killed a few days later. *(Welsh Guards Archives)*

This was confirmed by Captain R. V. J. Evans, who noted in his diary that not only was no opposition encountered but that the entry into Brussels was one of the most memorable incidents of the war:

> As I started into the outskirts now by myself I saw a fantastic sight, there lining the roads and advancing down the roads thirty or forty deep were thousands and thousands of people all absolutely crazy with joy. From this moment onwards there was chaos. All hope of seeing or moving tactically was finished, at least forty people climbed in the tank, men, women and children embracing me, bottles of wine being broken and poured all over me, boxes of 100 cigars ad lib. The tank was covered nearly 6 ft deep with every conceivable type of fruit and flowers, particularly some wonderful gladioli.[9]

For the record, the first Allied vehicle to enter Brussels was a tank belonging to No. 3 Troop, No. 1 Squadron, 2nd Armoured Reconnaissance Regiment, Welsh Guards. It was commanded by Lieutenant J. A. W. Dent and driven by Guardsman E. J. James; Lance Corporal E. K. Rees was the gunner, Guardsman Robert Beresford the hull gunner and Guardsman Ralph Beresford the wireless operator. As the leading tank approached the city centre, it destroyed a busload of Germans near the Avenue des Arts and then proceeded to the Arc de Cinquentaire where it knocked out an enemy tank before the Squadron halted for the night to laager with Prince of Wales Company from the 1st Battalion.

Peter Leuchars remembered that, although all his men were present at stand-to the following day, 'most were looking slightly the worse for wear!' More soberly, the Commanding Officer of the 2nd Battalion noted in his personal diary: 'The last time I had been in Brussels was in July 1940 as a fugitive escaping from the Germans. On that occasion I had entered the

Germans surrendering at Arras, 1 September 1944. From the Seine to Brussels, the Welsh Guards Group had to cope with an increasing numbers of prisoners. Met by the leading tanks, they would be quickly searched and passed on to following units. *(Welsh Guards Archives)*

city from the east in a tram. Today I entered it from the west in a tank.'[10] The revels continued for a further two days but the people of Brussels were not the only ones to be pleased by the achievement of the Welsh Guards; so too was XXX Corps commander Lieutenant General Brian Horrocks. Describing the capture of the city in a BBC *War Report* programme the following day, he said that it was all due to the fact that he 'had the pick of the British Army and nothing could stop them'.

To commemorate the rapid capture of the city and the part played by the Regiment, the well-known figure of Brussels's Mannekin Pis (literally 'Little Man Pee' in Marols, a Dutch dialect spoken in Brussels) was presented with a Welsh Guards uniform a year later. Also known as the 'Petit Julien', the 2-ft tall bronze statue stands above a fountain on the corner of Rue de l'Etuve and Rue des Grands Carmes and showers water into it in a most ingenious fashion. Many legends surround the statue, which was designed by Hiëronymus Duquesnoy the Elder in 1618 or 1619, but it has become the custom to dress the figure in different suits of clothes; each year on 2 September he finds himself wearing the bearskin and scarlet tunic of a Regimental Sergeant Major of the Welsh Guards in honour of the regiment which helped

WO2 (Drum Major) Percy Hitchcock with a liberated ham. DMaj Hitchcock commanded the anti-aircraft troop of Crusader tanks fitted with twin 20 mm Oerlikon cannons. He was Drum Major of the 2nd Battalion between September 1941 and August 1945, having transferred from the Coldstream Guards in 1941. *(Welsh Guards Archives)*

RSM Ivor Roberts of the 2nd Battalion dressing the Mannequin Pis, Brussels, 1944. RSM Roberts was known as 'Scratch' because of his tendency to scratch himself before giving words of command on parade. *(Welsh Guards Archives)*

LCpl Digville, a despatch rider, is shown the way to Brussels, September 1944. *(Welsh Guards Archives)*

A tank crew with their Cromwell IV. The white diamond insignia on its front shows that this is a Battalion HQ vehicle. *(Welsh Guards Archives)*

to liberate his city all those years ago. And the welcome was sincere from a population which had been under Nazi control for four years. As lorries of the 1st Battalion entered the city a young Belgian mother pleaded with the men for some soap for herself and her child. Charles Murrell responded by digging out a half-used bar which the woman accepted as if it were a diamond. For him and the others in the truck that made it all worthwhile:

> There are no doubting Thomases amongst us now concerning the rottenness of the German new order. The greatest cynics amongst us are impressed by this touching demonstration by the people of this liberated capital city. On our trucks they chalk *Merci aux nos liberatans* [sic]. Light-cream painted trams and trailers packed to overflowing with cheering folk – their bells incessantly ringing and the fainter toot of French motor horns – pink and yellow-tipped matches.[11]

The liberation of Brussels, September 1944. *(Welsh Guards Archives)*

A tank crew in front of their Cromwell. (All photographs show the 2nd Battalion). *(Welsh Guards Archives)*

3rd Battalion, Italy

While these stirring events were taking place in France and Belgium, the 3rd Battalion had already entered Europe as part of the invasion of mainland Italy that had begun in September 1943, following the successful capture of Sicily. Overall command for the

A returning patrol submits a report to Capt Sir Edmund Bedingfeld's Company HQ after the action near Le Bas Perrier, August 1944. *(Welsh Guards Archives)*

Group of 2nd Battalion officers with Rex Whistler on the extreme right. *(Welsh Guards Archives)*

operation was in the hands of General Sir Harold Alexander (Fifteenth Army Group), Lieutenant General Sir Oliver Leese commanded British Eighth Army which included 6th Armoured Division, the parent division of 3rd Battalion Welsh Guards, while Lieutenant General Mark Clark commanded US Fifth Army. The plan was for the British to land at Reggio Calabria on the Italian side of the Strait of Messina (Operation Baytown), while Clark's army landed south of Naples at Salerno (Operation Avalanche). Both invasions enjoyed mixed fortunes: the British landed unopposed and made good progress, but Clark's army encountered stubborn resistance from German land and air forces, and only the intervention of the firepower of the Royal Navy allowed the landings to be secured by the middle of the month. Although both armies then made progress in their advance northwards, lack of firm operational planning meant the campaign quickly degenerated into a remorseless slogging match, with bad winter weather playing havoc during the advance to the Garigliano and Sangro rivers. In January 1944 Allied troops landed at Anzio (Operation Shingle) in an attempt to outflank the Germans but this too ran into difficulties and made little progess inland.

In February 1944 the 3rd Battalion was shipped to Naples on board HMT *Moriarty*; not only was the ship full to capacity but it was a rough crossing. (An advance party had already arrived on board the Free French destroyer *Terrible*.) On arrival the weather was so bad that the only physical feature to stand out was Mount Vesuvius with its distinctive volcanic plume. Worse followed on disembarkation on 5 February 1944 when it was found that the Battalion's transport had not arrived. Even so, almost straight away, 3rd Battalion Welsh Guards was pitched into action. Their objective was Mount Cerasola, a thousand feet high, on the northern bank of the River Garigliano, and their task was to relieve 2nd/4th Hampshire Regiment, which had taken heavy casualties while holding a shoulder known as Mount Furlito. To get there, the Guardsmen had to walk in single file carrying all their equipment, in conditions which Major J. K. Cull, commanding No. 4 Company, remembered as 'pouring rain and bitter cold which was in marked contrast to the weather in North Africa a few days previously'.[12] As the relief had to be conducted in darkness and because daylight was short, on 10 February the Commanding Officer (Lieutenant Colonel Sir William V. Makins, Bt) took an orders group forward to reconnoitre the apex of the summit. This involved traversing a narrow path on the reverse slopes of a point known as Mount Fuga. Blizzard conditions meant that visibility was often down to twenty yards and the problem was exacerbated when the orders group found itself under enemy fire from the northern peaks of Ornito and Cerasola. At one stage the small party exchanged fire with the Germans after Makins gave the unorthodox order: 'O [orders] Group, load.' During the resulting fire-fight, Major R. L. Pattinson shot dead the leading German on the heights and the fighting came to a stop.

This allowed the party to make contact with the Hampshires' headquarters thus permitting the company commanders to make sense of the positions they were to occupy. As Major Kemmis Buckley remembered, this was not an easy task:

> It was extremely forbidding. The mountain was a confused mass of bare rocks with no soil or vegetation and rose like the rim of an amphitheatre to a crest on the western side. To the east the rocky ground fell gradually away towards the River Garigliano. Such protection as there was could be provided only by the building of sangars and as the battalion was of greater strength than the Hampshires, when dawn came many of the Guardsmen were unprotected.[13]

By then, 04.00, the rest of the Battalion had arrived under the command of the Adjutant, Captain David Gibson-Watt, and had taken shelter as best they could. At dawn the following day the Germans attacked the Welsh Guards' positions, with the main weight of the assault coming in on No. 2 Company. Already at a disadvantage because the lie of the land was unfamiliar, the Guardsmen also had to contend with the fact that the Germans had occupied a ridge above their position. To clear this, the officer commanding No. 2 Company, Captain D. P. G. Elliot, ordered a bayonet charge, which, though partially successful, resulted in him being killed. Command passed to Sergeant William Doyle until the arrival of Captain R. G. Barbour, the Battalion Intelligence Officer. Because the Germans had not been dislodged, Barbour ordered another bayonet charge, which succeeded in clearing the ridge. During the action he too was killed and No. 2 Company lost twenty-two dead and forty-nine wounded. The Battalion was further discomfited when the Commanding Officer had to be withdrawn due to illness; his place was taken by Major D. G. Davies-Scourfield, who managed to disguise the fact that he was suffering from a poisoned foot.

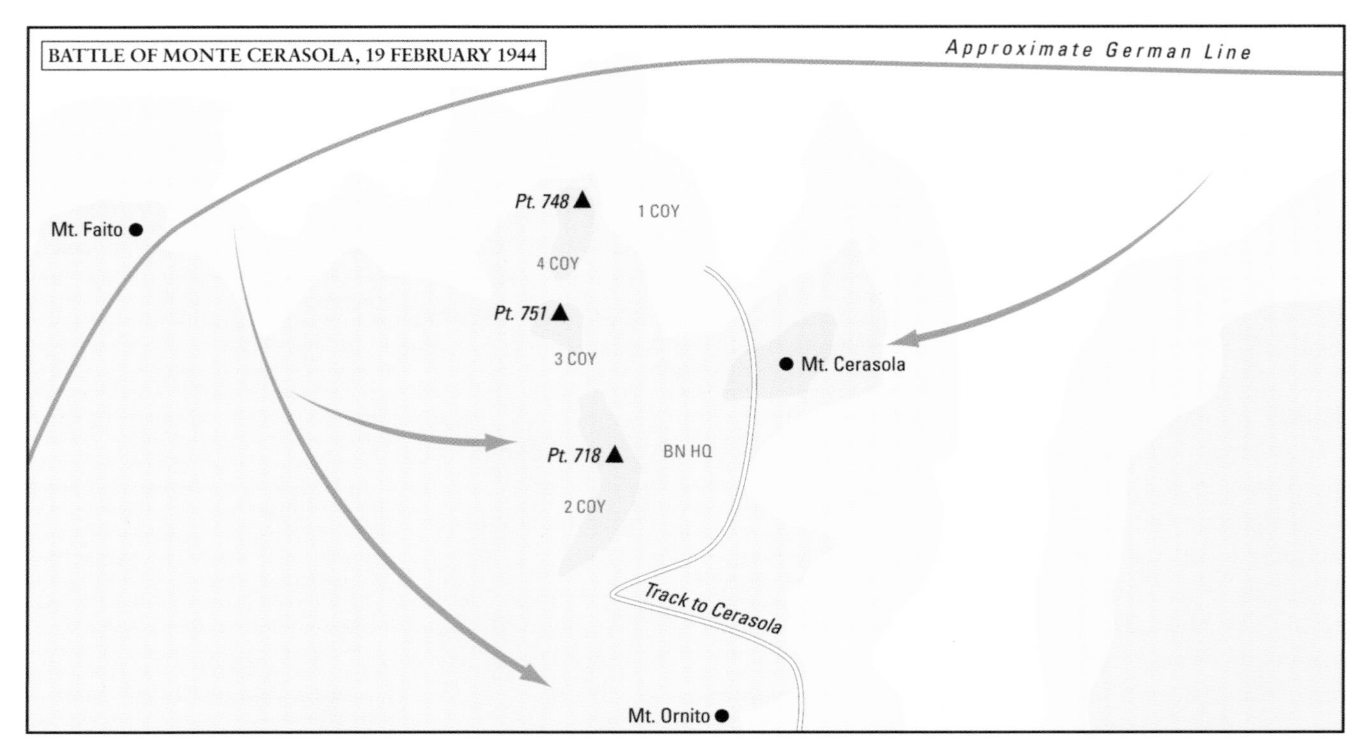

Battle of Cerasola, showing position on 11 February 1944 at the time of German counter-attacks.

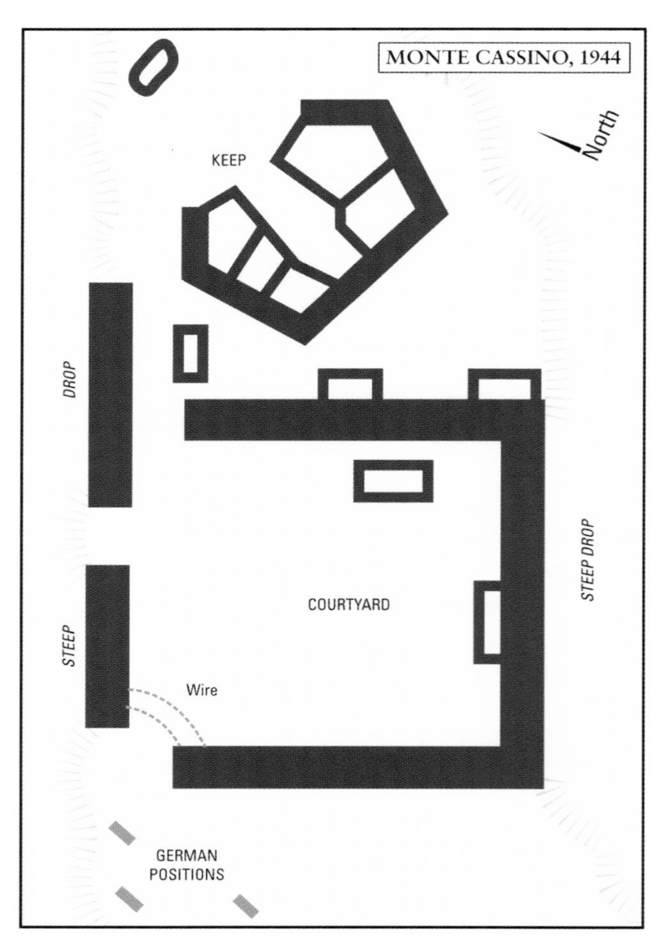

Map of the Castle at Cassino at the conclusion of the fighting, February 1944.

For the next week the Battalion remained in a precarious position on the heights of Cerasola. Not only were the conditions bleak – wet and bitterly cold – but contact with the enemy had to be maintained through aggressive patrolling of no-man's land, while casualties from enemy action and the atrocious weather took a steady toll. Despite the conditions, though, the Battalion remained in good heart, even when under heavy fire. Major Cull remembered seeing Sergeant Evans 98 standing on a ridge in full view of German fire while he surveyed their positions. Called back to safety, he looked at Cull 'slightly quizzically' and said, 'If only I could speak German I could get hundreds of them to surrender to me.'[14] The maintenance of morale was helped in a number of other ways. First, the Commanding Officer ordered that men were permitted to grow beards in order to conserve water; this was much appreciated but it resulted in the Battalion looking as if they were fighting in an earlier era. Secondly, the line of supply never faltered and much of the credit was due to the efforts of Regimental Quartermaster Sergeant H. P. N. Dunn, who had nineteen years' service in the Regiment.

Thirdly, the Battalion was lucky to have a superb Medical Officer in Captain David Morris who insisted on high levels of personal hygiene and intelligent man-management. This meant that the incidence of malaria in both North Africa and Italy was kept to 3.7 per cent over a two-year period and that cases of 'battle

Winston Churchill inspecting a Cromwell tank in March 1944 as the 2nd Battalion prepared for D-Day. To the right is Maj J. O. Spencer, who commanded No. 2 Sqn until he was killed at Hechtel in September 1944. Next to the hull-mounted 7.92 mm Besa machine gun, the face of Sgt T. Dredge, the vehicle driver, can just be seen behind the open visor. The markings on the front are (*L to R*) the number 45 painted in white on a green and blue square, which denotes that the vehicle belongs to an armoured reconnaissance regiment; the number 26 on a black circle is the bridge class (tanks were not allowed to cross bridges with a lower number); the white square shows that this is a No. 2 Sqn vehicle and the A within the square marks this as the squadron leader's tank; and, lastly, the divisional badge on the far right is that of the Guards Armoured Division, the ever open eye. *(Welsh Guards Archives)*

exhaustion' or 'anxiety neurosis' were never more than 2 per cent over the same period – 'amazingly low figures' according to a paper prepared by Dr Morris after the war. Venereal disease levels were also low, the overall incidence within the Battalion being 2.75 per cent over a nineteen-month period, which compared 'favourably with the incidence in some units of the US Army in the Mediterranean Theatre of Operations where an annual rate of 120 per cent was not uncommon'. The only concern came from the high incidence of hepatitis amongst the officers – eleven cases compared to forty-six Guardsmen. After the war Dr Morris discussed this with Sir Cyril Clarke, FRS, a distinguished physician and expert in preventative medicine, who told him that the figures were probably due to the difference in messing. Whereas the officers had a communal arrangement in which all crockery, cutlery and glasses were washed in the same receptacle with little or no water and no rinsing, the Guardsmen 'cleaned their own mess tins, mugs and so on and those with infective hepatitis did not pass it on to their comrades'.[15]

During the fighting the Germans asked several

The RSM of the 3rd Battalion with his soldier-servant, Italy, 1944. *(Welsh Guards Archives)*

ABOVE AND BELOW: Moving up to the front line. There were insufficient roads to serve the vast concentration of troops arriving at the Normandy bridgehead area. This meant that a journey of only few miles could easily take an hour or two. Motorcycle despatch riders, such as those seen in the picture below, were essential in getting through heavy traffic, June 1944. *(Welsh Guards Archives)*

times for truces so that their dead could be collected, but these were refused on the grounds that the enemy would be able to gain useful intelligence about the Battalion's dispositions. However, the brief ceasefires did allow the enemy to be seen at close-quarters and several Guardsmen commented on the fact that the German officers were wearing summer uniforms yet appeared not to suffer any ill effects. When asked, the Germans responded disdainfully that they had just returned from the Russian Front and that the conditions in Italy were not really cold!

The end came on 19 February when the Germans made one last effort to drive the Battalion from Mount Cerasola, but this was beaten off, allowing 3rd Battalion Welsh Guards to regroup before being replaced by 6th Battalion, York and Lancaster Regiment. Following a short spell in the line from Cascano to Purgatorio, west of the Garigliano River and covering a front of 4,000 yards, their next destination was Cassino – town, monastery and mountain – which was one of the most iconic battlefields of the war.[16] At that time, early in 1944, the Allied advance had been held up in the Liri Valley, south of Rome, where the enemy resistance centred on the monastery at Monte Cassino, the mother-house of the Benedictine Order. It stood on high ground outside the town of the same name, which had been razed to the ground and, being still partially occupied by German forces, was the scene of fierce fighting. Four assaults were made on this key position before it fell; 3rd Battalion Welsh Guards arrived on 7 April, following the third attack, only to find themselves living a 'troglodyte existence' amongst the wrecked buildings.

During the Battalion's two tours of duty in Cassino, the Commanding Officer ordered that an aggressive stance be maintained against the enemy, mortaring suspected positions and calling in artillery fire whenever possible. Concerned that the Germans were listening in to radio messages, the Signal Officer, Lieutenant M. J. B. Chinnery, decided 'to try a little Welsh' in all communications. This was a radical change of policy as Army regulations had banned Welsh-speaking soldiers from writing letters home in their native language because of problems over censorship.[17] It seemed to work, as twenty-four hours after the policy was introduced, to Chinnery's amusement the Battalion area was bombarded with German propaganda leaflets printed in Urdu. Recording the continuing struggle to discover the whereabouts of German positions, the War Diary noted that further amusement was generated by Lieutenant I. P. Bankier of No. 2 Company who 'dangled a portrait of Hitler on a fishing rod to attract fire. Succeeded.'[18] In fact, the offending portrait was of

A Cromwell tank wading ashore at Arromanches from a landing ship, tank (LST). All tanks were waterproofed to wade in salt water, up to the height of the turret if necessary. *(Welsh Guards Archives)*

a nineteenth-century Italian soldier which had been extracted from the ruins and reworked into an uncanny likeness of the Nazi leader. As the War Diary recorded, by firing at it the Germans revealed their own positions.

The Battalion's second tour lasted until 19 May when at long last the Cassino position fell into Allied hands at the fourth time of asking. When they emerged blinking from their sangars and other entrenched positions, the first thing the men noticed was the sickly smell of rotting corpses – there were seventy-four bodies of different regiments and nationalities lying unburied by the narrow path up to the castle summit, a sight which impinged itself on the senses of everyone who witnessed it, such as Philip Brutton who later described the scene facing the 3rd Battalion as it moved along the road to the north:

> At this point the stink of the town hit one forcibly – the sickly sweet smell of putrefying bodies which were left unburied under or in the rubble of the rocks and ruined buildings, too dangerous to approach and dig out for a decent burial; the stench of urine and human faeces, the first pungent because of the lack of rain, the second diminishing because of the daytime heat and fairly well controlled by latrine discipline; the nasty smell of our smoke shells and the nasty drifting waftage of cordite emanating from the armaments and explosions on both sides within the perimeter of battle.[19]

The fall of Monte Cassino opened up access to the valley of the River Liri and the coastal littoral, thus allowing the Allied infantry and armoured divisions to move towards Rome. This was a tipping point in the campaign. With the Germans in full retreat, the Allies pushed north-west towards the Italian capital. As happens so often in war, though, not everything went to plan and the 3rd Battalion had its fair share of frustration. On 26 May, 6th Armoured Division's advance stalled near the town of Arce, which straddled the strategically important Route 6 that led to Rome. Intelligence suggested that the town was in the process of being evacuated by the Germans, but this proved not to be the case. In fierce fighting to clear enemy

These three pictures show scenes which frequently occurred between the River Seine and Brussels. The top picture shows men from the Motor Transport Platoon taking cover in a ditch by the side of the road. Capt S. G. Holland is in the foreground on the bank. The centre picture shows LCpl Davis escorting a captured member of the *Herrenvolk*. At the bottom of the page, Welsh Guardsmen are watching the flank of their company while at the halt. The man in the centre is equipped with a 2-inch mortar. *(Welsh Guards Archives)*

Guardsmen from the 3rd Battalion Welsh Guards are passed by a Sherman tank of 26th Armoured Brigade as they move forward to attack Monte Piccolo outside Arce, Italy, 28 May 1944. *(Imperial War Museum)*

positions on the high ground at Monte Piccolo and Monte Grande, the Battalion suffered 112 casualties. Amongst the many acts of courage demonstrated by Welsh Guardsmen during this testing battle none was finer than the devotion to duty displayed by Lance Sergeant Frank Goodwin, in civilian life a bus conductor from Cefn, Denbighshire, who single-handedly attacked a German machine-gun position at the top of Monte Piccolo. When his section reached the position, they found that Goodwin had thrown himself onto the gun to silence it with his own body. Three days later the Battalion entered Arce unopposed, riding on the hulls of the tanks of 2nd Lothians and Border Horse.

Ahead lay the valley of the River Tiber – Rome was bypassed by the Battalion but the dome of St Peter's was clearly visible in the distance – and 1st Guards Brigade made relatively easy progress as it sped towards Perugia. Here the Germans had created an intermediate defensive line to stabilize their position before they pulled back again towards the Gothic Line in the northern Apennines. Alexander's plan was to concentrate his forces in the west in order to take Florence before pushing on towards Bologna. The battle for possession of Perugia began on 20 June. In the opening stages much of it involved fighting in built-up areas, never an easy task for infantrymen. It was a confused business: driving rain did not help matters and, as happened so often in Italy, there was a succession of wearisome fights to take possession of the hills and high ground to the north. It was not until the end of the month that the Germans were finally cleared from the Perugia area. After seven weeks of continuous fighting the Battalion badly needed to regroup and refit. Fortunately they were able to make use of a villa on the shores of Lake Trasimene which had been requisitioned on the orders of the new Commanding Officer, Lieutenant Colonel J. E. Gurney, DSO, MC. This welcome addition to the Battalion's rest and recreation resources was immediately christened the 'Cardiff Arms'.

The sense of respite continued into July, which was spent mainly in the valley of the River Arno, a delightful area which one young officer, John Retallack, characterized in terms of fond remembrance: 'If one must go to war, the Arno valley that summer was not a bad place to do it. The Battalion was not in continuous contact with the enemy … in the afternoons the local custom of the siesta was adopted.'[20] Not that it was all easy-street. The brigade used its battalions to leap-frog up the valley, providing a screen for the division's right flank as it advanced up Route 69, passing through Quarata, Renacci, Torre a Monte and San Ellero. As July gave way to August, the first tranches of leave in Rome were granted, the Battalion War Diary marking the occasion with the comment that Padre Richards

Montchamp painted by L. F. Ellis. This shows typical *bocage* country. In the first phase of the attack by the 1st Battalion on 4 August 1944, seven Germans were captured in the cottage seen in the foreground. Away behind Montchamp lies the ground near Estry. *(L. F. Ellis/Welsh Guards Archives)*

made 'a particularly virulent attack on sins of the flesh'.[21] But it was not all fun and games. As the advance continued up the valley, contact with the enemy was maintained by fighting patrols; a steady succession of German prisoners-of-war came into the hands of the Battalion, often with the help of local partisans. As a result the casualties sustained during this phase of the campaign, No. 2 Company was replaced by a company of Grenadiers.

At the end of August the Battalion reached the head of the Arno valley and, by mid-September, was within reach of the Gothic Line, the Germans' last redoubt in northern Italy which ran from La Spezia in the west to Pesaro in the east and made good use of the unbroken line of the Apennine mountain range. The Allied plan called for the British Eighth Army to attack up the Adriatic coast towards Pesaro and Rimini and draw in the German reserves from the centre of the country. At the same time, the US Fifth Army would then attack in the weakened central Apennines north of Florence towards Bologna, with British XIII Corps on the right wing of the attack fanning towards the coast to create a pincer with the Eighth Army advance. The latter advance would include 6th Armoured Division and 3rd Battalion Welsh Guards. Although the Germans were in the process of withdrawing they offered stiff resistance and the Battalion soon found that they had to contest every ridge in the mountains north of Florence.

The fighting began in earnest on 2 October in the area dominated by Monte Battaglia to the west of Posseggio, a high point (2,345 feet) at the end of a long narrow ridge, with a ruined castle at its summit where the Germans had created a defensive position. The mountain had been contested by US troops and the litter of unburied corpses gave some indication of the ferocity of the earlier fighting. In addition to the foul location, the weather deteriorated and subsequent memories are of constant rain, cold and high winds which combined to make life extremely unpleasant for everyone on the mountain. Guardsman Fred Burton from Newport later recalled the problems he faced while acting as a stretcher-bearer carrying casualties during a trip that could take up to four hours: 'The mud was so deep that the stretcher bearers would trip

The Welsh Guards Group had to cope with prisoners all the way from the Seine to Brussels. In a single day of action they took more than 2,000 prisioners. The top picture shows the Welsh Guards' Cromwells on Vimy Ridge. *(Welsh Guards Archives)*

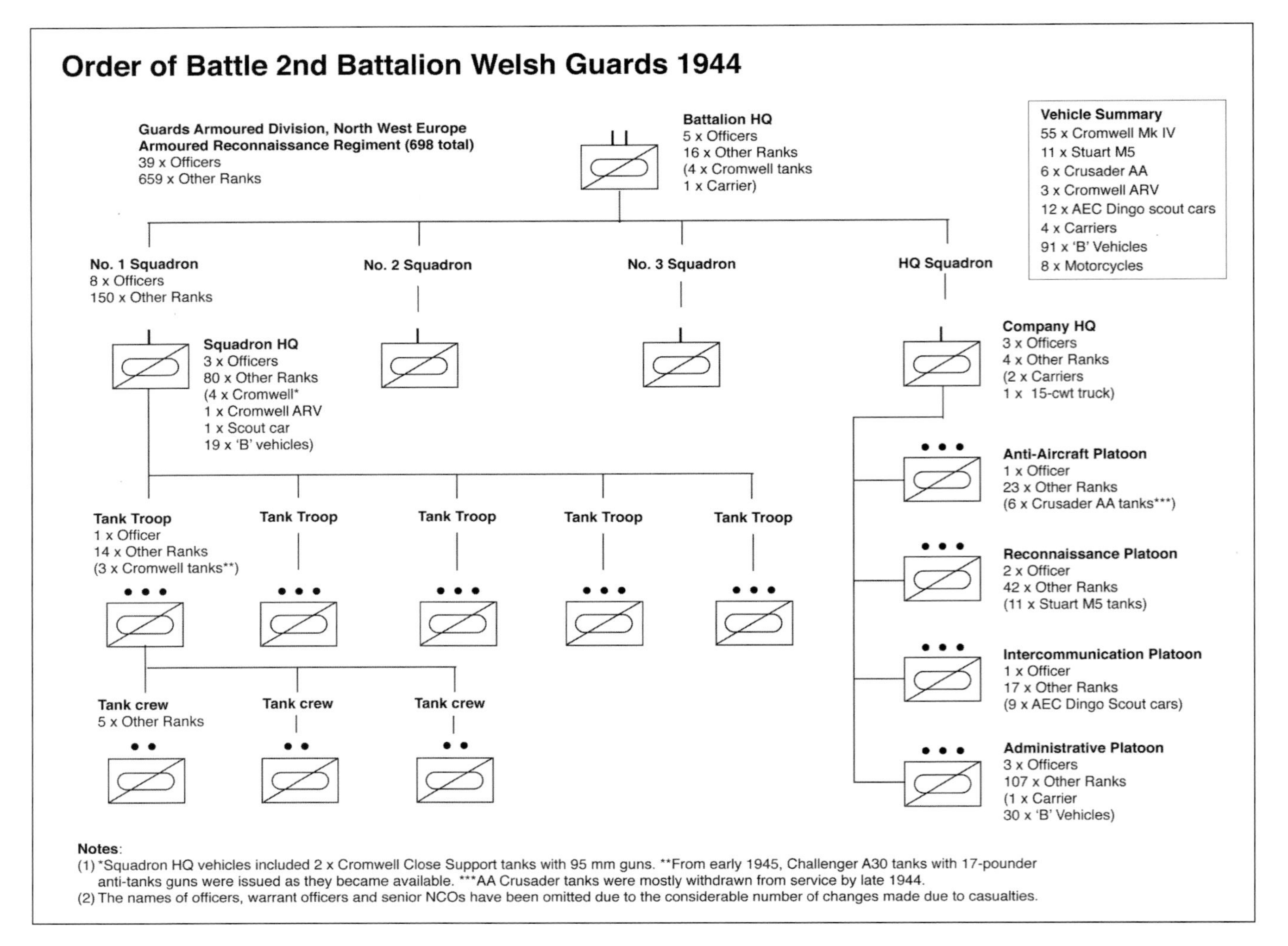

up and the bodies would roll off them back into the mud. We would have to pick them back up and start again each time.'[22] The Battalion remained in the Battaglia area until 25 October before moving into the Santerno valley and settling into a routine by which companies alternated between time spent in the town of Greve and time spent in positions in the mountains under fire, each time pushing further north. It was hard-going in vile weather and, although contact with the enemy was kept to a minimum, it was a test of the Guardsmen's morale and capacity to withstand hardship and danger. That they were able to overcome those conditions says much for natural Welsh tenacity, which may be summed up in an observation made by Andrew Gibson-Watt, younger brother of David, who joined the 3rd Battalion in May 1944:

ABOVE RIGHT: The bare mountain top of Monte Cerasola was held against repeated enemy attacks. RIGHT: Men of the 3rd Battalion wear leather jerkins and carry rolled-up blankets to protect them from the ice, wind and rain. *(Welsh Guards Archives)*

A scene from the cab of a 15 cwt truck on the road to Brussels, 3 September 1944, drawn by Sgt C. Murrell. *(Courtesy of the Murrell family.)*

> Being Welsh, our soldiers were very apt, when the going got really tough and miserable conditions prevailed, to start up a soft melancholy singing, usually Cwm Rhondda or another solemn Welsh hymn: when I heard my platoon doing that, for instance when they had staggered, dog-tired and wet through, into the back of a truck for a four-hour bumpy night drive back to our billets near Florence, I was happy, because I knew they were all right.[23]

Like many other young officers joining the Welsh Guards, Andrew Gibson-Watt quickly noticed that humour was also a vital ingredient in making hardships more bearable and keeping difficulties at bay. One story in particular seemed to sum this up: a Guardsman from Swansea complained that he was offended because a sergeant had insulted him and he wished to lodge a formal complaint. This was granted and he was marched into the Commanding Officer on Memoranda. The conversation went like this:

> Guardsman: Leave to speak, sir.
> Sergeant X called me a 'bugger'.
> Makins: A bugger?
> Guardsman: Yes, sir.
> Makins: Well, are you one?
> Guardsman (outraged Swansea accent): No, sir!
> Makins: Well then, march out.[24]

In the midst of that unpleasant winter one memory shines out in the history of 3rd Battalion Welsh Guards: the remarkable story of Lorna Twining and the 'tea and wads' she provided for the men of the Battalion in the Santerno, 'a deep gloomy valley in which it was usually raining or snowing and from time to time shells fell nearby'. Born Lorna Althea Ravenhill, she had married Pilot Officer Geoffrey Christopher Holt, RAF, but when he was killed in a flying accident early in the war she married Captain Richard Twining, 3rd Battalion Welsh Guards, who was killed in April 1943 at Fondouk. With the help of influential friends, she managed to make her way to North Africa to visit the cemetery at Enfidaville where her husband and fifty other Welsh Guardsmen were buried. By then she had determined to follow the Battalion on its travels and to share its difficulties and dangers. No sooner had the Battalion deployed to Italy than Lorna Twining was also there with a mobile canteen to supply the men with small necessities and refreshments. She was not alone in doing this – other mobile canteens were supplied by organizations such as the Women's Voluntary Service, the YMCA and the Church Army – but she was one of the very few to remain attached to a particular battalion and, as Captain M. C. Thursby-Pelham, the Adjutant, remembered, her loyalty and 'gentle kindness' were greatly respected by all who served with her:

Commanding Officer's Memoranda

'Inherited from the Grenadier Guards, this ritual was a parade for defaulters who had broken rules such as "slashing our peaks, commenting on the parentage of superiors" or "getting excused boots and blanco for no genuine reason". All cases were subject to a "Welsh Guards Scale of Punishments" and these guidelines determined the punishment for committing crimes as various as "bad order for duties" (i.e. having three articles dirty) or "untidy boot laces" which brought a punishment of 5–7 days CB [confined to barracks]. Other charges included "grumbling when warned for duty", "making a reply in the ranks" and "being in a verminous condition". Most minor charges were either dismissed or punished with an Extra Drill or Extra Parade. Guardsmen tended to concede that the ultimate sin was being found out.'

From 'Crime and Punishment', *Welsh Guards Magazine*, 1998–1999, pp. 37–8.

The Guardsmen thought the world of her and instinctively protected her from the unwelcome attention of soldiers from outside the Brigade who failed to treat her with what they regarded as proper respect. A shout of 'Hey Blondie! Give us a cup of cha' will ya?' was enough to detach two large men from the nearest group of Guardsmen who would walk over to have a quiet word with the offenders. 'Don't speak to our Mrs Twining like that. D'you see?'[25]

Although Lorna Twining survived the war and married Major Desmond Chichester, who had served with the Coldstream Guards, she died suddenly and unexpectedly a few hours after giving birth to their only son. The date was 9 April 1948, exactly five years after the death of her second husband at Fondouk.

Monte Cassino by L. F. Ellis. In the middle ground are the prison ruins, with the castle and the monastery behind. *(L. F. Ellis/Welsh Guards Archives)*

Scenes from the Normandy campaign and the road to Brussels, both painted by Sgt C. Murrell in 1944. *(Courtesy of the Murrell family)*

CHAPTER SEVEN

Victory in Europe, 1944–1945

1st and 2nd Battalions, Belgium, North-West Europe
3rd Battalion, Italy and Austria

After the fall of Brussels there was an understandable tendency to relax and to promote a feeling that the Germans were on the run and that it would be possible to enter the Ruhr and the Saar by the late autumn. Many believed that the war might be over by Christmas and this was not idle thinking. The Germans seemed to be on the back foot; their forces had been broken up and demoralized; the Allies held the upper hand and the argument for a rapid attack into Germany seemed irresistible. Thus was born the controversy of the 'single thrust' versus 'broad thrust' argument, with Montgomery arguing that a potent concentrated assault would penetrate German defences rapidly and decisively, while a broad-front approach would dissipate resources. By now, though, Montgomery was no longer in the ascendant, having handed over command of the Allied land forces to Eisenhower on 1 September, but his approach was still given the go-ahead a week later. Montgomery's plan was to leap over the Meuse and the Lower Rhine and to establish a foothold in the North German Plain. This 'race to the Rhine' would use an airborne assault along a fifty-mile corridor to seize bridgeheads at Eindhoven, Nijmegen and Arnhem while the armoured and motorized columns of Horrocks's XXX Corps would break out of Belgium to reinforce them. At one stroke the Netherlands would be cut in two, the German defences would be outflanked and the Allies

Tanks of the 2nd Battalion Welsh Guards carrying infantry of the 2nd Battalion Scots Guards during the advance into Germany. The code word 'magnum' was given over the radio whenever it was decided to mount infantry on tanks prior to an attack. A single Cromwell could carry a full section of ten men. Usually infantry were carried forward in trucks before dismounting and advancing on foot. *(Welsh Guards Archives)*

The Earl of Gowrie painted by Kenneth Green. Colonel of the Regiment, 1942–53; Commanding Officer, 1st Battalion, 1919–20; Regimental Lieutenant Colonel, 1920–4; Governor-General of Austrália, 1935–45. The second son of 7th Lord Ruthven of Freeland, he was created 1st Earl of Gowrie in 1945. *(Estate of Kenneth Green)*

would be established in the northern Ruhr.

Known as Operation Market-Garden, it was a bold plan, but from the outset it contained several flaws and it is clear that these were known to the planners. For the Allies there had never been an airborne operation of this size before and weather conditions in the Low Countries, even in September, were uncertain, but the greatest hazards were on the ground. Everything depended on the tanks of XXX Corps relieving the airborne forces and reaching Arnhem within forty-eight hours. Under any circumstances this would have posed difficulties, but in the Netherlands the terrain was totally unsuited to speed. The armoured formations would have to advance along a single main road as the surrounding countryside was unsuited to tank operations, being wooded and traversed by dykes and waterways. Also, by attacking on a narrow front, XXX Corps would pass the tactical advantage to the defenders and the timescale was perilously tight if the Germans responded with armour and artillery against the lightly armed airborne troops.

Almost immediately, Operation Market-Garden ran into difficulties. Although the British and US airborne forces completed their first drops on 17 September – US 82nd Airborne Division at Grave and Groesbeek, US 101st Airborne Division at Eindhoven and Vegel, British 1st Airborne Division at Arnhem – there was a steady accumulation of delays which held up British XXX Corps. Added to stout and unexpected German resistance (unknown to the Allies two SS panzer divisions (9th and 10th) were refitting in the area and were equipped with Mark IV tanks and assorted support weapons) the airborne forces were up against it from the outset. Ironically, their opponents were refitting with the intention of training to repel a landing by airborne forces. It did not help matters that the Germans came into possession of the opposition's order of battle and air plan which had been improperly carried into combat by an Allied officer. By the end of the first day of the operation, the German field commander, Field Marshal Walter Model, knew exactly what his opponents were going to do and he also knew what was expected of him – slow down the advance of XXX Corps and destroy the airborne forces on the ground. Several historians have noted that the first hours decided the outcome of Operation Market-Garden.

1st and 2nd Battalions, Welsh Guards Group

As part of the plan prior to the airborne assault, the Guards Armoured Division was ordered to cross the Albert Canal and to head towards the Meuse–Escaut Canal, which would be the start line for the race to the Rhine. The plans included the capture of the strategic villages of Helchteren and Hechtel, which stood in a terrain which lent itself to defence – sandy heath broken by small streams and patches of swamp. Here the Germans had dug themselves in, holding the crossroads and villages; having regrouped and reorganized themselves, they were prepared to fight with a fanaticism that surprised the Welsh Guards Group, whose men soon discovered that they would have to slog it out from street to street and house to house

The operation began on 6 September when the Welsh Guards Group led the Guards Armoured Division towards its first objectives, the bridges over the Albert Canal at Beeringen and Tessenderlo. At Beeringen, the first crossing point, it was found that the bridge had been partially destroyed and that the Germans were firmly entrenched on the other side; meanwhile at Tessenderlo the Grenadier and Coldstream Groups found the bridge there equally badly damaged. Heavy machine-gun fire from the

The Farewell to Armour Parade, Rotenburg Aerodrome, Germany, 9 June 1945. By this time the 2nd Battalion had been issued with a limited number of A30 Challenger tanks, shown here, mounting the 17-pounder gun. The Battalion's tanks were painted dark grey for the occasion using supplies of paint 'liberated' from a German naval yard. *(Welsh Guards Archives)*

opposite bank at Beeringen prevented any further progress until quite suddenly the Germans made an unexpected tactical withdrawal which allowed three rifle companies of 1st Battalion Welsh Guards to cross the canal. First on to the hastily repaired bridge was Prince of Wales Company, led by Major J. M. Miller, with No. 3 Company close behind. As Meirion Ellis remembered, clambering across the badly damaged structure could be a chancy business:

> Major Miller was urging everyone to get across, it was just what we were waiting for. We raced for the bridge in good section order, more like hares or mountain goats, a massive scramble of Welsh Guardsmen, we leapt from board to board doing exactly what Major Miller asked us to do. We were too preoccupied ensuring that we did not fall into the canal to worry about Germans – a slip would have meant certain death as the weight of our kit would have taken us to the bottom of the canal. Yes! Welsh Goats would have been a more appropriate word to describe us on that day.[1]

By then a team of Royal Engineers had arrived to repair the bridge and early the following morning the tanks of the 2nd Battalion were able to cross to the other side of the Albert Canal; by then, too, the Germans had returned to mount determined opposition with mortar fire and artillery fire from a number of self-propelled guns. One German battalion, according to the War Diary, had 'come all the way from the Rhine on bicycles'.[2] It was the beginning of five days of fierce fighting against an enemy who was determined to yield little ground and then only after putting up the stoutest resistance.

Having crossed the canal, the Welsh Guards columns pushed eastwards towards the village of Helchteren where they again came under concentrated fire, mainly from anti-tank guns. It was obvious that the Germans had managed to regroup their defences and were using the flat uncultivated countryside, with its hedges and woodlands, to good advantage. During this advance the Welsh Guards Group scored an easy if bloody victory over a German cyclist battalion which was resting at the roadside. (Later it was found to be a brigade headquarters defence battalion of the Hermann Göring Regiment.) To the Welsh Guardsmen, the orders of Lieutenant Colonel J. F. Gresham (commanding in place of the injured Heber-Percy) were clear: 'There is a battalion of infantry approaching us down the road. Go and destroy them.' As Meirion Ellis recorded in his memoir, the Welsh Guards did just that in an action which lasted barely three minutes:

> We threw grenades at them, fired our brens and rifles, grenades thrown by the armour were exploding as we passed, those of us firing could see bodies bouncing up in the air because of the impact of the bullets, those who could not or

The 2nd Battalion Rugby XV, 1945. *(Welsh Guards Archives)*

The Farewell to Armour Parade, Rotenburg Aerodrome, 1945. *(Welsh Guards Archives)*

would not fire quite simply looked with astonishment at this pure act of war and naked aggression.[3]

According to the War Diary, following the 'shooting and shouting', several hundred Germans had been killed or wounded, 150 were taken prisoner and their vehicles were left blazing on the roadside. In the days to come there would be no other victory so cheaply won.

From Helchteren, the heavily defended village of Hechtel stood four miles to the north – it lay on the axis of the Beeringen–Eindhoven road, which provided a direct corridor into the northern Netherlands – and the first attack on it was made by No. 3 Company, with the tanks of No. 1 Squadron in support. However, it came to naught and, after meeting heavy opposition, the Welsh Guardsmen were forced to withdraw at nightfall, the 1st Battalion War Diary recording that 'No. 3 Company had a nasty time getting back to the harbour area as the leading tank of their Group was shot up thereby blocking the road, and both their leading TCLs [troop-carrying lorries] got bazooka'd, killing two men,

A tank of the 2nd Battalion passes over Nijmegen bridge, September 1944. This bridge was located approximately ten miles from Arnhem and was described by General Eisenhower as a 'valuable prize'. It marked the entry point to the flat piece of land that divided the River Waal from the Lower Rhine, which was called 'the Island'. The exposed single road north often proved to be a bottleneck as vehicles were vulnerable to enemy fire. *(Welsh Guards Archives)*

No. 3 Squadron at Cuxhaven, VE-Day 1945. Maj C. A. la T. Leatham and Lt R. P. Farrer, MC, talking to German naval officers. *(Welsh Guards Archives)*

Hechtel, Germany, after the battle. Padre Payne can be seen walking towards a Bren gun carrier on the left of the picture, September 1944. *(Welsh Guards Archives)*

one of whom was 4573 Guardsman Burton, who had all along done invaluable service as a stretcher bearer, and wounding several others.'[4] During the evening the Irish Guards Group relieved the Welsh Guards at Beeringen, allowing the Group to concentrate behind Helchteren prior to the morning's attack on Hechtel. 'We got into the outskirts the first night and couldn't get any further, so we withdrew,' remembered Peter Leuchars of No. 3 Company. 'We then tried a second attack but that got held up and we had to pull out. That was very unpleasant. Meanwhile, every night that village was being reinforced.'[5] Facing the Welsh Guards at Hechtel was a reinforced German group made up of 1st Hermann Göring Regiment and 10th Gramsel Regiment, consisting of experienced parachute troops as well as Hitler Youth elements, who fought hard but then wept boyish tears of mortification on surrendering – much to the embarrassment of the Welsh Guardsmen who led them into captivity. There was also a serious side to this kind of bloody-minded resistance. In a letter to his father, Major J. D. A Syrett, commanding No. 4 Company, described the German forces as including large numbers of eastern Europeans who were usually anxious to surrender, unlike the Hitler Youth elements, who simply refused to give up:

> The Germans are mostly very young – boys of sixteen, seventeen and eighteen, nearly all fanatical Nazis who still believe Germany has some tremendous surprise up her sleeve and will win in the end. Their ignorance of the real state of affairs is astonishing, and they all believe we shall shoot them on capture; that, no doubt, is why most of them fight so desperately.[6]

At 08.30 hours on 8 September the Welsh Guards Group resumed its attack on Hechtel with Prince of Wales Company and X Company Scots Guards (attached to 1st Battalion Welsh Guards since August and commanded by Major Steuart Fotheringham), plus the tanks of No. 2 Squadron. Both infantry companies encountered determined opposition as they attempted to occupy their objectives – X Company the north-eastern side of the village and Prince of Wales Company the western side; by evening they had dug in and entered into a truce which allowed both sides to evacuate their wounded. It was a scene of utter chaos:

> And so night came over Hechtel. In the Kloosterstraat the house of Henri Snoek was on fire and in the Kirkstraat the flames of the house of Hubert Wuyts threw a red colour over the surroundings. In the flickering light of the flames phantoms could be seen of people who tried to save what they could of their possessions and of soldiers who were pulling away the wounded.[7]

During the night both companies had to fight off German attacks on their positions and suffered accordingly during hand-to-hand fighting. Some idea of the nature of the fighting which typified the Hechtel battle can be found in the experience of a lance sergeant who bumped into an equally astonished German soldier while they were searching a house; each let the other go – a short personal truce in the midst of a hard-fought battle. Even the use of the 2nd Battalion's tanks failed to break the deadlock, and the fighting continued in the same pattern for the next two days as the rifle companies, supported by armour, tried unsuccessfully to move forward into the heavily defended village. 'Everyone is tired,' noted Captain Evans in his diary on 11 September. 'We have all seen enough and heard enough of this bloody village called Hechtel – may it rot in hell and the Germans in it.'[8] That same day No. 4 Company attacked from the south of the village and linked up with No. 3 Company, but could get no further.

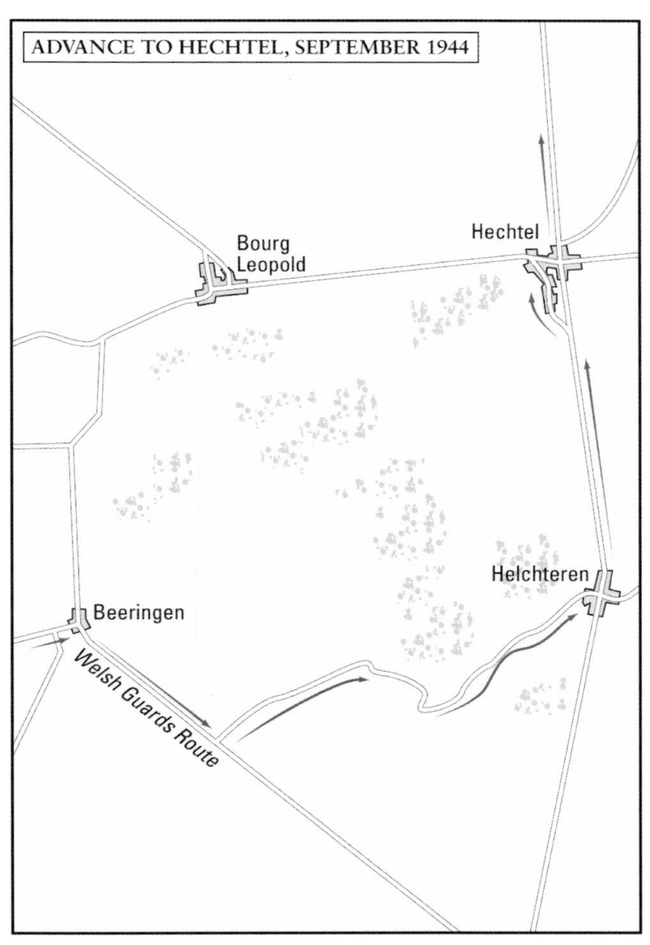

Map showing the route of the 2nd Battalion as it advanced towards Hechtel, 1944.

Brigadier the Earl of Gowrie, VC talking to DSgt John, MM, during a visit to Scotland. The 1st Battalion returned to the United Kingdom in March 1945 having sustained heavy casualties. *(Welsh Guards Archives)*

The stalemate was eventually broken on 12 September after Hechtel had been bypassed during a forward move by Guards Armoured Division towards the De Groot Barrier, the bridgehead over the Escaut Canal. Even so, although the village had been bypassed and surrounded, the Germans refused to surrender in the face of an all-out assault by the Welsh Guards Group. Backed by mortars and medium artillery this began at 08.00 hours. Peter Leuchars remembered that the weight of the attack knocked the stuffing from the defenders:

> The Germans were very professional and I'm full of admiration for the people we fought against. It's a very sandy area there and you dig slit trenches in about two minutes, but of course if a shell falls anywhere near them then the whole lot falls in. Once we got medium artillery involved then the young Germans – who weren't really properly trained and it was jolly bad luck putting them in there – were terrified and went and hid in the cellars. So on the last day it became not too difficult a battle.[8]

The five days of fighting took a heavy toll in terms of lives lost and property destroyed. The casualties in the Guards Armoured Division were sixty-two killed and many more wounded, while the Germans suffered around 150 casualties. A further 500 went into captivity, largely in the hands of the Welsh Guards Group. The fall of Hechtel and the role played by the Welsh Guards secured a vital part of the corridor which led towards Eindhoven. A large German force

LSgt Joyce winning the MM, Germany, 1945, as depicted in *The Victor* comic. He was later commissioned. *(The Victor, D. C. Thomson & Co.)*

had been defeated and a substantial quantity of German equipment had been destroyed. In a message to the Welsh Guards Group, Lieutenant General Sir Richard O'Connor, commanding VII Corps, expressed his thanks in the following way: 'It is one thing to gain ground but it is quite another to destroy a complete enemy battalion.'[10]

Ahead lay the race to the bridges at Nijmegen and Arnhem, the culmination of Operation Market-Garden. The story of this lost opportunity is well enough known and the narrative of the operation need not be told in its entirety. With the passing of the years it is possible to see where the mistakes were made at strategic and operational levels. It was wrong to drop the forces so far from their intended targets and equally erroneous not to fly two lifts on the first day. Once on the ground at Arnhem, British 1st Airborne Division was sluggish in getting to its target. Faulty or inaccurate intelligence gave insufficient information about the opposition, especially the presence of the two panzer divisions. Lieutenant General Frederick 'Boy' Browning, deputy commander of the Allied Airborne Army, failed to give US 82nd Airborne Division priority at Nijmegen and the ground forces showed a woeful lack of initiative. Horrocks's XXX Corps had been ordered by Montgomery to ensure that the attack should be 'rapid and violent, without regard to what is happening on the flanks', but this did not happen. As the authors of the British Official History described the outcome of Operation Market-Garden: '[It] accomplished much of what it had been designed to accomplish. Nevertheless, by the merciless logic of war, Market-Garden was a failure.'[11]

In the first stage of the ground assault, the Guards Armoured Division was led by the Irish Guards Group and later by the Grenadier Guards Group. To the Welsh Guards Group fell the task of covering the area around Grave, with its bridge over the River Maas carrying the road to Gelderland. This was a reasonably mobile phase of the battle, with key points being taken, but already the forward progress was getting depressingly slow due to delays on the road and stubborn German resistance. By the time that Arnhem's church spires were in sight, British airborne forces had been forced to surrender the bridge, allowing German armour and artillery to cross it. At this point, Heber-Percy returned to command the 1st Battalion after being wounded at Montchamps during the operations in Normandy. On 21 September the Welsh Guards Group moved forward to assist in the push from Nijmegen to Arnhem, but as they moved into the last stretch between the Maas and the Rhine the advance faltered.

By then both Battalions were ensconced in a fen-like landscape of reclaimed marshland whose only road was raised above the dykes, making all vehicles perfect targets for enemy attack either from the ground or from the air. As a result the advance frequently stalled, leaving tank crews as sitting ducks. One tank commander in the 2nd Battalion recorded his fears as he stopped 'stationary on the bridge … it is slightly uncomfortable as we are a perfect, silhouetted target for German bombers, if any'.[12] The flat lands were an uncanny place, much disliked by the Guardsmen, and the area was quickly christened 'the Island'. Bad weather, the bane of all infantrymen in open country, also added to the sense of pervading gloom, leaving Charles Murrell to complain to his diary on 23 September:

> Only two hours' kip – so rather tired. The night kept fine but after stand-down over came the low grey clouds and heavy rain – our concentration of gunfire during the night very heavy and almost incessant. Had breakfast in the downpour – a very miserable meal – one's fags too soaked to smoke. The dead cattle are humming quite a bit and some dead Germans lying around don't smell too pleasant either.[13]

The operation wound down on 25 September when the British airborne forces in Arnhem started assembling their remaining personnel for withdrawal and the Island was stabilized as the extent of the XXX Corps' advance. As a result, the Welsh Guards Group was not pulled out of the area until 4 November. Their next stop was Veulen where the 1st Battalion spent the fifth anniversary of its first arrival in continental Europe – 'a long time', noted Murrell, 'very few of that fine Battalion with us today. We did not think then that this would last five years.'[14] Christmas and New Year were celebrated in bitter winter weather at Namur in the flat lands of Belgium with still no inkling that the end of the

NAAFI break during the race to close the Falaise Gap. A well-deserved beer is enjoyed using a Nazi flag as a picnic blanket. Gdsm Holvey of Blaenavon reads the *Monmouth Free Press*. *(Welsh Guards Archives)*

RSM Rees, 2nd Battalion, 1943. *(Welsh Guards Archives)*

No. 3 Platoon, No. 1 Company, of the 3rd Battalion, at Lake Trasimeno, Italy, 1944. *(Welsh Guards Archives)*

war might yet be in sight. For some Welsh Guardsmen that moment could not come soon enough. With commendable understatement the 1st Battalion's War Diary recorded a narrow escape for Lance Sergeant E. T. Williams who 'had a very close call when a Spandau bullet went through his cap and parted his hair for him'.[15] On a less serious note a lunch was held on 1 December to celebrate the award of the Military Cross to Padre Payne, although, as the War Diary also recorded, the day almost ended in disaster:

> After lunch he [Payne] went round Coys and received a rapturous welcome wherever he went. He was given two chickens in one place which had to be hastily stowed away in his car as a Civil Affairs Officer made a sudden and unwelcome appearance.[16]

By the end of 1944, the summer dash out of Normandy was but a warm memory for the Allied armies as they faced the coming winter in the increas-

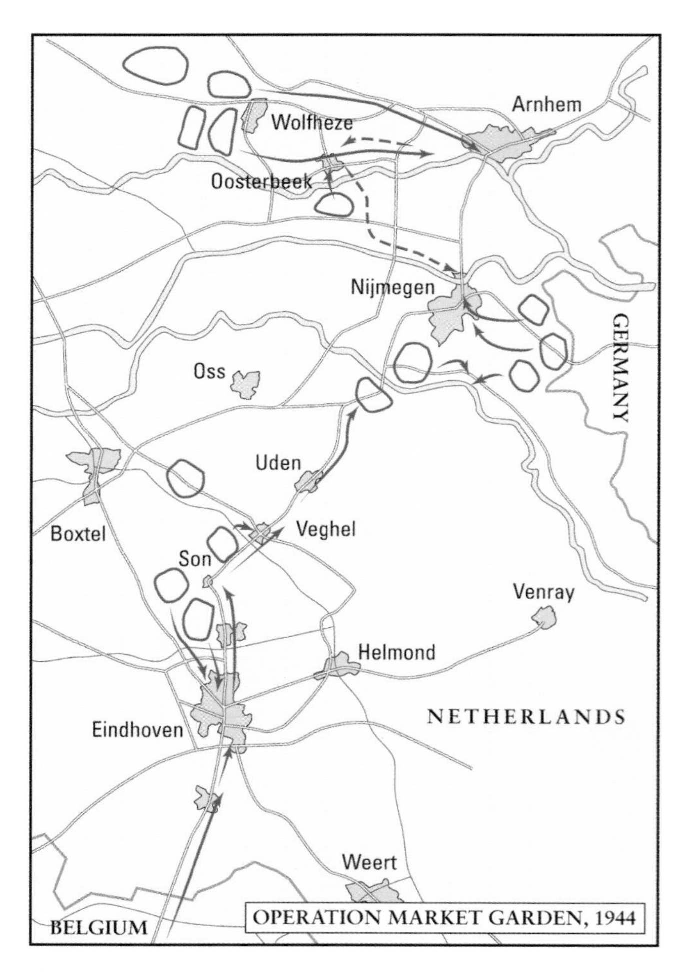

Operation Market-Garden, the 'bridge too far'. XXX Corps attack towards Arnhem, 17–25 September 1944.

They Were Not Divided

No account of the Welsh Guards' participation in the race to the Rhine would be complete without mentioning the film *They Were Not Divided* (1950). Written and directed by Terence Young, who had served in the Irish Guards during Operation Market-Garden, the film followed the fortunes of three recruits to the Welsh Guards from their basic training at Caterham to their service with Guards Armoured Division in Europe. The characters were an Englishman, an Irishman and an American, all played by relatively unknown actors – Edward Underdown, Ralph Clanton and Michael Brennan respectively – but the film also used real soldiers including Regimental Sergeant Major Ronald Brittain of the Coldstream Guards, known as 'The Voice' and a well-known character both inside and outside the Army. Other roles were played by actors who went on to enjoy successful careers, including Christopher Lee and Desmond Llewellyn. The film was notable for its realistic battle scenes, which included German armoured vehicles, and gave a spirited account of life in a close-knit regiment during the campaign in North-West Europe. Later in his career, in the early 1960s, Young directed the first James Bond films.

Poster from the 1950 film *They Were Not Divided. (ITV/Rex Features)*

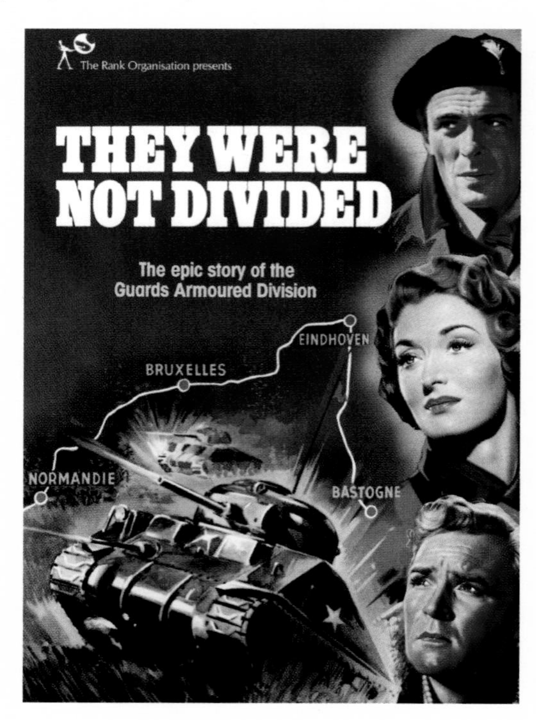

The Heber-Percy Family

The Heber-Percys of Hodnet Hall in Shropshire were one of the best-known hunting families in England during the 1930s and had a close connection with the Foot Guards. Four sons were born to Algernon Hugh Heber-Percy (1869–1941): the first, also Algernon (1904–61), commanded 3rd Battalion Grenadier Guards in Italy during the Second World War, while Cyril Hugh Reginald (1905–89) and Alan Charles (1907–34) served together in the Welsh Guards. A fourth son, Robert Vernon (1911–87), had a dubious reputation as a socialite and was widely known by his nickname of 'Mad Boy'. Early in life he became the constant companion of Lord Berners and helped to manage his estate at Faringdon. All four brothers were prodigious horsemen, one obituary of Robert noting that they all 'enjoyed the rude sports of a country gentleman'.

While serving in the 1st Battalion in the 1930s both Cyril and Alan built up reputations as madcap riders who pushed their luck to the limits and frequently fell foul of the authorities. On one occasion they hired an aeroplane to attend a steeplechase in the north of England and when they arrived back late at Pirbright they flew it low over the ranges so that Alan could blow the 'gone away' call on his hunting horn. Tragically Alan was later killed in a accident in a National Hunt steeplechase, but there was a serious side to their soldiering. Although Cyril left the Army to become Joint Master of the Cotswold Hunt, he re-joined the Regiment at the beginning of the war and served with the 2nd Battalion in the operations at the Hook of Holland and Boulogne. As Lieutenant Colonel C. H. R. Heber-Percy, DSO, MC, he commanded the 1st Battalion between October 1944 and August 1945. Like his father he was a noted gardener and wrote an affectionate portrait of his country childhood in *Us Four* (1963).

Lt Col C. H. R. Heber-Percy, DSO, MC (*right*). *(Welsh Guards Archives)*

ingly bitter weather conditions of North-West Europe. By then, too, the war of movement had given way to a war of attrition across the long front from the Channel to the Swiss border and it took its toll on men and equipment. At that stage in the campaign Eisenhower had at his disposal 73 divisions, the majority of which were American (49 US, 12 British, 8 French, 3 Canadian and 1 Polish), and he planned to use this overwhelming strength to advance on a broad front into Germany. Ahead lay the obstacle of the River Rhine; Eisenhower planned to address it with a pincer movement to the north which would roll up Army Group B, using Montgomery's 21st Army Group reinforced by the US Ninth Army, while Bradley's forces would advance towards the river between Cologne and Koblenz.

As the year began, the air was thick with rumours about a forthcoming offensive, with plans being put forward on an hourly basis and just as quickly being dropped, but as the 1st Battalion's War Diary noted, 'nothing passes the time quicker than taking over a part of the line and preparing for a major operation simultaneously'.[17] The next battle, Operation Veritable, opened on 8 February 1945 with one of the biggest artillery barrages of the war, with Canadian First Army attacking on the left towards the Rhine and Second British Army on the right towards the Reichswald. The 1st Battalion joined the attack on 14 February while the 2nd Battalion supported first the Irish Guards, then the Coldstream Guards. During the course of the operation, 2nd Battalion Scots Guards joined the Guards Armoured Division with the aim of replacing 1st Battalion Welsh Guards, which was under-strength and fought its last action on 23 March

No. 1 Platoon, No. 1 Company, of the 3rd Battalion, at Lake Trasimeno, near Perugia, Italy, 1945. *(Welsh Guards Archives)*

before returning to England. This move allowed the creation of a Scots-Welsh Group with 2nd Battalion Scots Guards taking over the motorized infantry role, having served previously in North Africa and Italy.

This was the formation which crossed the Rhine at Rees on 30 March, but instead of making a rapid gallop into Germany they found their way blocked by lightly armed troops who proved to be experts at delaying tactics – blowing bridges, blocking main roads and using anti-tank artillery to good effect. Even so, it took the Group just ten days to clear the first hundred miles into enemy territory. One of the highlights was a daring decision made by Windsor-Lewis to push his tanks rapidly forward by night from Nordholt to Lengerich, a move which took the Germans by surprise because at that time armoured formations were not expected to move in hours of darkness. This allowed them to strike northwards into the country between the ports of Bremen and Hamburg. On 27 April the last action was fought at Kirchtimke during which seven tanks were lost. On the credit side, 2nd Battalion Welsh Guards was able to liberate a nearby prisoner-of-war camp at Westertimke which held a Guardsman who had been taken captive with Windsor-Lewis at Boulogne and who had the pleasure of being freed by his old company commander. From there the Battalion moved north to Stade, a fine old Saxon town which astonishingly had been untouched by the war, a fitting end to a long campaign. Losses in the Regiment throughout the war were 633 killed and 1,306 wounded.

There remained one last duty. With the war over there was no longer any military necessity to maintain a large number of armoured divisions and it was clear that the days of the specially converted armoured battalions were numbered and that they would all revert to the infantry role. For the Guards Armoured Division the end came on a fine summer morning, 9 June 1945, when the entire Division was drawn up on an airfield at Rotenburg in Lower Saxony. On the instructions of the divisional commander, Major General Allan Adair, the resultant 'Farwell to Armour' parade was an impressive spectacle. The Division's Cromwell and Sherman tanks were specially painted battleship grey for the occasion, the paint having been liberated from the German Navy dockyards at Cuxhaven, and were drawn up in ranks with polished barrels raised in the morning sunlight. Distinguished guests arrived by motor vehicles and by Auster aircraft, and the parade was inspected by Field Marshal Montgomery, who praised the Division for setting 'a standard that will be difficult for those that come after to reach'. Then, on the orders of Brigadier G. W. Gwatkin, commanding 5th Guards Brigade, the tanks moved off in four battalion columns, turning their turrets and dipping their guns as they passed the saluting base and then moving out of sight in great clouds of dust. With the bands of the Scots and Welsh Guards playing 'Auld Lang Syne', it was an emotional moment.

3rd Battalion, Northern Italy and Austria

For the men of the 3rd Battalion who had pushed up to the Gothic Line, the winter of 1944–5 life reduced itself to a well-worn and familiar routine in the Battaglia sector. Up to five days at a time were spent on operational duty in the mountains above the Salerno valley where there was a programme of active patrolling interspersed with occasional short sharp actions. It is difficult to disagree with the finding of the Regiment's war historian when he noted that, 'It would be tedious to recount in detail their moves backwards and forwards and there is not enough of military interest in this period to give importance to a full record.'[18] The weather was uniformly awful, although the supply of winter-warfare kit, including string vests and heavy white socks, helped to make the cold bearable, and conditions were generally tough and unyielding. Had it not been for the services of the mules and Indian Army muleteers who brought up supplies on a regular basis, life could have been pretty grim for the men in their mountain fastness. The officer in charge of the mule operation, Captain J. J. Gurney, provided a vivid description of the trials and tribulations involved in bringing up rations and equipment over rocky and difficult ground, often in darkness and foul weather conditions:

> All the kit was ferried forward by jeep and taken over by the mule base company representatives, who laid everything out by loads in two parallel lines ready for the mules. Each company representative was told how many mules he would be allotted: when they arrived he was to collect this number, lead them between the two parallel lines and load up … the Indians made little effort to help, but the Guardsmen worked brilliantly, cutting the [signal] wire into lengths and tying up the wet and muddy loads by the light of a few torches. All this in pouring rain and bitter cold.[19]

If any reminder were needed that Italy had been no picnic, Peter Leuchars received it when he arrived in Naples in the spring of 1945 on transferring from the 1st Battalion to the 3rd Battalion. Before he left to drive north, a fellow soldier said to him: 'During your journey up to Rome count the number of rivers you cross. And then remember that every one had to be fought for.'[20] Although Leuchars found this 'a sobering thought', he was pleasantly surprised when he reached the Battalion to join No. 4 Company under the command of Major David Gibson-Watt. By way of contrast with normal wartime conditions, the Battalion was out of the line and was based in the pleasant town of Greve in the centre of the main Chianti wine-producing area to the south of Florence. A process of rotation allowed officers and men to lead 'a double life, dividing their time between slit trenches in the barren mountain tops and Florence drawing rooms and restaurants'.[21] During the winter the Battalion even managed to hold a ball in Florence to repay local families who had provided hospitality. The advent of the New Year was celebrated by a clandestine partridge

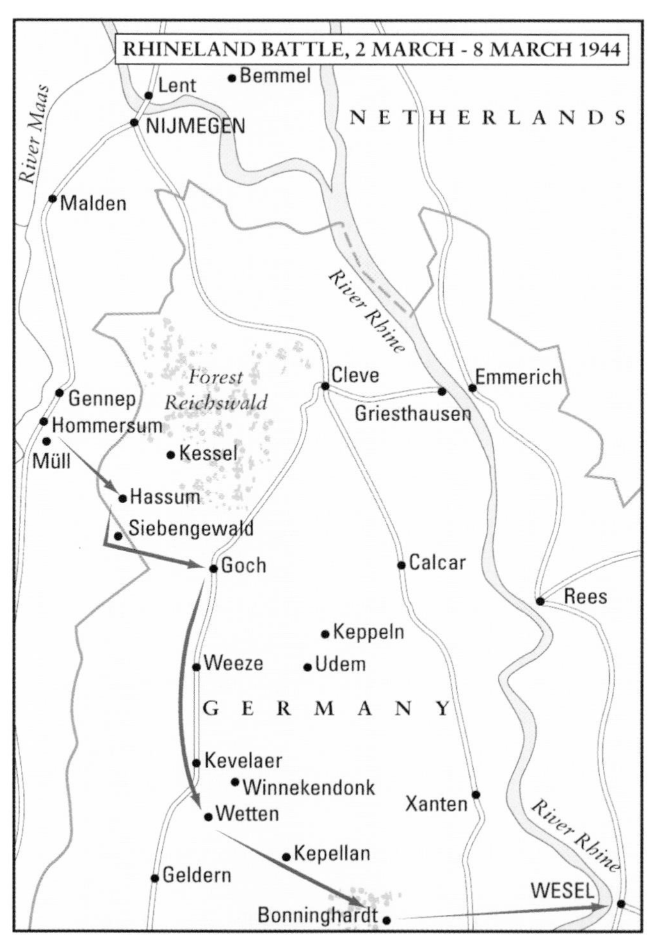

The Battle for the Rhineland which began on 8 February 1944. The 2nd Battalion crossed the Rhine at Rees at the end of the month.

MQMS and RSM of the 2nd Battalion, WO1 Ivor Roberts (*right*), Germany, 1945. *(Welsh Guards Archives)*

shoot and by the Germans firing tracer rounds into the night sky to form a huge 'V' sign. The Allies responded with one round of gunfire from their own artillery.

On 17 February 1945, as winter began to loosen its grip on the land, the Battalion left the mountains north of Florence for the last time and prepared to join the Allied advance, which would begin on 9 April, towards the Rivers Po and Adige, the Allied plan being to trap and destroy German Army Group C in a pincer movement involving the US Fifth Army and British Eighth Army. The former would make its attack north to the west of Bologna while the British forces would seize the Argenta Gap between the River Reno and Lake Comacchio. This move took the Battalion to the historic Adriatic seaside resort of Porto San Giorgio, which the War Diary described as 'a delightful fishing town. Modern but pleasantly situated overlooking a large flat stretch of beach … 2 Coy had the best billets occupying a delightful house on a hill a few miles inland.'[22] Ahead lay a period of hard fighting for the Battalion as it provided a flank guard for 26th Armoured Brigade as it advanced northwards towards the River Po along a route which showed increasing signs that the enemy was losing heart – abandoned fighting vehicles, white flags flying from surrounding farms and clusters of spiked artillery pieces. There was still some resistance at contested sites, such as the crossing of the Combalina Canal, but it was difficult to avoid the impression that the opposition was slowly collapsing. On 21 April the 3rd Battalion War Diary recorded that Sergeant Shiers of the Carrier Platoon came across a German Mark IV tank 'rammed tightly into a barn; this was found to be in perfect working order in every respect and it was brought back to Bn HQ in triumph'. On the following day Lieutenant Colonel R. C. Rose Price, DSO, was appointed to take over command of the 3rd Battalion; he was the son of Colonel T. R. C. Price, Regimental Lieutenant Colonel 1924–8, and the brother of the actor Dennis Price.

As Price took command, the fighting in Italy was reaching a fitful climax. Owing to losses in the Coldstream Guards they had been replaced in 1st Guards Brigade by 1st Battalion Welch Regiment, and the two Welsh formations were involved in successfully repulsing a determined German counter-attack the following day when six Tiger tanks attempted to dislodge the newly-gained bridgehead on the Combalina. Then came the momentous news that the leading element of the brigade had reached the River Po and that contact had been established with the US Fifth Army to complete the encirclement of the enemy. As the Allied amphibious vehicles began to arrive along the river banks, an unnamed officer envisaged the scene as 'like the Thames at Henley in regatta week as seen by H. G. Wells'. A platoon attack was made by men of No. 4 Company in which Sergeant M. G. G. Chatwin successfully took all objectives through 'a combination of skilful leadership and unskilful map-reading'. The day ended with an attempt to take a bridge over the Adige at Castel Guglielmo, which was only foiled by the onset of darkness and the realization that another crossing had been found a few miles away. During the operation the Battalion fired their last shots of the war on 27 April when a party of a dozen German soldiers led by two NCOs on horseback attempted to cross the bridge and surrendered after coming under fire. The following day saw Operation Ferreting, a tongue-in-cheek exercise involving 3rd Battalion Welsh Guards, 3rd Battalion Grenadier Guards and 1st Battalion Welch Regiment, which was intended to round up prisoners and stragglers or, as Brigade orders put it, to 'liquidate enemy tps and suspicious characters in this area'. After the carriers cleared the start line, the two Guards' battalions began the beats, while the Welch acted as long stop, the adjudication being stated in the following sporting terms:

> The bn sending the biggest bag to Bde HQ will be adjudged the most successful. Entries close 24.00 hours 29 Apr. Marks will be deducted for innocent Italians on the scale of 2 marks against 1 for every genuine prisoner. Extra marks will be given for good MI [military intelligence] sent to Bde HQ.[23]

The final bag for 3rd Battalion Welsh Guards was seven Italian prisoners-of-war dressed in plain clothes, but this was bettered by the Grenadiers who bagged nine.

With that light-hearted foray the shooting war ended for the 3rd Battalion Welsh Guards, but ahead lay their unanticipated involvement in one of the biggest tragedies of the immediate post-war period – dealing with assorted anti-Communist Croats and Slovenes and pro-German Russian Cossacks, White Russians and Ukrainians, many accompanied by women and children, who converged on the Austrian province of Carinthia hoping to surrender to the Western Allies before the advancing Red Army arrived. The first inkling that the Battalion had of the scale of the problem came on 2 May at Conegliano when the Commanding Officer, Adjutant and Padre were visiting No. 1 Company and came across 'a weird sight consisting of thousands of hungry Germans wrapped in blankets huddling forlornly around railway sleeper wood fires – with the rain pouring down – the acrid smell of tar smoke and icy scuds blowing off the Alps'.[24] These were the remnants of the German 65th Division and 1st Parachute Division who were waiting to be processed as prisoners-of-war, the usual fate awaiting a defeated army, the Germans in Italy having surrendered the preceding day. But the real extent of the problems that lay ahead became apparent two days later as the Battalion moved further northwards towards Villach and Rosegg in southern Carinthia. Here they encountered a 'continuous stream of approx 6,000 Chetnik soldiers [who] passed us trundling along with waggonettes, camp followers, bedraggled females and other appendages that go with an ill-organized and ill-directed army'.[25]

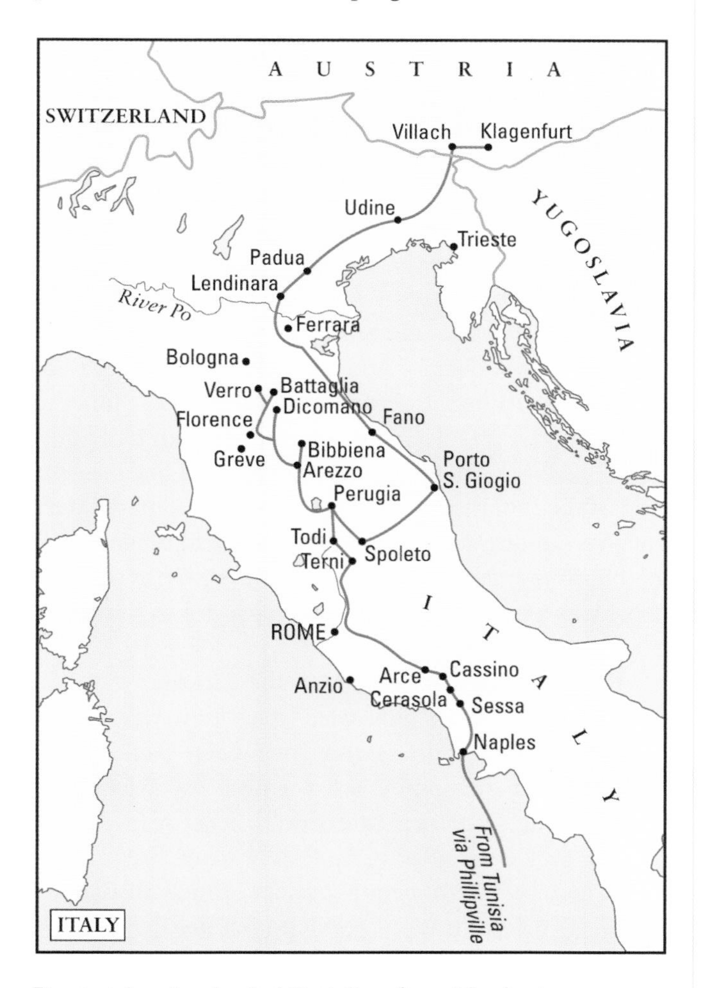

Route taken by the 3rd Battalion from Naples to Klagenfurt, 1944–1945

These Serb and Croat pro-monarchist nationalist forces, numbering some 27,000, had fought on the Axis side throughout the war and had become infamous for their use of terror tactics against sectors of the population in the Balkans which had been pro-Communist or supported the aims of Josip Broz Tito's Partisans. As Nazi collaborators they were associated with the enemy, but it was another matter how to deal with them now that the war was over. To complicate the situation in Carinthia, where British, Soviet and pro-Tito forces had begun arriving in the summer of 1945, it was necessary to handle around half a million German prisoners-of-war at a time when there was a danger that Tito's victorious Partisan army might advance into Italy and occupy the province of Venezia Giulia. To counter this threat, Field Marshal Alexander, now the Allied Commander-in-Chief, Mediterranean, was ordered to take the necessary action to stop this happening. His response was that he could do nothing until he was allowed to 'clear the decks' by handing over the Chetniks and other pro-Nazi Yugoslavs to the Soviet authorities.[26]

Unfortunately, owing to the tenor of the times and the fraught situation in Carinthia, this piece of *Realpolitik* involved force and deception and, as an entry in the War Diary for 19 May made clear, it also involved the men of the 3rd Battalion in the process:

The Price Family

This father and son each had the distinction of commanding a battalion of the Welsh Guards in war. Thomas Rose Caradoc Price (1875–1949) was born at Vellore, India on 2 August 1875. Following school and university in Melbourne, Australia, he was commissioned in the Royal West Kent Regiment in 1893 before serving in India, where he transferred to the Indian Cavalry in 1899. After seeing operational service in North-West Frontier Province, he was appointed to the staff of 2nd Indian Cavalry Division. He transferred to the Welsh Guards in December 1915 and saw service with the 1st Battalion on the Western Front where he was mentioned in despatches on three occasions and awarded the DSO. Between 1 October 1924 and 30 September 1928 he was Regimental Lieutenant Colonel.

His son, Lieutenant Colonel Robert Caradoc Rose Price, DSO, OBE (1912–88), commanded the 3rd Battalion in Italy in 1945; another son was the well-known film and stage actor Dennis (Dennistoun John Franklin Rose) Price.

> Lovely day. Evacuation of Croats begun. Order of most sinister duplicity received i.e. to send Croats to the foes i.e. TITs [pro Tito forces] to Yugo Slavia under the impression that they were to go to Italy. TIT guards on trains hidden in guards van.
>
> 2,500 Croats evacuated.[27]

The orders came from V Corps headquarters and had to be carried out but they caused huge resentment within the 3rd Battalion. Watching the operation was the author and publisher Nigel Nicolson, who was serving as a captain in 3rd Battalion Grenadier Guards. Long after the war he recorded his disquiet in a lengthy interview with the Imperial War Museum:

> The Welsh Guards who had the major task of forcing these people into the trains – they were at one point almost on the verge of mutiny. They were saying to their officers, 'Is this what we fought the war for?' And when one company of the Welsh Guards was relieved by another, a sergeant in the first company who'd been through this just for one day said, 'Well, I'm very glad we haven't got to do it again because I couldn't answer for what the men would do if we were ordered to do this a second time.' So they rotated them, the companies: it was as bad as that – the feeling was running as high as that.[28]

Equally troubling was the treatment of some 40,000 'Cossacks' who in fact came from a variety of racial backgrounds, including Poles from the Baltic, White Russians from Yugoslavia, as well as genuine Cossacks from the Caucasus and western Ukraine. Under the terms of the Yalta Agreement signed on 11 February 1945, all Soviet citizens liberated by the Allies and all British subjects liberated by forces under Soviet command were to be handed over – with some exceptions including White Russian émigrés who had escaped in the aftermath of the Bolshevik revolution of 1917 but who had become embroiled on the Nazi side in the Balkans. At the same time, others were liable to be returned and, following a meeting at V Corps on 21 May, crude steps were taken to hand these men and their families over to the Soviets, even though the move was met with dismay and disgust at lower levels of command. Although little or nothing was done to screen those about to be repatriated and individual cases were not examined, the repatriations began at the end of the month and 3rd Battalion Welsh Guards were involved in one such action at Weitensfeld where cages had been prepared in preparation for the repatriation of Cossacks to Judenburg in the Soviet sector to the north:

> Rather trying day. Half of the Cossack offrs when they were informed that they were to embus refused to as they had been tactlessly informed that they were going to Russia. Thinking that their fate would be a desperate one they demanded to be shot or given firearms in lieu. By tactful negotiation and timely display of a Wasp [tracked vehicle-mounted] flame thrower they were induced to get into the TCVs [troop carrying vehicles].[29]

Brigadier The Earl of Gowrie, VC, PC, GCMG, CB, DSO, visits the 1st Battalion in Scotland shortly before its deployment to Palestine in 1945. Behind Lord Gowrie, speaking to a Guardsman, is Colonel Sir Alexander Stanier, Bt, DSO, MC, Regimental Lieutenant Colonel, and on Lord Gowrie's right is Major Sir Richard Powell, Bt, MC. *(Welsh Guards Archives)*

But just when it seemed that the tension could not be raised higher, there was a last-minute reprieve from V Corps when it was found that there were fifty White Russians in the party who had not lived in the Soviet Union since 1920 and were not liable for repatriation. They were immediately debussed and returned to the cages amidst 'widespread rejoicing as no one relished the idea of sending them to an uncertain fate in Russia'.

The episode involving the Cossacks was later the object of intense historical debate and resulted in a high-profile libel case in 1989, but the fact remains that the officers and men of 3rd Battalion Welsh Guards found themselves in an impossible situation, being forced into an action which involved 'a good number of white lies' and angered the Guardsmen who had to take part. Later in his life, Colonel Rose Price admitted that his men were 'only able to continue in this harrowing task by constantly reminding themselves that former events had proved that, had the boot been on the other leg, the kicks would have been no less savage'.[30] For the Battalion which had fought so hard and over so many months it could have been a dispiriting way in which to end the war, not least when orders arrived on 22 June ordering the men to be prepared for service in the Far East where the war against Japan was still being prosecuted. Fortunately that order was rescinded and, at the end of the month, the Battalion left Austria to return to Italy, taking with them memories of a brief period when it was possible to be in a beautiful country without the perils of fighting a bitter enemy. Following the move into Italy, the War Diary recorded the bucolic thoughts of the Battalion's Adjutant, Captain F. L. Egerton, a man who was clearly in tune with his surroundings:

> Bn moved at an early hour of 0300 hrs and moved west up to time. The drive was an

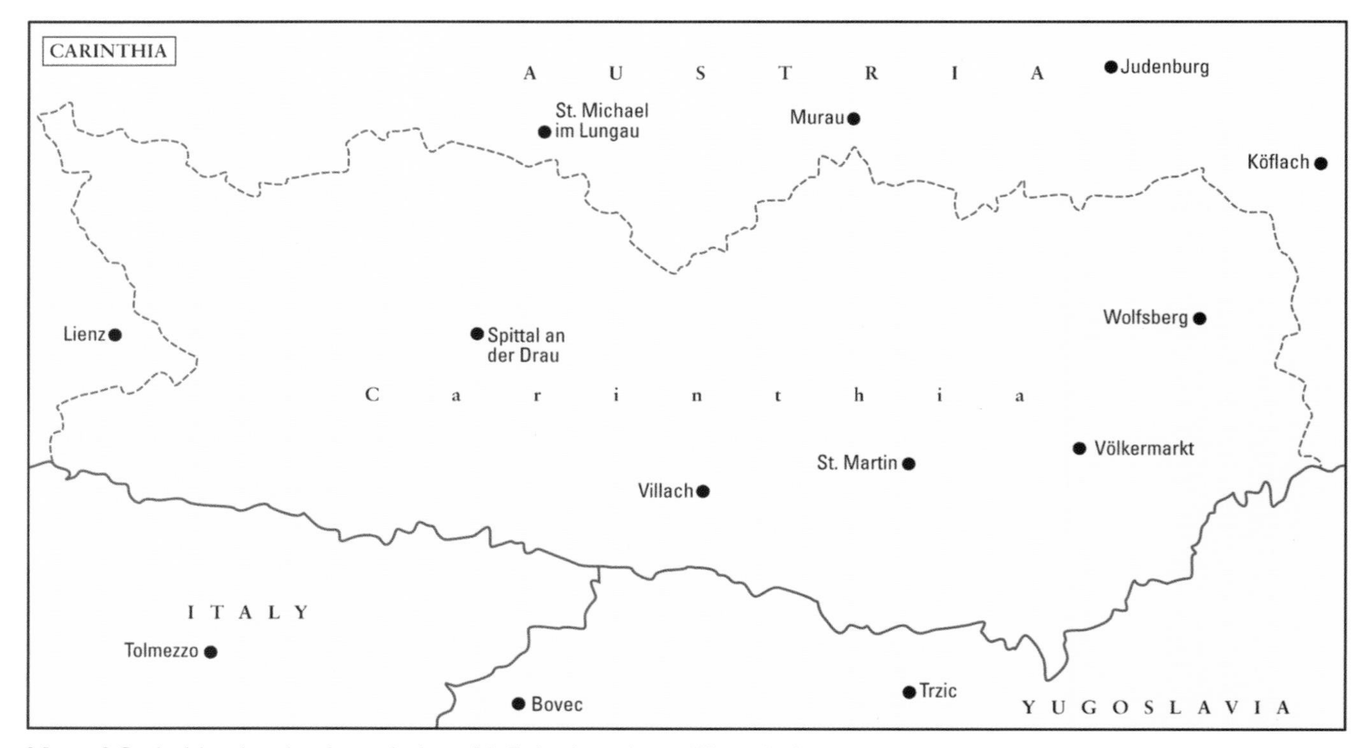

Map of Carinthia, showing boundaries with Italy, Austria and Yugoslavia.

> exquisite one, winding up over Alpine passes, first in darkness, then through a mere glimmer of dawn and then through the pine scented woods and alpine pasture.[31]

And so the war came to an end for 3rd Battalion, the self-styled 'Fighting Third'. As one of their officers, Sir Frederic Bolton, MC, said later, they were raised for the duration of hostilities and fought two major campaigns in North Africa and Italy. As a result, he wrote, 'We became much more self-contained: we were only visited twice, I seem to remember, by RHQ, which encouraged us to think of ourselves as being in some way unique – the same syndrome as MEF [Middle East Forces] standing for Men England Forgot or Burma being the province of the Forgotten Army, not the Fourteenth.'[32]

The Second World War had been a huge test for the nation, not least because it was an effort that involved everyone; very few people, even in the most remote or under-populated rural areas, had escaped the reality of modern total war. Just about every British family contributed members (male and female) to undertake some form of service under the various National Service Acts which harnessed the energies of the bulk of the population and this helped to foster a belief of everyone being in it together. It also meant that the Regiment was fully engaged not just through its three battalions on the battlefronts in North Africa, Italy, France and North-West Europe, but also on the home front. RHQ had to deal with the welfare of the Regiment as a whole, supervising everything from dealing with Guardsmen's families to helping with the organization of drafts. The Training Battalion at Sandown Park was also central to the cause of maintaining a steady stream of replacements for the three battalions, as was the Armoured Training Wing at Pirbright.

This meant that everyone connected with the Welsh Guards family was to a certain extent also on the front line. This was especially true for those who served or lived in the London area, who had to confront the dangers of modern warfare during the German bombing campaign. This began with the Blitz in 1940 and culminated in the last year of the war when the Germans introduced the V-weapons, the V1, a rudimentary cruise missile, and the V2, a supersonic rocket. Fired from launching sites mainly in the Netherlands, the weapons caused over 12,000 deaths and twice as many injuries, and were rightly feared by the civilian population in the south of England, being haphazard in their targeting. (To make matters worse the engine of the V1 cut out before falling to earth, adding to the terror by creating uncertainty.)

On a hot and sultry summer's day, 30 June 1944, a V1 fell on Imber Court, the Metropolitan Police

Recreation Centre in East Molesey where an athletics meeting was going on involving men from the Welsh Guards Training Battalion. It exploded in the centre of the athletics track where seconds earlier athletic events had been taking place and the band had been playing; in the subsequent inferno eighteen Welsh Guardsmen and two other service personnel were killed. A further 105 were injured, many of them seriously, making it one of the worst incidents involving flying bombs. As the shocked survivors made their way back to barracks, one of their number, Dick Fletcher, recalled that on arrival at Sandown Park members of Headquarter Company were treated to some instant 'counselling' by CSM George Treble, a Cockney:

> He paraded the any-order company in the MT area and 'chased' us for 45 minutes without a second of respite up and down the temporary drill square. At 'Dismiss' we hated George more than Hitler. I remember his red face, snag-teeth and with a half-smile he said, 'We have to do sumfink!' He was a marvellous man.[33]

Due to the demands of wartime censorship, the incident was not fully reported at the time but half a century later the victims were remembered at a service at the modern Imber Court, and in 2001 the Regiment inaugurated a memorial stone made of Welsh slate and engraved by a stonemason in Llanelli. It was a fitting monument to one of the lesser-known but no less human tragedies of a war in which 633 Welsh Guardsmen were killed and another 1,306 wounded.

Assault crossing of the River Po, Italy, spring 1945. *(Welsh Guards Archives)*

Guardsmen enjoying the distractions of a French bordello, Autumn 1944. Painted by Sgt C. Murrell. *(Courtesy of the Murrell family)*

Welsh Guards Prisoners-of-War, 1939–1945

The first Welsh Guards prisoners-of-war were captured by the Germans at the conclusion of the Battle of Boulogne in May 1940. Through no fault of its own the 2nd Battalion was forced to retire towards the harbour area for re-embarkation on ships of the Royal Navy. During this process a small force under the command of Major J. C. Windsor-Lewis was cut off and left behind. It consisted mainly of men from Nos. 2 and 4 Companies, some Irish Guardsmen, about 150 refugees, around twenty French troops and other assorted soldiers with a variety of cap badges.

With no reasonable option but surrender, Windsor-Lewis went into captivity but later escaped (see separate entry). Amongst those less fortunate was Guardsman Sydney Pritchard from Aberdare who spent the rest of the war in Stalag XXA at Thorn in Poland. At the end of the war he was repatriated and wrote a memoir of his experiences, *Life in the Welsh Guards 1939–46* (2007). Captain Peter Hanbury later privately published a personal account of the action at Boulogne and his subsequent capture by the Germans.

Other Welsh Guardsmen were taken prisoner during the retreat to Dunkirk. Amongst them was Major A. W. A. 'Billy' Malcolm, who was captured near Dunkirk whilst attempting to sort out the traffic chaos created by the German advance. Although he insisted that 'at least I was still alive', he managed to escape under difficult and dangerous circumstances towards the end of the war. In the post-war world Billy Malcolm had the distinction of commanding all three Battalions, including the 1st Battalion at the Sovereign's Birthday Parade in 1949.

Cromwell tanks, winter of 1944. A drawing by Sgt C. Murrell. *(Courtesy of the Murrell family)*

Imber Court

Welsh Guards casualties, 30 June 1944

2nd Lieutenant G. A. M. Barker
3769857 Guardsman C. W. Barstow
2nd Lieutenant J. A. L Crofts
2739165 Guardsman A. Fernihough
2739388 Guardsman J. F. Fernyhough
6145541 Guardsman C. C. Field
2739223 Guardsman I. G. Glen
3778234 Guardsman G. H. Green
2734026 Sergeant T. G. Griffiths
2737388 Guardsman A. G. Hill
14372843 Guardsman J. T. Hughes
2738924 Guardsman S. E. Jones
2731946 Guardsman C. H. Lang
14295068 Guardsman A. E. Lemon
Lieutenant W. F. Moss
6468155 Lance Corporal C. Richardson
2739424 Guardsman A. F. Street
2739402 Guardsman H. Wheeler

Imber Court – A Survivor's Story

'It was a sultry afternoon, a band was playing and everybody was having a good time. I'd just taken part in the running relay and was watching the 100 yards sprint when I heard someone shout that a flying bomb was on its way straight towards us.

Nobody heard it coming because of the band, so no one had any time to get out of the way. It landed precisely in the centre of the oval running track. It was carnage, I saw things which were unutterable, men and women were slaughtered and over 100 suffered terrible injuries. I was lucky because even though I was in the middle of it I only got shrapnel wounds and was later operated on at a hospital at Chelsea.

Some of the dead were never found and yet not a line of the incident made it into the press.'

As recounted by Sergeant Islwyn Evans of Betws-y-Coed in the *North Wales Weekly News*, 23 August 2007.

Officers studying mosaics and defensive overprints, prior to Operation Veritable (the advance to the Rhine), at Vught, Holland, February 1945. Drawn by Sgt C. Murrell. *(Courtesy of the Murrell family)*

CHAPTER EIGHT

Brave New World, 1945–1956

UK, Palestine, West Germany, Egypt

BRITAIN ENDED THE WAR with armed forces numbering just over five million men and women in uniform and with the same strategic obligations that it had possessed in 1939, plus the need to garrison Germany and Austria. The Army alone had forty divisions but it very quickly became clear that the cost of maintaining those forces was beyond the reach of a country whose economy had been devastated by the war. Many things remained the same in the armed forces but the next twenty years were to witness a sea-change as their size was decreased and Britain's overseas colonial holdings were dramatically reduced. Fighting the Second World War had drained Britain financially, and Clement Attlee's post-war Labour

Prince of Wales Company, Coronation Day, 1953. *(L to R) Front row*: LCpl Deag, LCpl Rowlands 36, LSgt Roberts 36, LSgt Burnell, LSgt Newell, 2Lt I. L. Mackeson-Sandbach, WO2 (CSM) Payton, Maj M. C. Thursby-Pelham, CSgt (CQMS) Williams, 2Lt J. W. R. Malcolm, 2Lt B. D. Stanier, LSgt Jones 56, LSgt Whiteman, LCpl Williams 85, LCpl Williams 75. *Second row*: Gdsm D. Richards, Gdsm F. Wells, Gdsm W. Evans 64, Gdsm D. Lovell, Gdsm R. Clarke 400, LCpl D. Powell, LCpl G. Davies 63, LCpl J. Jones 66, LCpl D. Evans 84, LCpl J. Redfern, LCpl R. Taylor 39, LCpl D. Sprake, Gdsm J. Martin, Gdsm A. Lee, Gdsm J. Stone, Gdsm R. Lawman. *Third row*: Gdsm T. Newman, Gdsm D. Powell, Gdsm T. Turner, Gdsm J. Kelly, Gdsm B. Harris, Gdsm C. Penn, Gdsm G. Scott, Gdsm I. Marshall, Gdsm D. Hobbs, Gdsm J. Bamsey, Gdsm J. Clark, Gdsm R. Williams 43, Gdsm T. Evans 51, Gdsm E. Astley, Gdsm A. Welsh. *Fourth row*: Gdsm J. Clayton, Gdsm G. Evans 19, Gdsm T. Jones 63, Gdsm R. Ford, Gdsm P. Hughes 75, Gdsm R. Brown, Gdsm P. Bayliss, Gdsm G. Canham, Gdsm L. Griffiths 90, Gdsm H. Teckoe, Gdsm M. Roberts 78, Gdsm R. McGlade, Gdsm W. Parsons, Gdsm W. Nicholson. *Fifth row*: Gdsm T. Robinson, Gdsm R. Gill, Gdsm N. Bennett, Gdsm T. Harding, Gdsm W. Rowland 22, Gdsm Padmore, Gdsm P. Davies 80, Gdsm K. Williams 96, Gdsm R. Ball, Gdsm R. Hill. Both the CSM and the CQMS went on to become RSMs of the 1st Battalion Welsh Guards. The photograph was taken outside the Officers' Mess, Victoria Barracks, Windsor, where the battalion was stationed for Coronation Day, prior to its embarkation for Egypt. *(Welsh Guards Archives)*

Brigadier Sir Alexander Stanier, Bt, DSO, MC, with his son 2Lt B. D. Stanier, Coronation Day, 1953. *(Welsh Guards Archives)*

government found itself having to grapple with the problems of recession, shortages and financial restrictions imposed by the shattered economy. In a world which saw Britain negotiating a loan of $3.75 billion from the United States, followed by harsh measures to restrict domestic expenditure, huge armed forces were a luxury the country could ill afford. Second World War equipment was not replaced in any quantity until the 1950s, forcing Field Marshal Viscount Montgomery of Alamein, Chief of the Imperial General Staff between 1946 and 1948, to remark that 'the Army was in a parlous condition, and was in a complete state of unreadiness and unpreparedness for war'.[1]

On the credit side, demobilization was carried out quickly and efficiently so that time-served and war-service personnel could return reasonably quickly to civilian life. To replace them, conscription was kept in being. Under a succession of post-war National Service Acts it became the law of the land for every male citizen to register at his local branch of the Ministry of Labour and National Service as soon as he became eighteen. Information about the relevant age-groups and clear-cut instructions were published in

Major Arthur Rees, MBE

Born in Ogmore Vale on 17 May 1916, Arthur Rees joined the Welsh Guards at the age of seventeen and within six months of leaving the Depot had been promoted lance corporal. Further promotion was rapid: by the outbreak of the Second World War he had risen to the rank of WO2, serving in that rank with the 2nd Battalion throughout the war. In August 1945 he was posted to the 1st Battalion as Regimental Sergeant Major and remained in that post during the tour of Palestine. His knowledge and experience were essential when the 1st Battalion returned to London to take up public duties, notably in preparing for the 1949 Birthday Parade. When the Army produced a recruiting poster with the slogan 'Pride of the Principality', Rees was the model wearing the uniform of the Welsh Guards. In June 1956 he received his commission and returned to the 1st Battalion as Lieutenant-Quartermaster, a post he held until 1961. He retired from the Army on 2 March 1968 in the rank of major, but worked as a retired officer at Headquarters, London District, for many years. Rees excelled at most sports, notably heavyweight boxing, and it was said of him that 'few of his opponents entered the ring with any degree of confidence of a successful outcome'. Rightly revered as an outstanding Welsh Guardsman, he died on 14 October 1991 while visiting his son in the USA.

Welsh Guards uniforms, post-war period. When the Second World War ended, the Regiment was scattered far and wide. 1st Battalion was in Scotland training young soldiers, and providing public duties detachments in London. In July 1945 it was ordered to move to the US to start training with 1st Guards Brigade for assault landings on Japan. The 2nd Battalion witnessed the German surrender on the North Sea at Cuxhaven in June 1945, and said goodbye to its tanks at the 'Farewell to Armour' Parade of the Guards Armoured Division. The 3rd Battalion ended the war in Austria, after a long haul from Naples through Cassino, to Monte Piccolo and the Gothic Line. The 3rd Battalion was ordered home in August 1945 and disbanded in Scotland just after St David's Day, 1946. *(From illustrations by Charles C. Stadden and Eric J. Collings commissioned by the Regiment)*

1946, Palestine, Guard Order. After the sudden capitulation of Japan in August 1945, 1st Guards Brigade was ordered to Palestine. It remained there in a peace-keeping role until the emergence of the modern state of Israel three years later. High standards of turnout were maintained by Battalion Headquarters guards at the Elizabeth Hotel, Tiberias. A new rifle was issued to the 1st Battalion while it was in Palestine. This was the .303 No. 5 Mark 1 Jungle Carbine, a cut-down version of the standard No. 4 Lee-Enfield rifle. Originally developed for the Far East, it was lighter with a shorter barrel, a flash hider and a rubber butt plate to reduce recoil. Its reduced size made arms drill more difficult.

1950, Mounted Officer, Service Dress. Here, a lieutenant colonel is turned out in mounted order. His breeches of Bedford cord have inside strappings of buckskin. Soft brown field boots have eyeletted insteps, and are jacked at the tops, where leather garters are 'worn between second and third buttons of the breeches'. The spurs and chains of Foot Guards officers are gilt, except when the wearer holds a General Staff appointment.

1954, Egypt, Drill Order. From 1953 until March 1956 the 1st Battalion served with the security forces in the Canal Zone of Egypt. The Sam Browne leather sword belt has a long history. It was adopted by a one-armed Victorian general of the name about 1860, as a practical method of carrying both sword and revolver. In those days it had twin braces worn crossed at the back. Taken into service during the 1870s, it has since become a standard item in many armed forces, worn by commissioned and warrant ranks alike.

1951, Germany, Training Order. During the early 1950s battalion life was centred on service with BAOR rather than in London. Training order varied little from fighting order. In this illustration the soldier is armed with a Bren gun.

1953, Berlin, Change of Station Order. The illustration shows a CQMS. The cap has a gilt badge and three brass bands around its peak. The arm badge is three chevrons surmounted with a crown. The appointment of CQMS dates from a 1913 re-organization when the two senior company pay sergeants became a company sergeant major, and company quartermaster sergeant. The greatcoat is worn open at the neck to show the collar and tie which were adopted by other ranks in 1945. A red worsted sergeant's sash is worn over it. Web equipment is the abbreviated 1937 pattern equipment. The rifle is the .303 No. 4 Lee-Enfield.

the national newspapers and broadcast on BBC radio, and schools and employers also played their part in passing on the relevant official information to their young charges. Short of deliberately refusing to register, there was no way the call-up could be ignored and those who did try to avoid conscription were always traced through their National Health records. Between the end of the war and the discharge of the last conscript in 1963, 2.3 million men served as National Servicemen, the majority in the Army – there was one sailor for every twelve airmen and thirty-three soldiers. In its final form the period of conscription was two years (there had been earlier periods of twelve and eighteen months) and, like every other regiment in the British Army, the Welsh Guards benefited from the contribution made by men who were the first peacetime conscripts in British history. To the Regiment's credit, the Welsh Guards refused to regard their National Servicemen as being any different from the regular volunteers: according to Arthur Rees, Regimental Sergeant Major in the early days of National Service, the result was a fine soldier.

> They were Guardsman So-and-so and they were treated just the same as a normal regular soldier. I must say that a lot of them were very intelligent and became NCOs. I always had a very high regard for National Servicemen. If they had stayed in longer I'm certain that quite a few of them would have gone all the way through the ranks – but most of them went out after doing their two years. We never treated them as separate entities at all, they were just ordinary Welsh Guardsmen.[2]

The end of the war also meant that the Regiment was once again reduced to one battalion – the 1st Battalion which had returned to England in May 1945 and had been warned to prepare for service in the Far East as part of a re-formed 1st Guards Brigade. Later the Battalion moved up to Stobs Camp near Hawick in the Scottish Borders, before travelling south again to Kington Camp in Herefordshire. The plan was to move the British contingent to the USA for embarkation to Japan from San Diego in California as part of a British Commonwealth Division, but the deployment never took place after the dropping of atomic bombs on Hiroshima and Nagasaki. As Lieutenant Colonel Sir John Miller remembered later, within the Regiment the decision was met with mixed emotions:

Vehicle checkpoint, Palestine, 1947. *(Welsh Guards Archives)*

Capt Geoffrey Gibbon winning the Prix de Rotterdam, 1952. *(Welsh Guards Archives)*

> I do not think any of us looked forward to this in any way whatsoever, although I do not think we realised that it was going to be as terrible as it would have been if it had taken place. Ever since then I have been extremely grateful to President Truman for dropping two atomic bombs on Japan, because if that had not happened, I do not think that any of us who were going to go there would have been here today.[3]

As for the 2nd Battalion, it had ended the war in the Baltic coastal town of Lübeck in Schleswig-Holstein where it occupied Furness Barracks, a former German Army facility, in which the main blocks were redesignated Cardiff, Swansea, Harlech, Caernarfon and Wrexham. To the pleasant surprise of the Guardsmen, the accommodation was centrally heated and the windows were double-glazed, a huge benefit as the winter of 1946–7 was one of the coldest on record. Amongst the Battalion's duties was the responsibility for guarding a frontier post at the Soviet–British zonal border and dealing with the many Displaced Persons (DPs) who passed through it. This disheartening job was made more difficult by the rigours of the weather; the *Household Division Magazine* reported that 'The frontier post is situated in what must be one of the coldest and most depressing parts of Germany; those

Guardsman, drill order, Palestine, 1947. *(Welsh Guards Archives)*

Guardsman, fighting order, Palestine, 1947. *(Welsh Guards Archives)*

on guard do their best to defy the Baltic winds with enormous sheepskin greatcoats conveniently "re-allotted" to us from a former German Army Ordnance Depot.'[4] The cold weather also interrupted sport, particularly rugby. Even so, the 2nd Battalion's rugby XV, captained by Major A. A. Duncan, won the British Army of the Rhine (BAOR) title to qualify for the Army Cup final, losing to a XV representing the Royal Army Medical Corps Depot and Training Establishment which contained several international players, including the England winger J. A. Gregory. On 1 July 1947, following a short deployment in Cologne with BAOR, 2nd Battalion Welsh Guards was placed in suspended animation and the Colours were returned to King George VI for safekeeping following a ceremony at Windsor Castle. The Ensigns were 2nd Lieutenant G. D. Young and 2nd Lieutenant B. M. Rhys-Williams and the last Commanding Officer was Lieutenant Colonel Sir William V. Makins, Bt.

By then the 3rd Battalion had also disappeared. Having ended its war in northern Italy and having undertaken difficult repatriation duties in Austria, it had been thought that 3rd Battalion Welsh Guards, along with 3rd Grenadier Guards, would be deployed to the Far East as reinforcements for the invasion of Japa, but the end of the fighting in August put paid to that possibility.[5] Instead, as a formation that had been raised specifically for wartime service, 3rd Battalion Welsh Guards returned to the United Kingdom in May 1945 and, to everyone's dismay, was despatched north to Selkirk in the Scottish Borders. As Drill Sergeant W. B. Davies 08 remembered, 'We were a tough and proud Battalion and resented what was happening.'[6] Although there was a farewell parade in the presence of the Regimental Lieutenant Colonel, there were no Colours to hand over and there was general unhappiness at what was happening to a Battalion which had fought together for almost thirty months in conditions which ranged from the heat and dust of the North African desert to the rain and snow of the Italian mountains. Disaster almost overtook the final parade, which was held in April 1946. When the Commanding Officer mistakenly gave the order 'Stand at ease' with the parade standing at the slope, the situation was saved by Regimental Sergeant Major 'Snowy' Baker who led the senior NCOs in 'a hardly audible rippling whisper, "Stand still".' As Davies 08 recalled later, 'No one blinked an eyelid, not a foot, arm, leg, head or rifle as much as quivered [*sic*].'[7]

Although the global conflict had come to an end, the world was still an unquiet place. In October 1945, under the command of Lieutenant Colonel R. B. Hodgkinson, MC, the 1st Battalion moved to Palestine, at that time a British-mandated territory, which had been designated as a home for the Jewish people in 1917. By the end of the war it was clear that a confrontation was inevitable between them and the Arab population, who resented the steady arrival of European Jews from the 1930s, many of them refugees from Hitler's Germany. As a result of the growing friction, the Jewish resistance group the Haganah was strengthened and began to adopt the offensive ethos embedded in the philosophy of Ze'ev Jabotinsky, the founder of Revisionist Zionism, who believed that, if a Jewish state were created in Palestine, confrontation

HM King George VI, Colonel-in-Chief, Welsh Guards, 1936–52. *(Welsh Guards Archives)*

Escort to the Colour found by Prince of Wales Company, 1st Battalion Welsh Guards, 1949. This was the first sovereign's birthday parade in home service clothing since 1939. Field Officer in Brigade Waiting: Lt Col A. W. A. Malcolm; Major of the Parade: Major B. G. P. Eugster, DSO, MC, Irish Guards; Adjutant in Brigade Waiting: Capt N. S. Kearsley; Captain of the Escort to the Colour: Major the Hon. R. W. Pomeroy; Subaltern of the Escort to the Colour: Lt A. J. Gibson-Watt; Ensign of the Escort to the Colour: 2Lt N. Webb Bowen. *(Welsh Guards Archives)*

with the Arabs was inevitable and that the Jews would be forced to fight for, and then defend, their homeland 'behind an iron wall which they [the Arabs] will be powerless to break down',[8] Standing between the two sides were the 20,000 men of the Palestine Police, backed by 80,000 soldiers of the British Army's 1st Division, 6th Airborne Division and, from March 1947 when it was deployed in southern Palestine, 3rd Division. It was a deployment which Britain was hard-pressed to make and throughout the period the military units involved in internal security duties were severely stretched.

On 9 October 1945, 1st Battalion Welsh Guards left Liverpool bound for Palestine on board TS *Volendam*, a former Holland-America Line luxury liner which had been torpedoed but not sunk by a German U-boat in 1940. Following repairs, it had been converted as a troopship and conditions were so cramped that 'at night it was impossible to walk around the Troop Decks as men are sleeping on hammocks, on the floor and even on tables'.[9] The Battalion formed part of 1st Guards Brigade which itself was part of 1st Infantry Division, under the command of Major General Richard 'Windy' Gale. On deploying, the Battalion was spread across Galilee, with companies based at Safed on the Lebanon border, Rosh Pinna overlooking the Jordan valley, and Merulla close to the Syrian border in the lee of Mount Hermon. As briefly stated by the military historian Correlli Barnett, who served as a National Service conscript soldier in Palestine in 1946, the position faced by incoming battalions amounted to a siege in which 'Two British divisions and support troops, some 60,000 soldiers, were stuck in Palestine, adding to the balance of payments deficit, carrying out clumsy and ineffective sweeps against the Jewish terrorists who were murdering their comrades, and otherwise doing nothing but guard their own barbed wire.'[10]

Despite the difficulties, the first impressions were favourable, so much so that the Battalion War Diary recorded on 27 October that 'The Sjts Mess had its first big evening, it was so noisy that they frightened a Jewish settlement nearby.'[11] After war-torn Britain where bomb damage was still part of the landscape and food rationing remained in force, Palestine was clear of any sign of war and the Guardsmen were astonished to discover fruit in abundance – bananas, oranges and grapefruit, all not seen in Britain for many a year. Another bonus was that the sun seemed to shine every day. In fact, the first winter in Palestine turned out to be something of a time out of life for the Battalion and a first experience of what many of the older men remembered as 'peacetime soldiering'. For a few months life in Palestine went on as before, as Britain and the USA, now the main Zionist supporter, attempted to find a solution which would allow continued Jewish immigration and the eventual creation of a bi-national Arab–Jewish state along the lines of a pre-war agreement brokered in 1939. During this period of relative calm it was possible to take advantage of living in the biblical lands of Tiberias by the Sea of Galilee, 'brown hills climbing into mountains with small villages clinging to the shore with, far down below, a lake of clear, blue water'.[12] At the same time the Battalion's training facilities had to be placed on an operational basis with firing ranges constructed, the digging being carried out by Guardsmen often in trying physical conditions. Internal security training and exercises were also carried out, drill courses were run and there was even

Map of post-war Palestine. In May 1948 the state of Israel came into being.

HRH The Duke of Windsor visits the Battalion, Wuppertal, Germany, BAOR, 1951. *(Welsh Guards Archives)*

a chance to take leave in such exotic places as Luxor in Egypt, Cyprus and neighbouring Lebanon. In an effort to instil an atmosphere of controlled aggression, the Commanding Officer made boxing compulsory for all ranks under the age of twenty-five and even offered 'to take on any novice Gdsm of my own height on Coy PT pde for 2 one-minute rounds. If he knocks me out he can have a week's leave and a five pound grant from the Commanding Officer's Fund.' Helpfully, Colonel Hodgkinson added that he had not boxed since leaving Sandhurst when he had been knocked out in the first round of the novice's competition by the future Shah of Persia.[13]

It was in fact the calm before the storm. At the beginning of 1946 the joint UK–US commission on Palestine recommended the repeal of restrictions on land sales and the immediate admission of 100,000 Jewish settlers. Simultaneously, the Zionists in Palestine began preparing for a 'Hebrew revolt', with the Haganah joining in common cause with two other underground groups, the Lohamei Herut Yisrael (also known as the Stern Gang) and the Irgun Zvai Leumi. A crackdown on their activities was inevitable when the Jewish Agency sponsored a policy of encouraging illegal immigration of Jews from Europe. To begin

Officers shooting with HRH The Duke of Windsor, southern Germany. *(Welsh Guards Archives)*

with, the British placed all illegal immigrants in detention camps in Palestine but, in August 1946, it was agreed to deport all immigrants and forcibly prevent the ships, most of them unseaworthy and overloaded, from reaching Palestinian territorial waters. Illegal immigrants were taken to Haifa where they were transferred to British transport ships and shipped to camps in Cyprus or Mauritius. However, as these operations were carried out under the eyes of the international press, the sight of immigrants being detained and transported encouraged allegations that British forces were engaging in Nazi-like behaviour. In fact the operations were difficult and dangerous for both sides. British boarding parties were usually opposed: clubs, iron bars and hatpins were used in the resulting scuffles and there were casualties. The captains of the immigrant ships also manoeuvred violently or stopped their engines when approached by British warships, making it difficult for them to get alongside, and because they were often overcrowded there was the ever-present danger of capsizing. The result was an increase in tension in the relationship with the Zionist community, and the Battalion Diary noted that 'the friendly atmosphere between the Battalion and some of the local settlers, cultivated during the winter, started to change rapidly'.[14]

The first flashpoint came on 22 July when the Irgun attacked and bombed the King David Hotel in Jerusalem, killing ninety-one and injuring forty-six. The building was home to the British administrative and military headquarters in Palestine: the bomb, placed in the basement, resulted in the destruction of a large part of the hotel's southern wing. Inevitably the incident caused outrage in Britain and amongst the garrison in Palestine and led to an immediate military crackdown. Although the Battalion had managed to mount an impressive Birthday Parade in Tiberias on 13 June when HH Emir Talal of Transjordan took the salute, it was followed immediately by a succession of internal security duties including 'a large scale cordon and search operation at the Jewish settlements of Kefar Giladi and Tel Hai not far from Metulla which heralded the start of several such operations that summer'.[15] Later in the summer the whole of 1st Guards Brigade moved across the border into Transjordan where they pitched their camp in flat and stony desert off the road between Zerqua and Mafraq, the headquarters of the Arab Legion, the country's largely British-officered army. As the Battalion diary made clear, this was less about training for internal security duties and more about relearning the tenets of

CQMSs, Egypt, 1954. *(From L to R)* CSgts Williams, Fletcher, Brawn Richardson and Nicholson. *(Welsh Guards Archives)*

Signals Platoon, Palestine, 1947. *(Harold Graham Williams/Welsh Guards Reunited)*

Capt David Gibson on Klaxton, painted by Edward Archibald Brown. He won the Grand Military Gold Cup three times in a row riding this horse and a fourth time riding another horse. Gibson presented the cup to the Regiment when he retired.

conventional warfare and culminated in a Brigade Test Exercise. A highlight was a visit paid to the Battalion by HH King Abdullah of Transjordan, a staunch British ally who took a keen interest in military affairs.

> He watched Prince of Wales Company doing a field firing exercise. Knowing his love of 'bangs', Prince of Wales's Company were determined to do him proud in this respect and, judging by the noise, a large proportion of the ammunition stocks left in the Middle East after World War II were fired in his honour![16]

The exercises came to an end on 25 October when the Battalion moved back to Palestine, this time to northern Galilee where once again the companies were spread over a large area centred on Rosh Pinna, one of the first of the modern Jewish agricultural settlements. While things were relatively quiet, there was a problem with manpower within the Battalion as release took its toll on numbers – on 7 January 1947 No. 3 Company was placed in 'a temporary state of suspended animation' – and there were few opportunities for training or sport.[17] On 11 December 1946, Lieutenant Colonel R. B. Hodgkinson stood down as Commanding Officer and was replaced on 3 February 1947 by Lieutenant Colonel A. W. A. Malcolm. At the beginning of the new year of 1947 there was a further move to a new hutted camp at Sarafand outside Tel Aviv which was part of a larger military complex and included cinemas, shops and sports facilities. To the

HRH The Duke of Edinburgh, Colonel of the Regiment

His Royal Highness Prince Philip, Duke of Edinburgh, was appointed Colonel of the Regiment on 6 July 1953 in succession to Brigadier the Earl of Gowrie, VC. The announcement was made the day before the official royal visit to Wales, the first to take place since the coronation of HM Queen Elizabeth II on 2 June.

In 1956, a few years after after his appointment, Prince Philip introduced a shooting trophy known as The Duke of Edinburgh's Trophy which was to be competed for by all units of the Corps and Regiments of which His Royal Highness was Captain-General, Colonel-in-Chief, Colonel or Honorary Colonel. The teams were to consist of one captain, two subalterns, three non-commissioned officers above the rank of corporal, three corporals or lance corporals and three private soldiers. The Regiment first won the trophy in 1958 with a team under the captaincy of Captain B. D. Stanier, with a score of 1,983 points.

Prince Philip attended five St David's Day parades during the twenty-two years when he was Colonel, on the following occasions: 1 March 1955, 1 March 1960, 1 March 1964, 3 March 1968, 25 February 1973.

HRH The Duke of Edinburgh, Colonel Welsh Guards, 1952–75, painted by Barrie Linklater. *(Barrie Linklater)*

The Malcolm Family

The Malcolm 'Celtic Connection' goes back to 1915 and the foundation of the Welsh Guards when Major the Hon. A. G. A ('Sandy') Hore-Ruthven, VC, King's Dragoon Guards, became the first Second-in-Command of the 1st Battalion. In 1919 he was promoted Commanding Officer and, after a distinguished military career (see Chapter 1), was created the 1st Earl of Gowrie and became Colonel of the Welsh Guards in June 1942 until he handed over to HRH The Duke of Edinburgh in 1953.

In 1922, his nephew, A. W. A. ('Billy') Malcolm, joined the Welsh Guards, despite the fact that his father had served as a regular officer in the Scots Guards. Taken prisoner in 1940 after Dunkirk, he managed to escape three months before the end of the war and commanded the 3rd Battalion before going on to command both the 2nd Battalion and the 1st Battalion, the latter in Palestine. This was a unique occurrence within the Regiment. On returning to London, he commanded the King's Birthday Parade in 1949, the first parade in full dress after the war.

His elder son J. W. T. A. ('James') joined the regiment in 1948 for his National Service and stayed on as a regular, commanding the 1st Battalion 1970–2, including two short tours in Northern Ireland. He was Regimental Lieutenant Colonel from 1972 to 1976 and in 1973 in that role commanded the Queen's Birthday Parade. He retired from the Army in 1976. His brother John also served as a National Service officer and carried the Queen's Colour at the Coronation Parade in 1953.

James's elder son, A. J. E. ('Sandy') Malcolm, followed him into the Welsh Guards and commanded the 1st Battalion 1997–9. Not only was he following his father, grandfather and great-great uncle, but he was by then the only Welsh Guardsman whose family had a record of continuous service from the Regiment's formation in 1915. Moreover, by commanding the Queen's Birthday Parade in 1998, he probably created a Foot Guards' record by following his distinguished forebears in the same role.

Lt Col A. W. A. Malcolm. *(Welsh Guards Archives)*

Col Sir J. W. T. A. Malcolm, Bt, OBE. *(Welsh Guards Archives)*

Col Sir A. J. E. Malcolm, Bt, OBE. *(Welsh Guards Archives)*

Capt A. W. A. Malcolm wearing guard of honour order in 1923. The waist sash is of gold and crimson silk. He is wearing 'butcher boots' with brass spurs. These so-called gold spurs were only worn by mounted officers in the Foot Guards. *(Welsh Guards Archives)*

The 1st Battalion Welsh Guards parades in home service clothing, Chelsea Barracks, 1949. *(Welsh Guards Archives)*

pleasure of the Adjutant (Captain Peter Leuchars) and the Regimental Sergeant Major (Arthur Rees), there was also a large and well-appointed square, the first encountered in Palestine.

Operational duties meant that the square remained unused until St David's Day. Both Prince of Wales Company and No. 2 Company were moved almost immediately to the garden city of Tel Litwinsky, while the rest of the Battalion deployed to Petah Tiqva to help oversee the evacuation of non-essential British civilians and families to Egypt during Operation Polly. By then the security situation in Palestine had deteriorated drastically. Terrorist attacks on key points were carried out almost on a daily basis as the Jewish underground attempted to keep the Palestine problem on the front pages of the world's press. Both Arabs and Jews committed atrocities and for the British security forces this meant a steady diet of patrols, guards and sweeps in an attempt to find illegal weapons and prevent further outrages. In a letter to the Regimental Adjutant in London, Captain Leuchars gave some indication of the problems facing the Battalion and the need to maintain full vigilance: 'The Jews have been fairly peaceful up to date but Peter Kingsley Heath spotted a lot of women doing Bren Gun combined handling and unarmed combat the other day, so I think we shall be doing another search in a day or two.'[18] On another occasion, while searching a settlement south of Haifa, the Battalion enjoyed an unexpected success when a National Service Guardsman put his civilian training as an engineer to good use while searching the settlement's engine room.

> Examining the engine cursorily he noticed a large bar in the middle not normally in that engine. The top was square shaped and looked

as though it needed a lever to turn it. To his surprise a section of the floor in the corner slid away to reveal a flight of stairs leading to a large room beneath in which were stored racks of weapons, piles of ammunition, and even an armourer's bench and tools. To provide air to the room when armourers were working there, an ingenious device had been installed from the children's playground, so that when the swings were being used air was fed through into the underground room.[19]

For the Battalion it was a welcome discovery, but it also served to reinforce the determination of the Jewish population to do everything possible to ensure the expulsion of British forces from Palestine. Under sustained international pressure to resolve the problem, the British government finally admitted defeat in February 1947 and announced that Palestine would be handed over to the stewardship of the newly formed United Nations (UN). The result was the formation of the United Nations' Special Committee on Palestine (UNSCOP), which was established on 15 May 1947 and got down to work immediately thereafter, taking evidence from all the parties involved in Palestine. Three months later, at the end of August, UNSCOP submitted its report which came down on the side of arbitration and ending the British mandate as soon as possible. Palestine would be partitioned into Jewish and Arab areas, while Jerusalem would become an international zone safeguarded by the UN. It was a compromise but one which the Jewish Agency was prepared to accept, if with a heavy heart, because it safeguarded the concept of a Jewish state. This would comprise the fertile coastal strip between Haifa and Jaffa, the Plain of Esdraelon and eastern Galilee, while Palestinian Arabs would be granted western and central Galilee, Samaria, the Judean Hills and the Gaza Strip.

This marked the third phase of the Battalion's

The Commanding Officer salutes the Inspecting Officer, King's Birthday Parade, Wuppertal, 1952. *(Welsh Guards Archives)*

HM The Queen inspects a Guard of Honour in Swansea, 1953. *(Welsh Guards Archives)*

Guardsman Harold Williams on the radio in a Daimler Dingo scout car, Palestine, 1947. *(Harold Graham Williams/ Welsh Guards Reunited)*

involvement in Palestine; during the latter part of 1947 it provided train guards on the main railway line between Tel Aviv and Jerusalem, as well as taking part in 'flag marches' together with tanks of the 4th/7th Royal Dragoon Guards in the more remote Jewish settlements in the hilly country around Jerusalem. The final St David's Day Parade of the tour was celebrated at Beit Leid in 1948 with the leeks being presented by the High Commissioner, General Sir Alan Cunningham. But this was a difficult and trying time for all the British security forces. The mandate would come to an end on 15 May 1948, Britain having already stated that it would withdraw all its forces by that date, and at the same time the British government insisted that it would take no further responsibility for any political or military action which implied support for the policy of partition, a decision which left the troops on the ground, as Cunningham put it, 'holding the ring' between the two opposing sides. For the Battalion this could have become personal. As Palestine slumped into violence, it became clear that a wider war might break out as the surrounding Arab armies prepared to involve themselves in the conflict that was bound to break out between the Palestinians and the planned new state of Israel. Amongst them was the Transjordanian Arab Legion. If that happened, and if 1st Battalion Welsh Guards was forced to intervene, then it could have found itself in action against the Legion's 3rd Brigade at Ramallah which was commanded by Lieutenant Colonel 'Teal' Ashton, a Welsh Guardsman known to most men in the Battalion.[20]

Fortunately that did not happen. The rest of March was taken up with preparations for leaving Palestine and, by the end of the month, the only remaining elements in the country were Headquarter Company, parts of Support Company and thirty men of Prince of Wales Company. Even then, their final days were not without incident. As their convoy made its way to Haifa for embarkation on HMT *Scythia*, their road was blocked by a fire-fight as Arab gunmen attacked a nearby Jewish settlement. As the Adjutant recalled later, 'having no ammunition and no particular desire to intervene, the column debussed, sat down and watched the battle. When it was over everyone embussed again and drove on to Haifa.'[21] With that somewhat surprising finale, 1st Battalion Welsh Guards headed for home, having completed a tour of duty which was by turns dangerous and frustrating yet always fascinating, not least because, as the Battalion diary noted, 'Every step one took was treading on history.' Six weeks after the Battalion's departure, and following the creation of the new state of Israel, the region was plunged into the first Arab–Israeli war.

On return to the UK in March 1948, the Battalion was housed in the old Chelsea Barracks opposite the Royal Hospital and, after leave, its first task was to prepare for Trooping the Colour. This caused a number of problems. With only 246 officers and men, the Battalion was severely under-strength, there was a shortage of home service dress (as the scarlet tunic

HRH The Duke of Edinburgh, Colonel Welsh Guards 1952–75, addresses the Battalion at Chelsea Barracks during his first visit, 1952. *(Welsh Guards Archives)*

and bearskin are correctly termed) and after nine years' absence from public duties there was a dearth of expertise in drilling, allied to a certain and inevitable rustiness as probably 90 per cent of the Battalion had never done public duties. The first post-war birthday parade had been held on 12 June 1947 when the King's Colour of 2nd Battalion Coldstream Guards had been trooped with the Guardsmen wearing battledress, and there was a growing feeling within the Brigade that, economies notwithstanding, the time had come to return to scarlet tunics and bearskins. In the following year, 1948, permission was given to 2nd Battalion Scots Guards to Troop the Colour in home service dress, provided that they returned to battledress for public duties thereafter, as the Quartermaster-General, General Sir Sidney Kirkman, GCB, KBE, MC, had decreed that shortages of available uniforms meant that 'We cannot afford it.'[22]

The decision was never put to the test, though, as the parade was cancelled at the last minute due to inclement weather early in the morning, which later turned to bright sunshine. Although the Battalion provided the Army element in the Tri-Services' Guard of Honour for the opening of the Royal Tournament at Olympia later that afternoon, there was still huge disappointment for the many National Service Guardsmen who had missed their only opportunity of taking part in the parade. By then the Battalion had moved to Wellington Barracks and the Adjutant remembered that some Guardsmen 'burst into tears when they heard the news' as they were National Servicemen and would not get the chance again. Fortunately the weather was benign on Thursday, 9 June 1949, when 1st Battalion Welsh Guards Trooped the Colour – a new stand of Colours had been presented only twelve days earlier at Buckingham Palace – and the whole Parade wore home service dress with the exception of HRH Princess Elizabeth, who continued to be dressed as Colonel of the Grenadier Guards, wearing the blue uniform she had worn in 1947. The Ensign was 2nd Lieutenant Newton Webb-Bowen and the Commanding Officer, Lieutenant Colonel A. W. A. Malcolm, was Field Officer. A professional film record of the ceremony was made by the Crown Film Unit and later released under the title *Trooping the Colour 1949* (Terry Bishop, 1950).

Towards the end of the ceremony, when the men were formed up in divisions to move off, Regimental Sergeant Major Rees noticed that a man in Prince of Wales Company was starting to waver and looked as though he might fall over. Acting quickly and unobtrusively, Rees gripped him by his waist belt and gave him a shake with the result that he could say later that 'We didn't lose a man on the ground that day.' A further cause of pride for Rees was the presence in the Escort of Guardsman Morris 44 from Cilfynydd who allegedly had 'hands the size of a shovel'. Earlier in his Army career, Morris had broken his rifle in half whilst striking the underside of the butt during the 'slope arms', much to the glee of Rees who was highly amused at the sight of one of his men performing such a feat.[23]

Because of illness, HM King George VI rode in an open landau but took the salute from a dais that was rolled out during the inspection. Despite those changes and concerns about the King's health, the return to scarlet uniforms and bearskins was regarded

Battle Honours and Trooping the Colour

Battle Honours are awarded by the sovereign as a recognition of a regiment's involvement in or contribution to a battle, but in the early days of the British Army only victories were honoured. Had that policy continued, 'contributions' by thirty-one regiments at Dunkirk in 1940 and seven at Arnhem in 1944, would not have been recognized.

Honours have been awarded for famous battles like 'Waterloo 1815' (thirty-eight regiments honoured), and minor engagements like 'Fishguard 1797', awarded to the Cardigan Militia for repelling a French invasion in Pembrokeshire. It is the only honour for an action within the British Isles.

Just over a hundred Battle Honours were awarded in the Great War 1914–18 for actions in France and Flanders alone. The first awarded to the Welsh Guards was 'Loos 1915', an honour given to almost a hundred British and Empire regiments. The Regiment's most recent is 'Falkland Islands 1982'.

The Colour of the 1st Battalion has been Trooped at the Sovereign's Birthday Parade on the following occasions:

1928	Colonel T. R. C. Price	RSM L. Pownall
1949	Lt Col A. W. A. Malcolm	RSM A. Rees
1965	Lt Col P. J. N. Ward	RSM I. A. James
1973	Lt Col Sir James Malcolm	RSM B. D. Morgan
1981	Lt Col J. F. R. Rickett	RSM A. J. Davies
1990	Lt Col C. R. Watt	RSM N. Harvey
1998	Lt Col A. J. E. Malcolm	RSM F. K. Oultram
2006	Lt Col B. J. Bathurst	RSM A. F. Bowen.
2008	Lt Col R. J. A. Stanford	RSM M. Monaghan
2013	Lt Col D. L. W. Bossi	RSM M. Topps
2015	Lt Col G. R. Harris	RSM P. J. Dunn

The Colour of the 2nd Battalion, formed in 1939, and put in suspended animation in 1947, was never Trooped. The 3rd Battalion, formed in 1941 and disbanded in 1946, had no Colours presented to it.

as a wonderful fillip to the nation after the war years and the austerity of post-war Britain:

> On the parade the young Guardsmen and colour-starved spectators seemed to rise to the long awaited occasion and show the world that this was Britain getting back into its stride. Ration books and utility clothing were forgotten in the splendour of this return to something which belonged to us and to no one else.[24]

The parade was also a fitting interlude in the life of the Battalion as it came to terms with the very different demands and innovations of post-war soldiering. Some of the changes were for the better. At the beginning of the year there had been the novelty of pyjamas being supplied to the Guardsmen for the first time in the history of the British Army. This had been a vexed issue ever since the war and had grown into a pressing social problem in the post-war years, especially with the appearance of National Servicemen. In the past, other ranks' families had been in the habit of giving their menfolk sheets and pyjamas, neither of which the Army supplied. Rationing, coupled with shortages of cotton, had put an end to that, prompting a War Office investigation, which, on 15 April 1947, led Frederick Bellenger, Secretary for War and a wartime officer in the Royal Artillery, to announce in the House of Commons: 'Pyjamas will be provided for all British troops as soon as enough can be produced.'[25]

Change was also on the agenda in other and more profound ways. Unlike the previous post-war experience, the pace of life within the Battalion quickened as it settled into a system eventually known as the 'arms plot', by which infantry battalions were periodically rotated to various locations and trained for different roles. On 4 March 1950, 1st Battalion Welsh Guards began moving to Wuppertal in North-Rhine Westphalia, where it occupied Anglesey Barracks, a former German Army site built as Sagan Kaserne in 1936. In West Germany the Battalion formed part of 4th Guards Brigade, which itself was part of the British Army of the Rhine, the command formation for British troops in West Germany in the post-war world. With the creation of the North Atlantic Treaty

Spandau Prison in West Berlin was used to imprison senior Nazi war criminals after WW2, including Rudolf Hess, and was guarded by the Allies until his death, when it was demolished. The CSM marching out the Old Guard is WO2 Bobby Joyce, MM. He had been awarded the MM as a lance sergeant fighting with the 1st Battalion (see Chapter 7 for *The Victor* illustration of him in this action). *(Welsh Guards Archives)*

Lt (QM) A. Rees, MBE, presents HM Queen Elizabeth the Queen Mother with a leek, St David's Day, 1953, watched by Col D. G. Davies-Scourfield, MC. *(Welsh Guards Archives)*

Organization (NATO) in September 1949, the Northern Army Group in Germany (plus Norway and Denmark) was put under British command. This was Britain's contribution to the post-war defence of Western Europe and, in some respects, West Germany was to take over India's role in the affections of generations of post-war British servicemen as one of the homes of the British Army. In the first years the rationale for being in post-war Germany, by then divided into western and eastern sectors, was not always apparent to soldiers but, following the blockade of Berlin in 1948, the confrontation between NATO and the Soviet Union became increasingly bitter and belligerent. In time the period would be known as the Cold War; for the rest of the century West Germany was to be a second home for the Regiment.

For everyone in the Battalion the posting to Wuppertal was a bit of an eye-opener. Not only were the barracks well-appointed and modern but they were self-contained units with ample facilities for off-duty leisure activities. The Army Kinema Corporation provided recent releases of popular films; the British Forces Broadcasting Services, founded in 1942, was an acceptable alternative to the BBC World Service, with its British Forces Network Germany broadcasting a cosy mixture of record request programmes and military gossip; and the ubiquitous NAAFI made every posting a home-from-home. During the early months of peacetime the British had pursued a vigorous policy of non-fraternization and had issued a handbook giving specific warning about the Germans:

> You are about to meet a strange people in a strange enemy country. When you meet the Germans you will probably think they are very much like us. They look like us except there are fewer of the wiry type and more big fleshy types. But they are not really as much like us as they look.[26]

A command post, Palestine, 1947. *(Welsh Guards Archives)*

Now it was possible to meet Germans and to visit local bars, even though the required linguistic ability was usually little more than '*Zwei Biere, bitte*.' Still, it was a welcome change.

While underlying tensions had not entirely evaporated by the time the Battalion arrived in Wuppertal, the restrictions had been removed as they were more or less unenforceable and for most Welsh Guardsmen West Germany was largely a pleasant and trouble-free experience. For many of them – with the possible exception of the war-service men and those who had served in Palestine – it also provided a unique occurrence of living abroad at a time when overseas travel was difficult and expensive and anything 'foreign' was considered a novelty. (At that time, travel between the two countries was mainly confined to rail and ferry, commercial air flights being prohibitively expensive.) Highlights included a staging of the pantomime *Aladdin* in time for Christmas 1950 and the visit of HRH The Duke of Windsor, formerly the Regiment's Colonel-in-Chief and Colonel. During his four-day visit in October 1951, which was filmed by British Pathé, he saw 'the Battalion training, watched a rugby football game, had a day shooting, attended a cocktail party in his honour and on the eve of his departure was given a tremendous welcome by the Members of the Sergeants Mess'.[27] Before leaving the latter he was greatly moved when the Sergeants Mess rose to their feet and sang 'God Bless the Prince of Wales'.

There was a change of rhythm in March 1952 when the Battalion moved to West Berlin where it was housed in Brooke Barracks, another structure from the 1930s, which stood adjacent to Spandau Prison, the home to seven imprisoned members of the Nazi hierarchy, including the deputy leader Rudolf Hess. Under a four-power agreement, each of the Allies supplied guards for the prison, with the British Army being responsible for the months of January, May and September, and this was one of the Battalion's tasks while stationed in West Berlin. Another innovation was the decision of the Commanding Officer, Lieutenant Colonel D. G. Davies-Scourfield, MC, to hold a full-dress Eisteddfod in which he was crowned Archdruid and the prize for male soloist was won by Guardsman Davies 40, a noted bass-baritone. The event did much to underline the regiment's links with Wales and emphasized the fact that *Gwarchodlu Cymreig* (the 'Welsh Guards') was central to Welsh life and society.

In March 1953 the Battalion returned home to settle in Corunna Barracks in Aldershot. Almost immediately the Guardsmen were plunged into a busy routine of training for the Birthday Parade which was held on 11 June, with 1st Battalion Grenadier Guards Trooping their Queen's Colour. Just over a week earlier, HM Queen Elizabeth II had been crowned at Westminster Abbey; 600 Welsh Guardsmen took part in the Coronation ceremony under the command of Lieutenant Colonel A. C. W. Noel, who had become Commanding Officer in November 1952. Although the weather was uncertain – 2 June itself, Coronation Day, was a miserable, November-like day in London with dull skies, a chill wind and sporadic outbreaks of rain during the morning – it was remembered with gratitude by all who took part and the events associated with it. On 9 July, the day before the beginning of the Royal Coronation tour of Wales, the season was capped by the announcement that HRH The Duke of Edinburgh would become the Regiment's next Colonel in succession to the Earl of Gowrie.

By then it was already known that the Battalion's next tour of duty would be overseas, namely a thirty-month tour to Egypt which would be quite unlike the previous visit two decades earlier. In place of the faded imperial grandeur of that first tour in the carefree 1920s, the Battalion would find a very different set-up at its first location, Wolseley Camp at El Ballah, right alongside the canal to the north of Ismailia, within a protected area known as the Suez Canal Zone. First impressions were not good: the fly-blown, sand-strewn military bases in the Zone were forlorn and disagreeable, the climate was vile and the local population were usually hostile. Dick Fletcher caught the mood when he described El Ballah as 'a somewhat dilapidated collection of sunken marquees and small broken-down outhouses, the whole surrounded by barbed wire'. In that flat windswept landscape the temperature often climbed into the high nineties Fahrenheit and there

The Battalion Rugby XV. Winners of the Army (Egypt) Inter-Unit Competition, 1953–4. *(Welsh Guards Archives)*

Lt Gen Sir Francis Festing, KBE, CB, DSO, GOCinC Troops Egypt, presents the Egypt Rugby Cup to Major J. M. H. Roberts. *(Welsh Guards Archives)*

was little of interest other than the regular sight of ships creeping through the canal to the immediate east of the camp.[28]

Of all the postings on offer to the post-war serviceman this was the most unpopular and it left an indelible mark on all who served there. The Zone had been created during the Second World War and in March 1947 it had become a secure base for the British Army following Egyptian nationalist protests at the British military presence in the country. (The treaty granting Egypt independence in 1936 had allowed British forces to remain in Egypt as 'guests'.) By the time the Battalion arrived the British garrison was locked in the Zone, ostensibly guarding the Suez Canal but in effect protecting themselves and British possessions from outraged Egyptian nationalists, including the Bulak Nizam, a group of paramilitary auxiliaries who had mastered the hit-and-run tactics of guerrilla warfare. In 1952 the atmosphere had worsened when the pro-British Egyptian leader King Farouk was deposed and replaced with a military council headed by Major General Mohammed Neguib. Two years later there was further change when Colonel Gamal Abdul Nasser, a perfervid Egyptian nationalist, seized power and served notice on Britain to quit the Zone within twenty months. During their time in the Zone the 1st Battalion's duties centred on providing guards for installations and carrying out training exercises in the desert area.

The Battalion arrived at Port Said at the Mediterranean end of the Zone on 13 October 1953, having experienced an eleven-day-long voyage on board the troopship HMT *Empire Ken*, the former liner *Ubena* of the German East Africa Line which had served in the German Navy as a U-boat depot ship during the Second World War. However, for the younger men, most of whom had not been abroad before, this was an exciting moment and Port Said offered the lure of eastern promise with its bustle, heat and stinks, its *gully-gully* men conjuring chickens out of thin air and its shady houses in dark alleyways. There was even a short-lived hint of the exotic in the tented camp at El Ballah, with its large double-skinned marquees to take advantage of any available breeze, but the harsh reality was that this was going to be a tough stretch with only the Guardsmen's natural good humour to lighten what has been described as 'a boring and monotonous period'. This was also an unaccompanied tour – the families remained behind in Wales or at married quarters in Hounslow – and at the time there were few telephone facilities (and certainly no email) to enable Guardsmen to keep in regular contact with their loved ones. At the end of 1954 some lucky families arrived to be quartered in Pegasus Village, built by Airborne Engineers on the shores of Lake Timsah, but their presence only underlined the general predicament faced by the rest of the Battalion. Khaki-drill uniforms were in short supply and a non-stop war was waged against the local population of flies, Part One Orders regularly declaring: 'Waving them aside is no good! Kill them! With a swot, gym shoe or any convenient weapon!' Under those circumstances, for some of the men, thirty months could have seemed like an eternity.

But of course there was more to soldiering in Egypt than developing a tolerance to the heat and the dust, the flies and boredom and the absence of friends and family. Internal security duties were an everyday fact of operational life. While this meant that visits to towns in the Zone were impossible without the presence of armed guards and that Cairo was off-limits, patrolling was carried out in earnest and had to be because there were British losses to terrorists during the Battalion's

Welsh Guards uniforms, post-war period. *(From illustrations by Charles C. Stadden and Eric J. Collings commissioned by the Regiment)*

1948, Company Sergeant Major, Guard Order. After three years in Palestine, the Battalion returned to London for public duties. Home service clothing was reissued, but all non-essential items of the old Slade-Wallace equipment were now discontinued. In this picture the CSM wears the embroidered colour badge of rank worn by senior NCOs of the Foot Guards. The Welsh Guards colour badge incorporates the dragon of Wales above a motto scroll, flying over two crossed scimitars, and with the sovereign's crown surmounting all. For colour sergeants the same design was worn superimposed on three chevrons. CSMs and drill sergeants wear the badge on the upper sleeve.

1960, Drum Major. Drum Majors carry out most of their duties in regimental dress, as shown here. The tunic, with its embellished gold lace and embroidery, remains a solitary concession to past glories. The rank badge of four inverted chevrons is worn above the right cuff, and the narrow white sword belt has a gilt and silver locket.

1949, Mounted Field Officer. 1936 saw the first major change since 1902 in a field officer's mounted dress, when pantaloons and butcher boots were abolished in favour of overalls and Wellington boots. A buckle was also added to the stirrup leathers.

1965, Captain and Adjutant, Blue Frock Coat. On 'undress' occasions Foot Guards officers have long been distinguished by their blue frock coats adorned with elaborate black braiding and olivets across the chest. An Army Order of 1920 restricted the use of this coat to certain regimental staff, including the adjutant and commanding officer.

1967, Drill Sergeant, Guard Order. The appointment of drill sergeant was formerly that of a staff sergeant class I, but is now that of a warrant officer class II. On full dress tunics the colour badge is worn by WO2s on the upper right sleeve. In greatcoat order this badge is worn on the right cuff. No waist belt or sash is worn over the coat. The sword is carried underneath, with its hilt protruding from a slit behind the pocket.

tour. Despite the conditions there were also opportunities for sport and in 1954 the Battalion rugby team won the Army (Egypt) Cup under the captaincy of Major John Roberts while the boxing team was narrowly beaten by 2nd Battalion, Parachute Regiment, in the boxing cup, losing by one bout. Matters began to improve immeasurably for the Battalion in 1954 when it was moved from El Ballah to what was described as a 'luxurious camp at Moascar from which weekend visits were made to the Red Sea where there was excellent swimming and snorkelling'.[29] More to the point, the camp was closer to Ismailia and consisted of hutted and brick-built accommodation which at least helped to keep the sand at bay.

Inevitably a tour of this kind cannot be assessed purely in terms of black and white. While service in the Zone could be a succession of days enduring hot and difficult physical conditions, and while there was the ever-present danger of local disturbances, there were ample opportunities for the Battalion to engage in realistic training exercises in the desert, and sport, too, had a high premium. For members of Support Company under the command of Peter Leuchars, towards the end of 1955 there was the bonus of an attached posting to the port of Aqaba in neighbouring Transjordan where they came under the command of the Queen's Bays. It was the first time in the Regiment's history that a complete rifle company had been detached from the Battalion and placed under the operational command not only of another headquarters but one operating in a different country. The three-month visit proved to be a great success for all concerned. Given relative freedom, Leuchars was able to organize a variety of training exercises, many of them held in the nearby Negev desert, and there were opportunities to visit Petra, 'the rose-red city, half as old as Time' and further afield there were trips to Jerusalem and Bethlehem. Not even a sudden flash flood on Christmas Day 1955 could dampen spirits, leaving Support Company with 'a Christmas we shall all remember, and many were the stories passed back to the Battalion in Moascar who, as we expected, did not believe them at all'.[30]

By then the British occupation of the Suez Canal Zone was almost at an end. On 19 October 1954, Britain agreed to withdraw from a base which was becoming increasingly difficult and expensive to maintain – at its height the garrison numbered 70,000 troops – and to

The Regiment exercises its freedom to march through Aberystwyth with drums beating, colours flying and bayonets fixed, 1952. Lt I. L. Mackeson-Sandbach carries the Queen's Colour, and Lt J. W. R. Malcolm carries the Regimental Colour. *(Welsh Guards Archives)*

The Welsh Guards Polo Team versus the Life Guards, No. 3 Ground, Smith's Lawn, Windsor Great Park, painted by Lionel Edwards. Welsh Guards players: *(L to R)* Lt Col R. J. Watt (WG 1937–61) riding Sugar, HRH The Prince Philip, Duke of Edinburgh, Colonel of the Regiment, riding the thoroughbred mare Bet-a-way, Capt E. A. M. Fox-Pitt riding the club pony Corona, and Lt Col John Miller. The two grooms to the front and right of the picture are *(left)* Sgt K. Barrett, Stud Groom of the Guards Polo Club, and CSgt Sgt A. J. Winser, Stud Groom of the Household Brigade Saddle Club. The polo pony being led by a dismounted figure *(front left)* is Sanquanina, an Argentine mare, which was HRH The Prince of Wales's first polo pony. *(Estate of Lionel Edwards)*

complete the operation within twenty months, keeping only the right to move forces back should the freedom of the Canal ever be compromised. It fell to 1st Battalion Welsh Guards to be one of the last British regiments to leave this unloved part of the world in March 1956. It seemed to all concerned that after seventy-four years the British link with Egypt had finally ended but, as it turned out, the British Army returned that same autumn in an ill-starred invasion to regain possession of the Canal, which had been nationalized in July, and to depose Nasser. The failure of the intervention gave notice that Britain was no longer a world power and that its position was now subordinate to the interests of the new superpowers, the USA and the Soviet Union. In that sense, no other post-war posting brought home so strongly the message that Britain's withdrawal from Empire would be dictated by the vocabulary of sandbag, roadblock, barbed wire and rear-guard action.

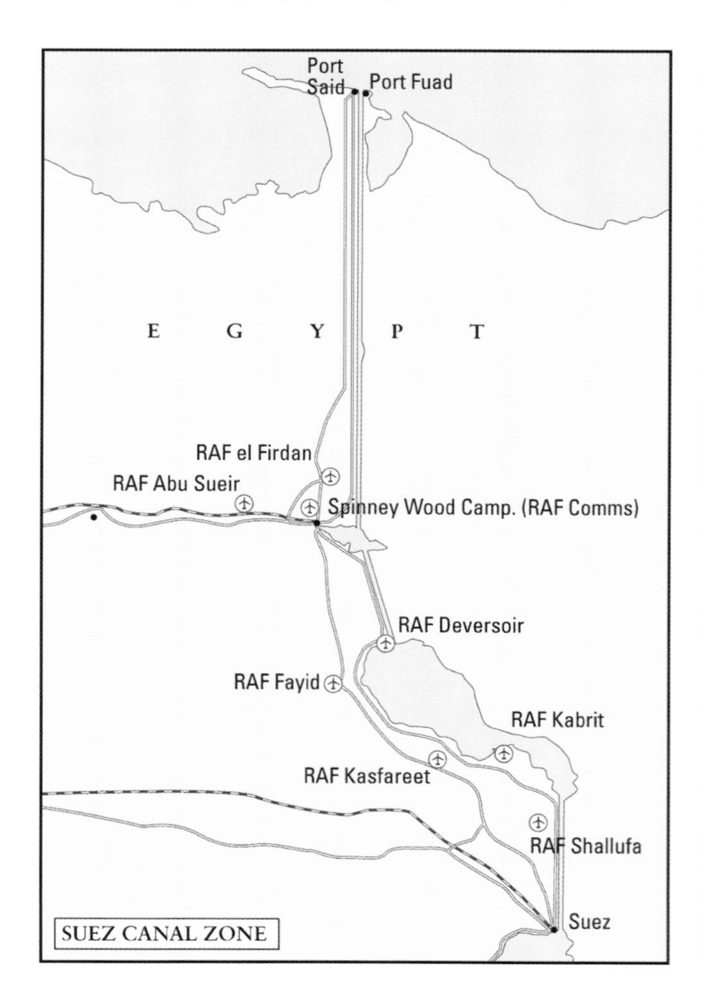

Map of Suez Canal Zone, 1956, showing main RAF bases.

Regimental Sergeant Major B. F. Hillier, DCM

Bert Frank Hillier joined the Welsh Guards on the 13 September 1934 and was Regimental Sergeant Major from September 1951 to September 1956. He served in the 3rd Battalion in North Africa and Italy. During operations at Monte Piccolo in May 1943, while serving as CSM with No. 2 Company, he was awarded the Distinguished Conduct Medal for leading a 'reconnaissance in force' which left him in charge of forty-six men after two officers had been killed and a third badly wounded. Between September 1944 and April 1945 he served as a drill sergeant with the Training Battalion.

Regimental Sergeant Major E. L. G. 'Tex' Richards, MM

2730511 Edward Leslie Gilbert Richards was only fifteen when he joined the Welsh Guards in 1920, but claimed he was eighteen years of age in order to draw a man's pay. Born in Madras, the son of a soldier, he was a drill sergeant in February 1935 and Regimental Sergeant Major of the 1st Battalion in 1938, with which he went to France at the outbreak of the Second World War.

During the operations around Arras he was Mentioned in Despatches and awarded the Military Medal for 'most valuable and gallant service in organizing the defence' and for 'playing a prominent part in organizing the defence and subsequent withdrawal'.

He remained with the last party but was severely wounded and taken captive, seeing out his war in Stalag 383 in Poland, the same camp that held CQMS Horace 'Phil' Philips (see separate entry). After being released he was again Mentioned in Despatches in 'recognition of Gallant and Distinguished Services as a Prisoner of War'. In the immediate post-war period he was appointed Regimental Sergeant Major of a Composite Battalion sent to Norway on internal security duties and, having been commissioned, was Camp Commandment at Pirbright, and 4th Guards Brigade in Germany. His final position was Quartermaster HQ Western Command. He retired from the Army in 1956 and died in October 1977.

He was always nicknamed 'Tex' after a country music singer and film actor called Tex Ritter.

Lieutenant Colonel (QM) Harold Humphries

The first Welsh Guards Quartermaster to reach the rank of lieutenant colonel, Harold Humphries joined the Regiment in August 1933 and the 1st Battalion in January 1934. In the same year he was posted to the Guards Depot as a clerk in the Orderly Room, where he eventually became a lance sergeant. He was promoted to sergeant and appointed Orderly Room Sergeant of the 2nd Battalion on its formation in 1939. He became ORQMS in September 1942 and held that appointment until July 1945 when he was appointed RQMS of the 2nd Battalion. In June 1946 he was posted to Regimental Headquarters as ORQMS and was appointed Superintending Clerk in December 1947. He was commissioned in December 1951 and appointed Quartermaster of the 1st Battalion, serving in BAOR, the UK (for the Coronation in 1953) and then the Canal Zone. In June 1956 he became Staff Captain at Headquarters, 1st Guards Brigade. He subsequently became Staff Captain PA 6 at the War Office in September 1958 and was then appointed Staff Quartermaster to the Director of Infantry in May 1961. He was promoted to lieutenant colonel on 31 May 1964 and in doing so became the first Quartermaster in the Regiment ever to attain that rank. He retired from the Army on 15 March 1968 and died in 2007.

CHAPTER NINE

Cold War, 1956–1966

UK, BAOR, Aden

ON ARRIVAL BACK FROM Egypt – again travelling on board the troopship HMT *Empire Ken* – 1st Battalion Welsh Guards moved into Shorncliffe Barracks at Folkestone under the command of Lieutenant Colonel C. A. la T. Leatham, who had replaced Lieutenant Colonel Archie Noel as Commanding Officer in November 1955. Despite the frequent tedium and discomfort of the months in the Suez Canal Zone there was a sense of anti-climax after the return to the United Kingdom and it was with some anticipation that the Battalion moved to Chelsea Barracks towards the end of 1956. Although the barracks remained the same 'walled, prisonlike structures of the nineteenth century', the Battalion was pleased to be back in London and to return to the measured certainties of public duties.[1] Other things, though, were changing within the Army. By the time 1st Battalion Welsh Guards had returned to London, the Suez crisis had come and gone. British and French forces had landed in Egypt on 5 November 1956; all Egyptian resistance had been quashed by the following day, but by then the 'war' was over before it had begun. Soviet belligerence and US financial pressure forced the government to call a cease-fire and to turn the

Presentation of new colours at Buckingham Palace. Behind Her Majesty is the Commanding Officer, Lt Col Colonel P. R. Leuchars. To her left is CSgt (CQMS) Guy holding the Company Colour of No. 2 Coy. *(Welsh Guards Archives)*

problem over to the United Nations. It was a humiliation which Britain had to accept; the Soviet Union was emerging as a superpower and Washington had refused to support Britain's application to the International Monetary Fund for a loan to prop up the falling pound unless those conditions were met.

In the aftermath of the action there was a rapid rethinking of British defence policy, caused not just by the Suez debacle but also by Britain's changing position in the world. This revisionist approach centred on the dangers of taking unilateral action, the relevance of maintaining large and ponderous conscript forces and the impact of nuclear weapons on modern warfare. Introduced by Defence Secretary Duncan Sandys, the resulting Defence White Paper of April 1957, *The Central Organisation for Defence*, put forward a new policy which would enhance Britain's nuclear deterrent, end National Service, create a new strategic reserve and replace the obsolescent Second World War equipment which had been so cruelly exposed at Suez. To make regular military service more attractive, a government advisory committee, chaired by Sir James Grigg, reported in the following year and recommended that service pay should be reviewed every two years, that housing should be improved and that officers should be drawn from a wider social background. Nevertheless, recruiting to the Army picked up only slowly and by 1962 the minimum requirement for 165,000 men had not been met, resulting in the retention of 9,000 National Servicemen to make good the deficit. It was not until the end of that year that the final National Service conscripts left 1st Battalion Welsh Guards, although quite a few had decided to remain in the Army after accepting three-year engagements.

For the most part, the ending of National Service was welcomed. By this time it had become unpopular with politicians and the general public alike and seemed to serve no practical military purpose. The transition was supposed to herald a return to professional volunteer armed forces, but it also meant that the Army in particular would have to work harder to find the next generations of recruits. As predicated by the 1957 defence review, spending on the armed forces was gradually reduced in real terms and the size of the Army fell, so that ten years after the end of National Service its establishment stood at 166,000, with 55,000 stationed in BAOR organized in one armoured and three infantry divisions (later four armoured divisions). The Army's strength may have declined in real terms but the aim was to build up

HRH The Duke of Edinburgh, Colonel of the Regiment, presenting Lt (QM) Arthur Rees, MBE, with the Long Service and Good Conduct Medal c. 1954. Rees had been Regimental Sergeant Major from August 1945 until September 1951. In the film *They Were Not Divided* he can be seen handing over the Colour on the 1949 King's Birthday Parade. He was also the Battalion's heavyweight boxing champion, a noted shot putter, discus thrower and rugby player. He was famous for having a size 8 cap. *(Welsh Guards Archives)*

skills and capabilities so that its units would be sufficiently flexible to operate anywhere in the world. For senior soldiers of that period, such as Major General John Strawson, the new slogan was that 'although small, the army led the world in experience'.[2] Against that strategic background the post-National Service Welsh Guards looked forward to a future in which, to borrow from Giuseppe di Lampedusa's famous dictum, things would have to change if people wanted them to remain much the same. That philosophy underpinned all that was to happen to the Regiment in the decade that lay ahead.

Whatever the impact of the new order, some things did remain the same, notably the ability to mix operational soldiering with public duties. Until 1959 all Guardsmen began their training at the Guards Depot at Caterham before moving to Pirbright where the Guards Training Battalion, formed in 1947,

WO1 (RSM) D. I. Williams and Gdsm H. Holland at Buckingham Palace in December 1966, where they were presented with the MBE and MM respectively. Holland won his medal in Aden. *(Welsh Guards Archives)*

Anti-tank Platoon with 3.5 inch rocket launcher, Aden, 1966. To the left is Guardsman G. Morgan, who later joined the police and was one of the detectives who arrested serial killer Fred West. *(Welsh Guards Archives)*

Anti-tank platoon with 120 mm BAT anti-tank gun, Aden, 1966. *(Welsh Guards Archives)*

LCpl D. 'Timber' Woods on a camel in a promotional shot for the documentary *All the Queen's Men*, Habilayn Camp, Aden, 1966. LCpl Woods later transferred to the Army Air Corps as a pilot and was subsequently commissioned. He is now President of Welsh Guards Reunited. *(David Woods/Welsh Guards Reunited)*

completed the field training of recruits. Now, the training course was changed in order to concentrate on an eighteen-week recruit course at Pirbright, but with a few exceptions to the familiar routines (no church parades, no pack drills), the regime would have been recognized by many an older pre-war Welsh Guardsman. In 1958 Alan Acreman from Bargoed was one of the last Welsh Guards recruits to pass through Caterham. His description of life during the first few weeks in Edward Block is worth quoting at length, simply because it recalls a bygone age and yet one that was central to the experience of being a Welsh Guardsman in the second decade after the end of the Second World War.

> We didn't have a Regimental Number yet, everyone was known by the letters NYA (Not Yet Accepted) and their surname. The Barrack Room was on the second floor of a three-storey block. There was a central staircase with a

Foot patrol, Habilayn, Aden, 1966. *(David Woods/Welsh Guards Reunited)*

Guardsman P. Young (*right*) pausing for water, Aden, 1966. *(David Woods/Welsh Guards Reunited)*

The local Arab who was in charge of all civilian staff in Aden. He was promoted to the unique rank of 'Number One' and given a sergeant's red sash, which he adored and never took off. *(David Woods/Welsh Guards Reunited)*

LCpl D. Woods controls mortar fire from a Sioux helicopter belonging to the Air Platoon, Aden, 1966. The pilot of this helicopter was Major Greville Edgecombe, known as Low-level Greville. *(David Woods/Welsh Guards Reunited)*

> Barrack Room each side. The ground floor was for the Stores and on the other side of the entrance was the Trained Soldier's room. They didn't have a rank except a star which they wore on the right sleeve of their uniform. Each squad had its own Trained Soldier who 'bunked' in the Barrack Room. Mine was Trained Soldier Carrol whom I loathed with such intensity; no man deserved to be loathed that way. His job was to turn us out to look like soldiers, sometimes he succeeded! Every evening we'd have a Shining Parade for a couple of hours. This entailed spreading your groundsheet on the bed to keep it clean, and polishing everything in sight. The boots were the very devil to get polished, when you got them from the stores they looked all right but before you could wear them they had to have a layer of polish put on them, then you 'burned' them with an electric iron until all the pimples had disappeared, then and only then could you start to polish them with a duster and water. The routine went something like this: take the duster and dip it in some polish (the forefinger having already been inserted), with another finger put a drop of water on the toecap, heel or side and work it until the desired effect has been acquired. You should end up with boots you could shave in, knife, fork and spoon you could match the Family Silver with.[3]

Acreman survived the experience and went on to serve for almost nineteen years in the Battalion, reaching the rank of WO2 and retiring from the Army in the mid-1970s. For the Regiment, the transition from post-war conscript army to a modern and professional all-volunteer force was helped immensely by the continuity provided by its officers and senior

Foot patrol cleaning weapons (7.62 mm L1A1 self-loading rifle [SLR] and 7.62 mm general purpose machine gun [GPMG]), Aden, 1966. The radio carried at the time was the Larkspur A41, which remained in service from the 1950s until the 1980s. *(Welsh Guards Reunited)*

WO1 (RSM) I. James talking to Guardsman A. Lumby, Aden, 1967. RSM James used to enjoy dropping in unannounced by helicopter, just to make sure everything was in order. *(Marilyn James/Welsh Guards Reunited)*

NCOs. In Tony Leatham the Battalion had a Commanding Officer steeped in the traditions of the Welsh Guards and possessed of a solid war record. Although he had a deserved reputation as a disciplinarian who demanded (and got) high standards from those under his command, he was no stick-in-the-mud traditionalist and 'took a keen interest in the catering trials then taking place in the Army. The improvement was immediate; from then onwards the standard of food changed for the better out of all recognition.'[4] It also helped that the redoubtable Arthur Rees was still with the Battalion, now commissioned and serving as Quartermaster. Having served throughout the Second World War and having acted as Regimental Sergeant Major for six years, he was in a good position to assess the changing scene as well as the merits of the Guardsmen under his command:

> Looking back at the Crimea, people said how wonderful the soldiers had been and then at the time of the Boer War they said, 'Ah, our men aren't as good as they were in the Crimea.' And the boys in the First World War weren't as good as the last lot. But we were and I think that the Welsh Guardsmen of today are as good as the men of 1939–45 and 1914–18. It's still a matter of discipline, pride and strict training which Welsh Guardsmen will happily adapt themselves to when they join the regiment.[5]

Other standards were also maintained. While off-duty in London, officers were supposed to wear a suit and a hat, no matter how senior they were. While he was a major, Peter Leuchars remembered 'being caught in Victoria Street by the Regimental Lieutenant Colonel' without a hat on: 'He asked me why I wasn't wearing one and when I said I had forgotten he reminded me that I had to wear a hat so that I could take it off when I saw him.' It was no different with the men. When Arthur Rees was Regimental Sergeant Major, he met a drummer escorting a girlfriend in Buckingham Palace Road and thought him improperly dressed, an unusual occurrence as at that time Guardsmen were supposed to be inspected before they walked out. As Rees passed the couple, he doffed his hat and said 'Good evening.' It was not the end of the matter. The next morning he had the drummer up in front of him, but instead of handing down a punishment, Rees told him in no uncertain terms that he had a nice girl and should show her more respect.[6]

In the midst of the normal round of public duties

HM The Queen, Colonel-in-Chief, attends the Welsh Guards Association Dinner, Cardiff, to mark the fiftieth anniversary of the Regiment, 1965. *(Welsh Guards Archives)*

there were other excitements. On St David's Day, 1957, leeks were presented by HM Queen Elizabeth, the Queen Mother, and this was followed a month later by another important event in the Regiment's connection to the City of Cardiff, which scarcely a year before had been officially recognized as the Capital of the Principality of Wales. The relationship between the Regiment and the city went back to the founding of the Welsh Guards in 1915 when Cardiff City Council had provided funds for a complete set of band instruments and had helped and offered encouragement with recruiting. It was fitting therefore that the newly elevated Welsh capital should mark its position by conferring on the Welsh Guards 'the privilege, honour and distinction of marching through the Streets of Cardiff, the Capital City of Wales, on ceremonial occasions with drums beating, bayonets fixed, colours flying and bands playing'. The resolution was passed on 15 March 1956 and a year later HRH The Duke of Edinburgh, Regimental Colonel, accepted the illuminated document granting such authority from the Lord Mayor, Alderman D. T. Williams, at a ceremony at Cardiff Castle on 27 April 1957. They then took the Salute as a Welsh Guards detachment with colour party, band and drums, marched past Cardiff City Hall with bayonets fixed and then through the city centre before enthusiastic cheering crowds. The granting of the same honour had already been conferred on the Regiment by Swansea in 1948 and by Aberystwyth in 1953, but it was generally agreed that 'Nowhere have the Regiment's achievements been acclaimed with greater zest than in the Capital City, where the first impetus was given to the formation of the Regiment and the first recruits enlisted.'[7]

At the end of 1957 there was a change of station when the Battalion left London for Pirbright in Surrey, the new home of the Guards Depot. Company training was the order of the day and for the first time it was possible to understand the need for the kind of versatility demanded by the modern post-National Service Army. In the summer of 1958 the Battalion concentrated on public duties and found the Guard of Honour at the Empire and Commonwealth Games, held in Cardiff, which was attended by 1,122 athletes from twenty-three countries. At the same time a composite rifle company under the command of Major Peter de Zulueta was sent to reinforce 1st Battalion Irish Guards, which had deployed on an emergency tour to Cyprus. The island had become a new battle-

Serving Welsh Guards Quartermasters, 1965. (*L to R*) Capt (QM) W. S. Phelps, MBE, Lt Col (QM) H. W. Humphries, MBE, Maj (QM) A. Rees, MBE, Lt (QM) A. P. Joyce, MM, BEM, Capt (QM) R. C. Williams. *(Welsh Guards Archives)*

Welsh Guards Freedoms

The conferring of the freedom of a city is an ancient honour granted to military organizations, allowing them the privilege to march into the city 'with drums beating, colours flying, and bayonets fixed'. The concept behind the grant stretches back into antiquity; today it is an entirely ceremonial honour, usually bestowed upon a Regiment with historic ties to the area, as a token of appreciation and esteem for long and dedicated service.

The first grant of the freedom of a city to the Welsh Guards was made on 2 October 1948 by the City of Swansea at a ceremony at St Helen's ground during the mayoralty of Councillor Sir William Jenkins, JP. The parchment scroll recording the presentation of the Freedom was received by Brigadier the Earl of Gowrie, VC, Colonel of the Regiment.

Since then the Regiment has been granted the freedom of the following cities, boroughs and towns:

- Freedom of Aberystwyth, 15 July 1953. The scroll was received by Colonel J. C. Windsor-Lewis, DSO, MC, the Regimental Lieutenant Colonel, and presented by the Mayor, Councillor W. G. Pryse, JP.
- Freedom of the City of Cardiff, 27 April 1957. The scroll was received by HRH Prince Philip, Colonel of the Regiment, and presented by the Lord Mayor, Alderman D. T. Williams, OBE, DL, JP.
- Freedom of the Royal Borough of Caernarvon, 10 May 1958. The scroll was received by Colonel H. C. L. Dimsdale, the Regimental Lieutenant Colonel, and presented by the Mayor, Councillor David Williams, JP.
- Freedom of the Royal Borough of Windsor, 4 May 1968, conferred on all regiments of the Guards Division.
- Freedom of the Town of Carmarthen, 30 April 1982, received by HRH Prince Charles, Colonel of the Regiment, and presented by the Mayor, Councillor S. David Thomas.
- Freedom of the Borough of Merthyr Tydfil, 30 July 1983, received by HRH Prince Charles, Colonel of the Regiment, and presented by the Mayor of Merthyr Tydfil, Councillor The Reverend W. B. Morgan.
- Freedom of the Borough of Taff-Ely, 8 August 1983. The scroll was received by Colonel D. R. P. Lewis, Regimental Lieutenant Colonel, and presented by the Mayor, Councillor Leslie L. Carter, JP.
- Freedom of Cynon Valley was conferred on 26 March 1996 in Aberdare and was accepted by Lieutenant Colonel C. F. B. Stephens, the Regimental Adjutant. The Regimental Adjutant was accompanied by the Mayor, Councillor David Barnsley.
- Freedom of the County Borough of the Vale of Glamorgan, 16 March 2006. The scroll was received by Lieutenant Colonel Ben Bathurst, Commanding Officer, 1st Battalion Welsh Guards, and presented by the Mayor, Councillor Margaret Alexander.
- Freedom of the County of Powys, 10 May 2011. The scroll was received by Colonel Tom Bonas, Regimental Adjutant, and presented by the Mayor, Councillor Garry Banks.
- Freedom of the County Borough of Bridgend, 11 May 2011. The scroll was received by Colonel Tom Bonas, Regimental Adjutant, and presented by the Mayor, Councillor Chris Michaelides.
- Freedom of the City of Birmingham, 21 March 2012. The scroll was received by the Major General on behalf of all five Regiments of Foot Guards and presented by the Lord Mayor, Councillor Anita Ward.
- Freedom of the County Borough of Rhondda Cynon Taf, 15 May 2013. The scroll was received by Colonel Tom Bonas and presented by the Mayor, Councillor Doug Williams.
- Freedom of the City of Bangor, 25 March 2014. The scroll was received by the Regimental Lieutenant Colonel, Brigadier Robert Talbot Rice, and presented by the Mayor, Councillor Douglas Madge.
- Freedom of the Town of Wrexham, 18 July 2014. The scroll was received by Major General Robert Talbot Rice and presented by the Mayor, Councillor Alan Edwards.

ground after Greek Cypriot guerrillas formed themselves into an underground army, Ethniki Organosis Kypriakou Agonos (EOKA, or National Organization of Greek Fighters), under the command of Colonel Georgios Grivas to fight for self-determination for the majority Greek-Cypriot population. The deployment provided some tough training as well as the opportunity to gain experience in counter-insurgency operations; it also rekindled the wartime policy of cross-posting of companies and platoons within the regiments of the Brigade of Guards.

The Company returned at the end of the summer to find further changes afoot. There was a new Commanding Officer, Lieutenant Colonel J. M.

Above: Lt Col P. J. N. Ward and Capt J. F. Rickett with local tribesman, Radfan, Aden, 1966. *(Welsh Guards Archives)*

Right: Gdsm Jones 64 and Gdsm Appleton patrolling an Aden street, 1966. *(Welsh Guards Archives)*

Miller, DSO, MC, another officer with a fine wartime record. Under his watch the Battalion converted to the new L1A1 Self-Loading Rifle (SLR) in February 1959. This came out of a requirement to find a replacement for the existing Lee-Enfield .303 rifle as part of the drive to replace equipment from the Second World War and to introduce standard rifle and machine-gun ammunition for common use within NATO. The first attempt was the experimental EM2 assault rifle with a 7 mm round, but this otherwise handy weapon was unacceptable to the US Army, which considered it to be under-powered; instead the choice fell on the Belgian FN self-loading rifle (SLR) with a 7.62 mm round.[8] In time it proved to be a useful replacement but its revolutionary shape meant that it could not be carried at the slope, necessitating a new arms drill. As part of the Army's round of testing new air-mobile equipment, the Battalion spent time attached to 16th Independent Parachute Brigade, which was part of the Army's strategic reserve and had seen recent operational service in Cyprus and during the landings at Suez. This period included a deployment to Northern Ireland to train in aircraft loading, working with helicopters and getting used to flying in the cavernous Blackburn Beverley heavy transport aircraft which had come into service with the RAF four years earlier.

In a sense, serving with 16th Independent Parachute Brigade was something of a home from home for the Battalion. After re-forming in 1948, the Brigade was home to 1st (Guards) Independent Parachute Company which acted as the formation's pathfinder force and was composed of serving soldiers of the Foot Guards or Household Cavalry Regiments, attached to the Parachute Regiment. At the time of the Battalion's deployment, the Officer Commanding the Guards Company was Major John Retallack who had served with 3rd Battalion Welsh Guards during the war and went on to write a concise history of the regiment in 1981.[9] For a time the Company had been billeted in the old spider huts at Pirbright and their presence was an obvious incentive to adventurous Guardsmen to apply for a posting. After the deployment in Northern Ireland, Guardsman Glyn Roberts did just that in 1961 – 'being a young seventeen and a half year old these Paras looked ten feet tall and just as broad, it was then I decided that's what I wanted to be' – and he stayed with them until 1967, before leaving the Army four years later. In 1971 Roberts rejoined the Welsh Guards and served again with the Guards Parachute Company before returning to the Battalion and finally retiring from the Army in 1984 in the rank of Colour Sergeant.[10]

In complete contrast, the Battalion also took part in an exercise in Norway, travelling in warships of the Royal Navy and taking part in Exercise Barfrost within the Arctic Circle. The objective of this combined operation was to test the capabilities of the Norwegian Army's Brigade North to resist an attack by Warsaw Pact troops in the Norwegian–Russian border area. All told, twenty-five ships took part in the exercise, including the recently re-commissioned aircraft carrier HMS *Victorious*, which supported the

Lieutenant Colonel Sir John Miller, GCVO, DSO, MC

Crown Equerry to HM The Queen between 1961 and 1987, John Miller was educated at Eton and Sandhurst and joined the Welsh Guards during the Second World War, serving with the 1st Battalion in France and North-West Europe. He was a courageous soldier, distinguishing himself in particular after the Allied landings in Normandy in 1944. As a company commander he rallied his men under intense shellfire and was awarded an MC. Only a month later he received the DSO for re-establishing two companies that had been dispersed during a fierce onslaught by enemy tanks. After the war Miller was ADC to Field Marshal Lord Wilson of Libya, head of the British Joint Staff Mission in Washington, 1945–7. An excellent horseman, injury ruled Miller out of Britain's eventing team for the 1952 Helsinki Olympics, but two decades later he was in the gold-medal-winning British team at the World Driving Championships in Germany; and in 1974 he was a member of the team that won a gold medal in Switzerland. He commanded 1st Battalion Welsh Guards from 1958 to 1961, thereafter leaving the Army to become a courtier, running the Royal Mews and all attendant equestrian and ceremonial affairs until 1987. On his retirement he was appointed an Extra Equerry to the Queen. He died in May 2006, aged eighty-seven.

Lt Col Sir John Miller with the Queen's horse, Burmese, at the Royal Mews, Buckingham Palace. This photograph was taken to commemorate the point at which Sir John and Burmese had accumulated 100 appearances between them on the sovereign's birthday parade. *(Royal Archives)*

Patrol in the Radfan, Aden, 1966. *(Welsh Guards Archives)*

airborne and amphibious landings, undertaken by 400 men of 16th Parachute Brigade and 200 of 1st Battalion Welsh Guards. By way of further contrast, in the following year, 1960, the Battalion joined 1st Guards Brigade on exercise in Libya, at that time an independent kingdom ruled by King Idris I, who encouraged close links with the US and UK. Although these were interesting and varied experiences that took the Battalion to differing environments, the received opinion was that it had been 'proved conclusively that most of the new equipment we were trialling did not work'.[11] At the end of the exercise there was a return to England, this time to Caterham in preparation for another deployment to BAOR in November 1960.

1st (Guards) Independent Company, Parachute Regiment

The first British paratroop force was formed on Churchill's orders in 1940 and, after various changes of name, became 1st Parachute Battalion in September 1941. It had originally boasted a strong Guards representation, which it retained throughout the Second World War. With the cessation of hostilities, the 1st Battalion was sent to Denmark with the rest of its brigade to oversee the surrender of German troops. In 1946, the battalion was re-designated the 1st (Guards) Parachute Battalion and despatched on a testing operational tour to Palestine. In 1948 it returned to the United Kingdom, shortly afterwards being disbanded, when the Airborne Division was reduced to a brigade, designated 16th Independent Parachute Brigade.

In July 1948 No. 1 Guards Independent Parachute Company was formed from officers and men of the 16th (Guards) Independent Pathfinder Company and in 1949 was renamed the 1st (Guards) Independent Company, Parachute Regiment. Initially tasked as pathfinders for the Brigade, parachuting into areas in advance of the main force, marking and defending planned dropping zones, they were subsequently given a reconnaissance role, first with Ferret scout cars and then, to meet changing operational needs, in four-man foot patrols.

During the early 1950s the Company moved to Egypt as part of the Brigade's deployment in the Canal Zone, being based in Ismailia. The Brigade returned to the United Kingdom in 1954 but was sent to Cyprus in mid-1956, as part of the initial response to the Suez Crisis. Their arrival in Cyprus coincided with mounting EOKA activity and they were immediately tasked with anti-terrorist operations. In the assault on Port Said itself, the airborne objectives were to seize Gamil airfield and the port, and a stick of the Company was dropped from French aircraft to capture a crucial bridge, the remainder immediately following by sea.

In December 1963, the Company was recalled from leave and deployed at short notice to Cyprus following the outbreak of fighting between the Turkish and Greek communities. However, on 2 February 1964 the Company was ordered to return to the United Kingdom and told to train for a further role alongside 22 SAS Regiment, dealing with incursions by Indonesian forces in Borneo. Following weeks of intensive specialist training, the Company deployed in mid-June 1964, remaining on operations as an independent squadron in support of 22 SAS until November, and returning to the United Kingdom in January 1965. Later that year the Company was recalled to Borneo on a second tour. In 1966 the Company Commander was informed that the SAS wished to form a Guards Squadron, 22 SAS. A Troop of the Independent Company began their training and so formed the nucleus of the new Squadron. They were joined by many of those in the Company when it was disbanded.

Three tours of Northern Ireland followed, until the final disbandment of the Company as part of the overall reduction in airborne forces. The farewell parade was held on 24 October 1976; a high proportion volunteered to join G Squadron, 22 SAS. Sadly, many of them were killed in a helicopter crash in the Falklands campaign.

Throughout its existence the Guards Parachute Company was made up of volunteers from the Foot Guards, and, from 1955, the Household Cavalry. All ranks had to pass selection at 'P' Company, undergoing tests laid down by the Director of Airborne Forces and, following that, completion of the parachute course. In addition to the Company, there has been a strong tradition of Household Division officers serving on the staff of parachute formations. Welsh Guardsmen have consistently provided their share of volunteers for airborne forces.

The tradition of Guardsmen in the Parachute Brigade is currently maintained by the Guards Parachute Platoon formed in May 2001 and attached to 3rd Battalion, Parachute Regiment. Commanded by an officer from one of the Foot Guards Regiments, the platoon has served in various theatres as part of the Battalion Group, including tours in Afghanistan during Operation Herrick.

The second three-year tour of West Germany took 1st Battalion Welsh Guards to Gort Barracks in Hubbelrath, now a suburb of the modern industrial conurbation of Düsseldorf, where it formed part of 4th Guards Brigade Group. For the last time on an overseas deployment, the Battalion travelled to its new home by train and by troopship from Harwich to Hook of Holland and thence on into West Germany; thereafter all trooping would be done by air. Once again the Battalion was occupying buildings which had been built in the 1930s for the pre-war German Wehrmacht as the home for the 64th Flak Regiment

Welsh Guardsmen on guard duty in a sangar, Aden. On the left is Guardsman H. Holland, MM. *(Welsh Guards Archives)*

(anti-aircraft artillery), although in this case the fabled German central heating system failed to live up to its reputation, being either too hot or prone to break down. One of the first training exercises took the Battalion to Leopoldsberg in Belgium which was close to Hechtel, scene of fierce fighting only sixteen years earlier – one of those who had taken part in the battle was the new Commanding Officer, Lieutenant Colonel V. G. Wallace, who was able to offer a first-hand account of the fighting.

By the time of this second tour, BAOR had settled down to a well-honed routine of training and co-operation with the armies of Britain's allies within NATO. Central to the defence thinking of the day was the creation in 1960 of the Allied Command Europe (ACE) Mobile Force–Land, one of many attempts to create a multinational formation which would possess sufficient flexibility and firepower to be despatched quickly to any part of ACE's command area, especially its flanks – from Northern Norway to eastern Turkey

2Lt M. Barnes takes off on a reconnaissance task, Aden, 1966. *(Welsh Guards Archives)*

– in order to demonstrate the solidarity of the alliance and to test its ability to resist any aggression against it. Originally envisaged as a brigade group with a strength of approximately 5,000, it comprised a headquarters structure designed to command three light infantry battalions supported by the appropriate combat support and combat service support elements.[12]

The concept was put into practice in October 1962 when the Battalion provided the British contribution to a Mobile Force–Land exercise in Greece, the other contributors being representatives of the armies of the USA, West Germany and Belgium. This got off to an unsettling start when the force landed at Larissa in heavy rain, making for an uncomfortable first night in the flooded staging camp. Matters got worse when the transport arrived – ageing former US Army M35 two-and-a-half ton trucks – to take them up into their defensive positions in the hills above Salonika along the anticipated invasion route from the north. Although the exercise tried everyone's patience, it was a useful introduction to the NATO concept of joint responsibility and the Battalion had the satisfaction of being the only one in the force to reach its assembly area within the given time limits.

Aden Air Platoon

When the Ministry of Defence announced that infantry battalions, on certain operational tours, would receive an air platoon consisting of three Sioux reconnaissance helicopters, there was a rush to get pilots trained for the 1st Battalion's tour in Aden. The aircraft arrived in early 1966 with the platoon commanded by Captain Garry Daintry (Irish Guards); the other two pilots were 2nd Lieutenant Micky Barnes (Welsh Guards) and Staff Sergeant Michael Cull (Royal Signals). Guardsmen were trained as ground crew and observers, with technical support being provided by REME. Much was learnt in the use of these aircraft, with reconnaissance, fire control, emergency resupply and general liaison duties being performed. Unfortunately, with later cuts to the Army, all aircraft were subsequently centralized at brigade level so that this was the only period that its own aircraft were an integral part of the Battalion. The aircraft and pilots stayed on in Aden when 1st Battalion Irish Guards took over in October 1966 and had some notable operational successes and awards.

LSgt R. Evans 87 (who later became Regimental Sergeant Major and then Lt Col Staff Quartermaster at the RMA Sandhurst) leads a patrol in Aden. *(Welsh Guards Archives)*

Capt J. B. B. Cockroft with the victorious Battalion Rugby XV, winners of the Army Rugby Cup, 1964. *(Welsh Guards Archives)*

It was also a useful interlude in a tour which by its very nature was circumscribed by the limits imposed by BAOR. This involved intensive training exercises at Sennelager, Haltern, Putlos and Vogelsang 'in weather ranging from the arctic to sub-tropical'. The following year, 1963, saw a signals training exercise in mid-summer in Norway and in August came a complete change of scene and temperature in southern France as guests of the French Army. As if that were not enough of a contrast, the year ended with a wet and extremely cold exercise on the Baltic coast. At the end of the year the Battalion returned to London under the command of Lieutenant Colonel Peter Leuchars and took up residence at the newly rebuilt Chelsea Barracks whose centrepiece was provided by two slightly startling thirteen-storey tower blocks. In 2008, the site was sold for development and the redundant futuristic tower blocks were demolished, having been described by the Ministry of Defence as 'a 1960s' eyesore'.[13]

The posting back to London involved a return to public duties, but in June 1964, there was a welcome change of pace and scene when the Battalion moved to Canada for seven weeks' training with the Canadian Army at Camp Wainwright in Alberta. Operation Pond Jump IV was something of a novelty for the Battalion as the creation of the British Army Training Unit Suffield (BATUS) was still almost a decade away – the vast armoured warfare training area came into being in 1971 following the accession of President Muammar Gaddafi in Libya and his decision to close down British military installations at El Adem and Tobruk. For a start, Operation Pond Jump IV was the first time that the Welsh Guards had been in North America and it was also the first time that the Battalion had moved by air over such a long distance, travelling from RAF Lyneham in Bristol Britannia turboprop airliners of RAF Transport Command. For everyone it was a unique experience which gave further evidence of the range of activities available to the soldier in the modern age. Everyone was impressed, even those who had seen it all before:

> RSM David Williams sat in his padded seat of the Bristol Britannia en route to England from Canada, and sweated slightly in his 1960 pattern Special Fitting No. 2 Dress. Refusing the fourth cup of coffee offered by the steward, he looked back on the past seven weeks in that vast country … and turned his thoughts to his

Welsh Guards Uniforms 1960s. *(From illustrations by Charles C. Stadden and Eric J. Collings commissioned by the Regiment)*

1949, Regimental Sergeant Major, Guard Order. In May 1949 HM King George VI presented new colours and in June of the same year the Battalion found the Escort and No. 2 Guard for the first sovereign's birthday parade in home service clothing since 1939. For guard mounting, the RSM does not wear a bearskin cap, but rather his dark blue forage cap. Its badge and buttons are gilt and the peak has five gold braids. He wears the distinctive large royal coat of arms worn only by RSMs and Superintending Clerks of Foot Guards battalions. Sword belt, sash and pace stick complete his attire.

1959, Guardsman, Guard Order. The newly issued 7.62 mm self-loading rifle (SLR) led to major changes in arms drill. Instead of the sloping arms, the weapon was now carried at the shoulder. At the same time, sentries at Buckingham Palace were moved behind the railings to protect them from growing numbers of tourists.

1964, Foul Weather, Guard Order. In exceptionally cold weather capes are worn over the Atholl grey greatcoats. Both garments have wide stand and fall collars. The greatcoat is fastened with five buttons, the cape with three. No insignia appear on the greatcoat, other than badges of rank. These are in dark blue, edged with red. They are worn on the right cuff and not on the upper sleeve.

1965, Musician, Regimental Band. Post-WW2 austerity delayed the return of the spectacular band tunics of pre-1939. When tunics did reappear, the only pre-1939 details that were modified were the shoulder wings. Forage caps were edged with gold cord, rather than brass.

1966, Lance Sergeant, Mess Dress. On promotion to lance sergeant, a Foot Guards NCO becomes a member of the sergeants' mess. It is an honoured institution. The RSM is the president and social arbiter. The mess dress jacket is similar to that worn by officers, with roll collar and pointed cuffs in the regimental facing colour. Leek badges are of silver wire. Around the waist of the full dress trousers is the dark blue cummerbund authorized for the Foot Guards and later replaced by a blue cloth waistcoat.

Major General Sir Philip Ward, KCVO, CBE

The first Welsh Guardsman to hold the position of Major General commanding the Household Division and General Officer Commanding London District (1973–6), Philip John Newling Ward was born on 10 July 1924. He was commissioned in the Welsh Guards in 1943 and served with the 2nd Battalion in the fighting in Normandy and North-West Europe. In the post-war period he served as Adjutant of the Eaton Hall Infantry Officer Cadet School (1950–2) and of the Royal Military Academy Sandhurst (1960–2), before being appointed Brigade Major of the Household Division and London District in 1962.

In 1965–6 he was the Commanding Officer of the 1st Battalion during the operational tour of Aden and the Western Aden Protectorate. He was a keen gardener and created a remarkable garden at the Officers' Mess in Salerno Camp that was the envy of all who saw it. When the government announced during the latter part of the tour the intention to withdraw from Aden, he was determined that operations did not warrant the life of a single Welsh Guardsman and in this he succeeded, despite some very active operations particularly in the jebels. Four year later, in the rank of brigadier, he returned to the area as Commander Land Forces, Gulf, with his headquarters in Bahrain. This was a demanding period in the UK's relationship with the Arab world. Aden had been abandoned as a colony and military base and the British government had announced an intention to withdraw from the Gulf by 1971. Both decisions caused unease in the Gulf states, but Ward steered a careful course, notably by seconding British officers and other ranks to assist with the gradual build-up of the indigenous armed forces.

In 1976 he was appointed Commandant of the Royal Military Academy Sandhurst and retired from the Army three years later thereafter holding a number of public positions, including serving as Lord Lieutenant of West Sussex, 1994–9. He died on 6 January 2003.

> waiting family in its sky-scraper flat in Chelsea Barracks – a Chelsea Barracks so changed from its former gaunt Victorian old self, a Chelsea Barracks of concrete and glass of straight lines and no cellars …[14]

Before leaving Alberta the Battalion had attended the annual Calgary Stampede and Rodeo and had also had the opportunity of visiting Edmonton and Banff in the scenic Canadian Rockies.

The return to London could have been a let-down but the following year, 1965, was both memorable and extremely busy. Not only was it the fiftieth anniversary of the founding of the Regiment but ceremonial duties predominated in the first three-quarters of the year. St David's Day was marked in style by commemorating the Regiment's beginnings. The salute was taken by Major General W. A. F. L. Fox-Pitt, one of the original founding officers, and the leeks were presented by the widow of William Stevenson, the first Regimental Sergeant Major, who had died four years earlier. In May, new Colours were presented to the 1st Battalion by Her Majesty the Queen at a ceremony in the gardens of Buckingham Palace and the new Colours were trooped a month later on the Birthday Parade which was commanded by the Regimental Lieutenant Colonel, Colonel M. C. Thursby-Pelham. For the first time, the Queen wore the uniform of the Regiment in her role as Colonel-in-Chief and was accompanied on to Horse Guards by HRH The Duke of Edinburgh as Colonel. Once again, as had happened in 1948, wet weather almost interfered with the proceedings, but the Major General (John Nelson) was loath to cancel the parade. Instead, he relied on a RAF forecast that the rain would last all day but would cease for a short period in the middle of the morning. As he remembered, that is exactly what happened: 'At 10.50, as the Queen came out of the Palace to mount her charger, the rain stopped and did not start again until she had dismounted two hours later. Weather forecasting had improved since 1948.'[15] The other great state occasion during the year was the funeral of

Sgt E. Pridham *(left)* on patrol in Aden 1967. He went on to become Regimental Sergeant Major and then Major (QM) E. Pridham, MBE. *(Welsh Guards Archives)*

Sir Winston Churchill on 30 January in which the Regiment played a part in the procession from Westminster Hall to St Paul's Cathedral.

The year also brought another tour of duty overseas under a new Commanding Officer, Lieutenant Colonel Philip Ward, who was later to be the Welsh Guards' first Major General. The Adjutant was Captain Charles (later Field Marshal the Lord) Guthrie and the Regimental Sergeant Major was David ('Ginger Will') Williams, who was succeeded in March 1966 by Idris 'Jesse' James. This time the deployment was to the Crown Colony of Aden to relieve 2nd Battalion Coldstream Guards at Salerno Lines in Little Aden, home of the BP oil refinery; as had been the case in Egypt ten years earlier, it was an unaccompanied tour. During the steam age, the port of Aden had been a vital coaling station on the sea route to India through the Suez Canal and, in the wake of the Suez operations, it became a major British military base and the headquarters of Middle East Command in 1960.

Aden consisted of two main parts. The Colony itself was relatively small, an area of some seventy square miles of rock and sand on the shores of the Red Sea. It consisted of the port, the airfield at Khormaksar, the BP oil refinery, Little Aden and the Crater district, all encircling the large natural harbour situated in an extinct volcano – the feature which had made it such a prized possession in the years of imperial sea travel. Surrounding it was the Protectorate, an area about the size of England and Wales, which ran towards the borders of Yemen and Saudi Arabia and which contained the main routes, little more than rough tracks, to the towns of Dhala and Taiz. The territory provided hard-going with steep-sided mountains and a generally inhospitable terrain.

By the early 1960s, though, Britain's only presence east of Suez was becoming a liability and preparations were being made to hand over Aden and its protectorates to a responsible government which would represent the Federation of South Arabia (FSA) – a coalition of the colony of Aden and the various sheikhdoms and emirates of the Yemeni interior which had been created in April 1962. The timetable for granting independence had been set during the defence review of 1964, when it was agreed that there would be a transfer of power within the next four years. At the same time Britain would retain its military and naval presence in Aden. Far from appeasing Arab opinion, though, the decision simply stirred up nationalist ire in the immediate area and throughout the Middle East, especially in Egypt where anti-British feelings were still running high. In 1962 Muhammad al-Badr, the king of neighbouring Yemen, was ousted by rebels backed by President Nasser of Egypt, an implacable enemy of the British presence, and the following year the National Liberation Front (NLF) was founded in Taiz in Yemen to launch attacks on targets in Aden. When Nasser visited Yemen in 1964, he announced his intention to 'kick the British right out of the Arab world', while the leader of the Soviet Union, Nikita Khrushchev, used a visit to Cairo to voice his support for the local 'demands to liquidate foreign military bases' in the Middle East. Due to Cold War rivalries, Soviet opposition to vestiges of British colonialism was taken seriously by the British government.

A guerrilla war was also fomented in the Radfan, the wild and mountainous area to the north whose indigenous Qutaibi tribes resented any external interference in their affairs and who were considered 'a xenophobic lot, equipped from boyhood with rifles, who regarded the arrival of the British Army in their mountains as the chance for a bit of target practice'.[16] The situation deteriorated and a state of emergency was declared in December 1963 following a dissident

Welsh Guards Uniforms 1960–75. *(From illustrations by Charles C. Stadden and Eric J. Collings commissioned by the Regiment)*

1966, Aden, Fighting Order. The order of dress for 'up country' patrols consisted of bush hat, shirt and shorts, socks and desert boots. For town patrols the order of dress was well-pressed khaki shorts, worn with brigade hose tops, and either a beret or forage cap. The weapon being carried here is the 7.62 mm GPMG.

1960, Pirbright, Training Order. The Guardsman shown here is dressed in training order, ready to fire his annual personal weapons Test (APWT). A khaki green combat suit has replaced denim battle dress. He wears 1958 pattern web equipment with blued metal parts.

1962, Guards Parachute Company. During WW2 the Guards Independent Parachute Company fought with the Airborne Division in Europe. After a tour of Palestine it re-formed in 1948 as No. 1 (Guards) Independent Parachute Company, Parachute Regiment. The Guards Parachute Company had its own colours – maroon with the Household Division Star and Pegasus sign of the Airborne Forces.

1972, Northern Ireland, Internal Security. The Guardsman here is shown wearing a black anodised leek in the newly issued brown beret. His body armour was known as a 'flak jacket', with a pocket for a two-way radio. He has a riot baton hanging from his belt, and is carrying his SLR at the high port.

1975, London, Drill Order. In this illustration a Guardsman is turned out in No. 2 Dress drill order. He carries an SLR, which now has plastic, rather than wooden, stock and grips.

attack on the British High Commissioner, Sir Kennedy Trevaskis, as he and FSA ministers gathered at Khormaksar airport for a flight to London to discuss constitutional issues with the Foreign Office. Direct rule from London quickly followed. As the Radfan tribes became bolder, they were encouraged by the NLF to attack government posts and to cut the main thoroughfare from Aden to Dhala, a traditional pilgrim route along the road to Mecca. Soon the Radfan became a no-go area and most of 1964 was spent in bringing the tribes to heel with two punitive operations involving 2,500 FSA and British troops, backed by armoured cars, field artillery and air power. This worked and, as the trouble began to die down at the end of the year, the NLF began to turn its attention to targets in Aden itself and to intimidating those in the

220,000-strong population who had dealings with the British. A rival group, the Front for the Liberation of South Yemen (FLOSY), was soon set up with a strong trade union influence and, as ever during that period, with support from Cairo. It was against that troublesome background that the Welsh Guards began their second half-century.

On 25 September the Battalion's advance party flew out of Gatwick on an exhausting 22-hour flight to Khormaksar where on landing they embussed on four-tonners for the trip to Salerno Camp. Initial signs were encouraging. The accommodation was relatively new and consisted mainly of prefabricated aluminium Twynham huts which were supposed to have ducted air conditioning, but the system was soon found not to be working and, as Guardsman J. M. Jones 23523201 remembered, that was not all:

> Neither were the toilets! These consisted of oil drums cut down to size, with a wooden seat and hessian partition, not the most hygienic place to be in the searing heat. Night time without a torch was also a bit dodgy as you couldn't see how full they were, but at least they were not covered in flies. This was not rectified until half way through our tour, when finally flush toilets were connected to running water.[17]

There were other potential drawbacks. From the outset it was made clear to all Guardsmen that anyone found unfit for duty due to sunburn would be fined. Inevitably there were defaulters as the desire to get a tan was strong – no one wanted to remain a 'Whitey from Blighty' – but on one occasion Lance Sergeant David Evans 13 fell asleep on the beach and had to be 'revived with a jerrican of calamine lotion and a rollicking at Memoranda when fit to hobble in'.[18] As daily temperatures could reach well over 30 degrees Centigrade and rainfall is rare in Aden, the avoidance of sunburn was a sensible precaution. The advance party began arriving at the beginning of October and soon thereafter the Battalion was involved in internal security duties, guarding the pipelines and patrolling within the colony itself, especially areas such as the Ma'alla district between Steamer Point and Crater where many British families lived.

As the months passed and the dissidents began targeting the built-up areas in Aden, this became a particularly dangerous task, but on 24 October the Battalion's operational area was moved up, country towards the Radfan to take over from 45 Commando, Royal Marines, for the first experience of fighting against the Qutaibi tribesmen. Battalion headquarters was at a tented camp at Habilayn, which had an airstrip large enough to accommodate Beverley transport aircraft, but the real operational work took place in the surrounding rural areas. As was noted at the time, this was very much a 'junior leaders' campaign' with companies and platoons dispersed over the Battalion area which lay some eighty miles north of Aden. Briefly stated, these were the Battalion's dispositions: Habilayn and a position known as H10 (Hotel Ten) overlooking the route north were the responsibility of No. 2 Company (Major Peter Williams), while further to the north were positions at Dhala which was home to No. 3 Company (Major Sam Gaussen followed by Major Dennis Stewart); Monk's Field, somewhat to the south-east, which was home to Prince of Wales Company (Major Barney Cockcroft followed by Major Charles Dawnay). To the east of Monk's Field was a platoon position known as the Cap Badge, a feature which dominated the surrounding countryside and divided two reasonably populous areas called the Wadi Taym

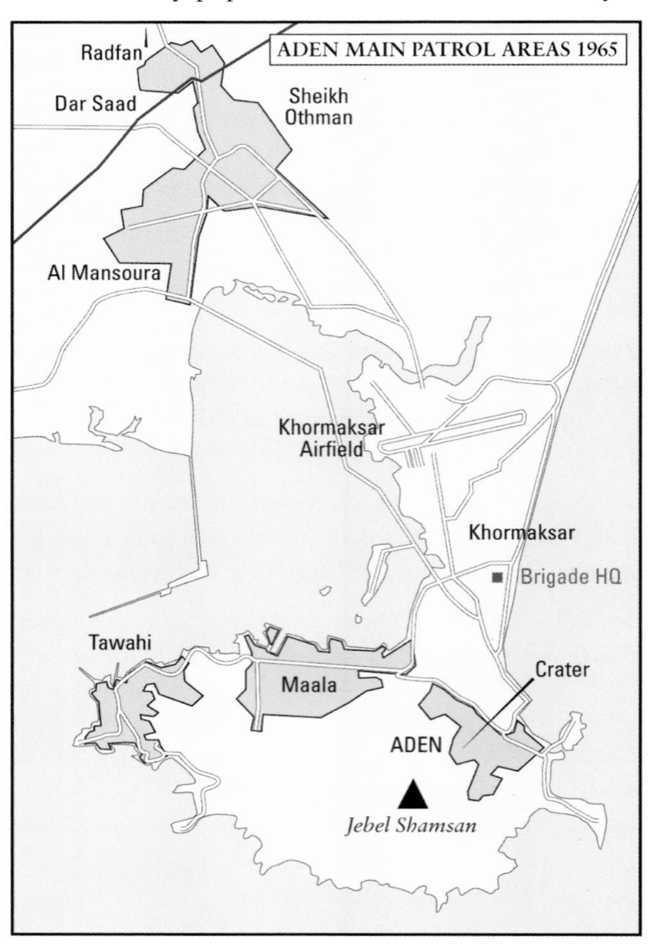

Map of the Crown Colony of Aden, 1965.

Regimental Headquarters, Welsh Guards

RHQ Staff, Welsh Guards, Wellington Barracks, 1964. Before major reorganizations of the Army in the 1980s and 1990s, RHQ Welsh Guards was established for one colonel (the Regimental Lieutenant Colonel), one major (the Regimental Adjutant), one WO1 (the Superintending Clerk), one WO2 (the RQMS); one sergeant (the Records Office) and several junior ranks. Maintaining other ranks' records was an important duty of RHQ until all regimental records offices were amalgamated and moved first to York and finally to Glasgow. Officers' records were kept at RHQ as well as centrally by the War Office/MoD at Stanmore, before also moving to Glasgow.

RHQ Staff, Welsh Guards, Wellington Barracks, 1964. *Rear row*: (*L to R*) Gdsm Watts, Gdsm D. Llewellyn, LCpl Jones 01, Gdsm Williams, LCpl Payne, LSgt Newport (Royal Army Pay Corps), Gdsm Woolley. *Centre row*: Gdsm Howen, Gdsm Davies 13, Gdsm Evans 34, LSgt Lewis 28, Gdsm Edwards, Gdsm Greenow, Gdsm Ganley, LSgt Evans 13, Gdsm Adams. *Front row*: Mr Dinnis, Sgt James, ORQMS Wilcox, Maj J. W. T. A. Malcolm, RSM Lewis, Col C. A. La T. Leatham, RQMS Jones, Capt D. R. P. Lewis, Sgt Patten, Sgt Ayton, Mr R. Robertson.

and the Danaba basin. Cap Badge had been identified as a known tribal stronghold and was certain to be well defended. To keep everyone on their toes and to maintain a sense of continuity, companies were rotated between their various tasks. Mention should also be made of the Air Platoon, which the Battalion operated during the deployment and which consisted of three Sioux helicopters. The intention was that there would be regimental pilots and observers/ground crew with REME providing technical support.

No sooner had the Battalion deployed than the trouble started. On 26 October Number Nine Platoon was ambushed in the proximity of the Cap Badge; although there were no casualties, the attack gave a good indication of the tip and run tactics being used by the tribesmen. To counter the threat, the Battalion engaged in a type of fighting which would have been recognized by British soldiers in the pre-war North-West Frontier Province in what is now Pakistan. To meet the threat posed by the Pathan tribesmen, the British Army evolved a policy which was known as 'butcher and bolt'. Basically this meant keeping its infantry battalions in the plains during the hot season and only entering the mountainous tribal areas when there was trouble to be put down. When the tribes became too aggressive and started breaking the peace, a punitive expedition would be mounted, the rebellious behaviour would be tamed by killing as many men as possible and then truces would be enforced which involved the surrender of weapons or the payment of a substantial fine. It was understood by both sides that no promises were binding and it was also recognized that violence would inevitably break out again after a suitable period of calm.

As the Battalion soon discovered during their deployment in the Radfan, history had a habit of

Capt J. de Salis with the Battalion team participating in the Nijmegen Marches, early 1960s. *(Welsh Guards Archives)*

repeating itself as the scattered rifle companies found themselves in the middle of a real shooting war against the Qutaibi tribesmen. The response to all tribal aggression was to pursue the opposition to their remote fastnesses in the hills and to show them that there would be no peace and quiet as long as they insisted on attacking the British positions. Exchanges of fire were a common occurrence and intelligence reports regularly noted 'fire returned, bloodstains later found'. Then, on 3 December, there was a large explosion in Dhala at the house of a member of the federal parliament which killed an FSA soldier as well as a number of civilians, including three small girls. By the time the up-country deployment ended just before Christmas 1965 no one had been left in any doubt about the serious nature of the local situation.

The return to Aden proper should have brought some relief, but by the beginning of 1966 it was becoming increasingly difficult to police the back streets and crowded alleys of Sheikh Othman, Tawahi, Ma'alla and Crater. The dissidents were also becoming more audacious, attacking not just military positions but also civilian areas and killing randomly, with fragmentation grenades being their weapon of choice. Often these were given to local youths in exchange for a bribe and ended up being exploded harmlessly on open wasteland, but all too frequently they were used in crowded areas with predictable results when thrown by the NLF's so-called 'Cairo Grenadiers'. And everyone knew about (but never witnessed) the incident in which a grenade pin was thrown from a taxi, which then exploded seconds later, killing the would-be assassin and his driver.

Even so, casualties were still being kept at fairly modest levels – 13 dead and 325 injured – but all that changed in February 1966 when the new British government, a Labour administration under Prime Minister Harold Wilson, announced that the timetable for withdrawal from Aden would be maintained and there would be no continuation of the British military presence. While this was greeted with enthusiasm by Arab nationalists, it caused consternation within the colony. Not only was it seen as a betrayal by those who had supported the FSA, especially the sheikhs who relied on British forces to underpin their position, but it also encouraged the unaligned to throw in their lot with the various dissident groups. It was at this time that FLOSY emerged to take advantage of the growing mayhem in the urban area. As this new grouping vied with the NLF for public support, it made life increasingly difficult for the security forces: 'We, unfortunately, were in the middle trying to keep them apart, so we never knew who our enemies were.'[19]

As the situation deteriorated in the early months of 1966, the Battalion returned to its positions in the Radfan and almost immediately it became clear that things were getting hotter and not just on the weather front. On 16 March, on Picquet Hill, a section of Prince of Wales Company came under heavy attack by dissidents armed with small arms and Blindicide unguided anti-tank missiles. During the engagement both NCOs were wounded and command of the section was taken over by Guardsman Harry Holland from Llangefni on Anglesey who restored order and made sure that the wounded were tended. For his bravery and calm handling of a potentially dangerous situation, Holland was awarded the Military Medal. Then, a week later, No. 2 Company came under determined dissident attack during what their Company Commander described as a 'coat-trailing exercise' designed to bring the opposition to battle. Although the Company suffered a number of casualties wounded, they returned fire and killed six of their assailants, 'as was the intention'. Before the month ended there were further incidents, one of which resulted in the award of a Military Medal to Sergeant Tom Edwards 59 who demonstrated coolness and good judgement while leading his platoon under enemy fire. GOC's Commendations were also awarded to Drum Major Bill Felkin, Lance Corporal T. Brindley, Guardsman E. Whitehurst and Guardsman J. Dodd.

Although there was a *Beau Geste* quality to the tour, especially up-country with its jagged peaks and wide-open spaces, Aden was a dangerous posting. In 1966 five soldiers were killed and 218 wounded; there were regular rocket attacks on the Battalion Headquarters at Habilayn and at Dhala; and in Aden itself regular strikes closed the airport and the harbour as teenage mobs went on the rampage. In a filmed television interview, Colonel Ward summed up the problems which the Battalion had faced during their year-long tour of duty in one of the most intractable and demanding of Britain's imperial holdings.

Reconnaissance Platoon preparing for Aden, with Lt R. F. Powell to the left, 1965. *(Welsh Guards Archives)*

> You can say it's offensive/defensive but it's basically a defensive war, it must be. The troops that are here are too few in number to sally forth into this infinitely remote uncharted territory in search of dissidents who in fact you don't really know if they're dissident or not because every man here carries a rifle and it's quite impossible to tell who is friend or foe. I'm not suggesting we enjoy being here because we enjoy fighting or killing, very far from it, but I think a little spice can be added to the training where you know that unless you do your job properly you will get killed.[20]

In September 1966 the Battalion entered the final stages of the tour and started preparing to leave Aden. It went home with mixed feelings: glad that the tour was over and thankful that casualties had been kept to a minimum. It was also a timely exit. In the following year the situation in Aden imploded as violence escalated and the security forces lost control of Crater. With the colony descending into chaos and civil war raging between the rival factions of the NLF and FLOSY, service families and civilian employees were evacuated and, although Crater was retaken by 1st Battalion Argyll and Sutherland Highlanders, in July, British forces finally pulled out of Aden on 29 November 1967.

By then the Battalion was resident at Victoria Barracks, Windsor, where it was involved in a busy round of public duties. Perhaps the happiest of those occasions was a parade on 4 May 1968 in which all the regiments of Foot Guards received the Freedom of the Royal Borough of Windsor. It was one of the very few occasions when all the regiments of Foot Guards received the Queen in line, with each one providing a Colour Party and a large marching contingent. By then the Household Brigade had been officially re-titled the Household Division, while the Brigade of Guards became the Guards Division, the Army Board having decided that the existing groupings of infantry battalions into brigades and large regiments was too unwieldy. The new Guards Division consisted of the eight regular battalions of Foot Guards.

CHAPTER TEN

The Long War, 1968–1982

UK, West Germany, Northern Ireland

FOR A REGIMENT SO firmly rooted in Wales yet based so frequently outside its borders, either in London District or on overseas tours, it was entirely fitting that the Welsh Guards should enter their second half-century by spending some time in Wales where they took part in two Searchlight Tattoos held in the grounds of Cardiff Castle. This event, featuring marching bands and military displays, was extremely popular in the 1960s and 1970s and always attracted enthusiastic audiences. It was also good for the profile of the armed forces and the presence of the Welsh Guards was always appreciated. (The tattoo came to an end in the late 1980s due to lack of finance; attempts to revive the event, the last as recently as 2012, floundered for the same reason.) The second involvement with Wales was more sonorous. On 1 July 1969 the Regiment took part in the rarely experienced Investiture of the Prince of Wales in the grounds of Caernarfon Castle. Although the Prince had received the title in 1958 by Letters Patent, the Investiture took place at a time when the twenty-year-old Prince was able to understand its significance both to himself and to the people of Wales. It is a ceremony steeped in history and had not been mounted since 1911 when

Welsh Guards Ski Team 1973, United Kingdom Land Forces Alpine Skiing Winners. (*L to R*) Capt M. Barnes, Gdsm Millington, Gdsm Stevens, LCpl A. Powell, Sgt Humphries, Gdsm Jackson, LCpl Robbins, Gdsm Jones. *(Welsh Guards Archives)*

HRH The Prince of Wales, Colonel of the Regiment

Appointed Colonel in 1975 in succession to his father, he is the fifth holder of the post and 2015 will mark his fortieth year as Colonel of the Regiment. As Colonel, he has been unstinting in his support to the Regiment, spending a considerable time with the Battalion, the Association and families. He has attended innumerable regimental events in his time as Colonel, including St David's Day, Remembrance Sunday, and many other special events such as presentation of new Colours and the Queen's Birthday Parade.

As Colonel he has paid particular attention to supporting the Battalion on operations, including Northern Ireland, the Falklands, the Balkans, and more recently Iraq and Afghanistan. In this context he has frequently visited predeployment training, and has also visited the Battalion in operational theatres. His personal contact with bereaved families and the wounded has been a particular focus in recent years, and has been much appreciated by families trying to come to terms with their losses. He has also generously supported and given considerable assistance to various regimental appeals, including the Welsh Guards Afghanistan Appeal launched in 2009. His Royal Highness has also reviewed his Company, The Prince of Wales's Company, on three occasions: 1980 and 1991 at Windsor Castle, and at Clarence House in 2008.

The Regiment is very fortunate to have had him as its Colonel for the past forty years.

HRH The Prince of Wales, Colonel of the Welsh Guards. *(Hugo Burnand)*

Welsh Guards rugby team in action, 1973. *(Welsh Guards Archives)*

WO2 (DSgt) John, QGM, with his father Tommy John, MM, the last RSM of the 2nd Battalion. *(Welsh Guards Archives)*

the future King Edward VIII was invested Prince of Wales in the same location.

The ceremony originated in a decision taken by King Edward I in 1301 to placate the people of Wales by creating his eldest son Prince of Wales following the defeat of Llewellyn ap Gruffydd, the last prince of an independent Wales. In a ceremony with many historic echoes, directed largely by the Constable of the Castle, Lord Snowdon, the Queen invested the Prince of Wales with the insignia of his Principality and Earldom of Chester: a sword, coronet, mantle, gold ring and gold rod. Over 4,000 guests attended the ceremony, many more gathered outside the castle and an audience of millions watched the Investiture on television. Although there was some opposition to the ceremony, mainly fanned by Welsh nationalists – on the eve of the Investiture two men were killed whilst placing a bomb outside government offices in Abergele – the Investiture cemented the role of the Prince of Wales within the Principality. It was entirely fitting that in 1975 he was appointed Colonel of the Welsh Guards in succession to his father, HRH The Duke of Edinburgh.

In addition to these pleasant tasks, which had the bonus of taking place in Wales, the Battalion had its share of operational duties. In the summer of 1968 it supplied a rifle company to train with 2nd Battalion Scots Guards at their base in Waterloo Barracks, Münster, where they formed part of 4th Guards Brigade. This was a particularly useful experience as 2nd Battalion Scots Guards was in the process of retraining to equip with the newly introduced FV432 tracked fighting vehicle, capable of carrying a section into battle. During this same period a platoon was attached to 1st Battalion Grenadier Guards in Sharjah in the Trucial States where the terrain and the weather conditions were not very different from those experienced in Aden. The British presence was intended to reassure the Gulf state rulers in advance of the creation of the United Arab Emirates, at a time when Britain planned to withdraw completely from the Middle East by 1970. That autumn the Battalion moved to Cyprus for a short tour under the command of Lieutenant Colonel Newton Webb-Bowen. They were based in the Sovereign Base Area at Dhekelia. By and large it was a relatively routine tour, with 'opportunities for training and recreation' in reasonably pleasant surroundings.[1] On return to the UK later in the year, the Battalion moved into Elizabeth Barracks at Pirbright.

In 1969 the overseas deployments continued when the Battalion returned to Norway to train again with the Norwegian Brigade North, whose task in the event of the outbreak of war was to advance to the Finnmark border to confront Soviet forces advancing from Murmansk. Its commander was well known to the Regiment and provided a suitably warm welcome. At the beginning of the Second World Ivar Froystad was

HRH The Prince of Wales presents a leek to Lt T. C. S. Bonas, St David's Day, 1975. *(Welsh Guards Archives)*

HRH The Prince of Wales presents a leek to WO2 (DSgt) Graham Pugh. The Regimental Lieutenant Colonel, Colonel J. W. T. A. Malcolm, is watching. DSgt Pugh was one of the all-time great Welsh Guards sportsmen. He played basketball for Wales, rugby for the Regiment, winning an Army Cup Medal, and cricket for the Army. He was also an outstanding athlete in field and track events and a member of the Battalion Shooting Team. *(Welsh Guards Archives)*

Inspection of reliefs from the Buckingham Palace detachment of the Queen's Guard, c. 1980. *(Welsh Guards Archives)*

a cavalry officer cadet in the Norwegian Army and managed to escape to Britain through Russia, Iraq and India. Initially he trained for sabotage operations with the Norwegian army-in-exile but, in the summer of 1944, Froystad joined the 2nd Battalion Welsh Guards in Yorkshire ahead of the Normandy invasion. While serving with them in North-West Europe, Froystad was awarded the Military Cross and, although he returned to the Norwegian Army at the conclusion of hostilities, he clearly valued his time with the Regiment, which featured affectionately in his memoirs.[2]

A different kind of deployment took place in October with an exchange involving 1st US Armored Division, based at Fort Hood in Texas. This was 'a great success … Guardsmen always seem to get on with the American Army'.[3] In March 1970 the 1st Battalion returned to BAOR, this time under the command of Lieutenant Colonel J. W. T. A. Malcolm, whose father had also commanded the Battalion. Its home was Waterloo Barracks in Münster, which had just been vacated by 2nd Battalion Scots Guards. Built on the site of a wartime Luftwaffe airfield called Loddenheide, the accommodation consisted mainly of single-storied prefabricated huts which had been constructed in 1953, allegedly with a ten-year life span, and were in need of refurbishment. The married quarters were also somewhat dilapidated and run down. Fortunately Münster itself was not an unattractive place and most of the wartime damage, particularly in the historic Old Town, had been repaired or restored. Fortunately, too, the Battalion was kept extremely busy as it transferred to its new role as a mechanized infantry battalion operating the recently introduced FV432 armoured personnel carrier (APC). With a crew of two (or three when fitted with the L37 machine-gun turret), it was able to carry a full section and was a big step forward in armoured warfare but it did have drawbacks. Unless fitted with the L37 turret, the only available weapon was the General Purpose Machine Gun situated by the commander's hatch, which required the commander to expose his head and shoulders in order to fire. Another drawback was that time spent in a closed-down APC could lead to disorientation when soldiers debussed to go into action.

These factors combined to make training a challenging experience. For the first time since 1941, when the 2nd Battalion was converted to fight in tanks, Welsh Guardsmen had to master training on unfamiliar tracked equipment as well as mastering the intricacies of the vastly different tactics of armoured warfare. Practical training was done on the three main BAOR ranges and manoeuvre areas at Sennelager, Soltau and Vogelsang, but by the late 1960s additional problems were created when the German authorities began placing restrictions on the use of non-military land and the movement of military convoys. Nevertheless, as their forefathers had done, the Battalion worked hard at converting to tracked fighting vehicles. Soon the ninety APCs were regarded as familiar friends and not as large unwieldy beasts.

The New Guard, found by Prince of Wales Company, marches into the forecourt of Buckingham Palace, January 1973. *(Welsh Guards Archives)*

Baruki Sangar, Crossmaglen, South Armagh, 1979. *(Charlie Carty)*

Foot patrol, Prince of Wales Company, South Armagh, 1979. *(Charlie Carty)*

That settled way of life did not last long. During the winter of 1970–71 the Battalion received notice of an impending move to Northern Ireland where British troops had begun supporting the Royal Ulster Constabulary (RUC) after a period of civil unrest in 1969. In the years following the political settlement of 1921, which had created the Irish Free State as part of a divided Ireland, the situation in the six counties in the north had always been volatile. Partly this was due to the emergence of the Irish Republican Army (IRA) as the military arm of the Sinn Féin nationalist party and partly it was because of a backlash from the province's Protestant community who wanted to remain part of the United Kingdom. Throughout that period there had always been tensions between the two communities and there is little doubt that this bad feeling was perpetuated by the segregated social and educational systems in Northern Ireland. As a result, a civil rights movement had come into being in 1968 and its emergence prompted a number of protest marches, which in turn encouraged rioting and outbreaks of violence throughout the winter of that year. In April 1969, in separate clashes, seventy-nine civilians were injured as well as over 200 members of the RUC. While this was happening there was a decisive split in the IRA, with the breakaway wing becoming the Provisional IRA (PIRA, Provisionals or Provos) which sought to use violence to end British rule and produce a united Ireland of all thirty-two counties. This was followed by even more serious rioting in Belfast, in which eight people were killed and houses were set alight. By the middle of August the RUC had lost control and the British government decided to despatch additional troops to Northern Ireland to help restore order. At first they were welcomed by the Catholic community, but that mood did not last long and soon PIRA gunmen were targeting British troops as well as the RUC. At the beginning of 1970 the situation had deteriorated to such an extent that the GOC Northern Ireland, General Sir Ian Freeland, announced that anyone seen with a weapon or throwing a petrol bomb was liable to be shot.

A 'Brick' of Prince of Wales Company pose in front of a Saracen APC, South Armagh, 1979. *(Charlie Carty)*

By the following year the communal hostility had reached new levels of violence, the first casualties had been sustained by the British Army and it was clear that the Provisionals had launched a full-scale terrorist war which required further troop reinforcements. Amongst them would be 1st Battalion Welsh Guards. The order arrived on the day that the rugby team was contesting the Army Cup at Aldershot against their old rivals 1st Battalion Royal Regiment of Wales. Tied at 3–3, the game had gone into extra time when, just as Guardsman Williams 66 was about to take a crucial penalty kick, the Commanding Officer received news that the Battalion was to deploy to Belfast in two days' time – on 25 March 1971. The kick at goal was successful, leaving the Battalion narrow winners, repeating their 18–6 victory of the previous year, over the same opponents.

The move to Northern Ireland was unlike anything the Battalion had experienced in the past. Barrack space was at a premium and improvisation was necessary: the Battalion's home was in a factory in the north Belfast suburb of Carnmoney, known locally for

Queen's Birthday Parade, 1981. Escort to the Colour found by The Prince of Wales's Company marches past HM The Queen in slow time. Field Officer in Brigade Waiting in Command of the Parade: Colonel S. C. C. Gaussen, Welsh Guards; Major of the Parade: Major S. O. Dwyer, Irish Guards; Adjutant in Brigade Waiting: Major J. L. Goodridge; RSM: WO1 A. Davies 22; Capt of the Escort: Maj G. N. R. Sayle; Subaltern of the Escort: Lt A. W. Ballard; Ensign of the Escort: 2Lt N. J. C. Drummond. *(MoD)*

HM The Queen, Colonel-in-Chief Welsh Guards, rides onto Horse Guards, Queen's Birthday Parade, 1981. *(MoD)*

its large and impressive cemetery. In contrast to later deployments, the Guardsmen were relatively lightly armed and the base at Carnmoney was only roughly protected by a wall of sandbags against small arms fire and blast bombs. While this was happening, No. 2 Company was attached to 3rd Battalion, Parachute Regiment, in the largely Catholic area of Andersonstown. During this deployment, which lasted until the end of July, the Battalion had no specific area of responsibility but provided Guardsmen as and when they were needed to support the RUC. It was a difficult and enervating period that required a good deal of patience and determination on the Battalion's part as they found themselves operating in a hostile environment which was nevertheless still part of the United Kingdom. By the summer, violence in the Province had increased exponentially: bombings were an almost daily occurrence, mobs roamed the streets causing mayhem and, by the beginning of August, twelve soldiers, two policemen and sixteen civilians had been killed. With extreme reluctance, the government introduced internment and, predictably, when it came into force on 9 August, it sparked further outrages across the Province.

As the Battalion ended its first emergency tour, it was clear that the situation in Northern Ireland was not susceptible to easy answers and it was with a certain sense of relief that the Battalion returned to the regularity of life in BAOR. Operation Banner was destined to be the longest ever active-service deployment by the British Army – thirty-eight years – and it put huge strains on the regiments and battalions (not to say their wives, partners and families) which served in the Province. For the Army this entailed four-month (later six) roulement tours, which were unaccompanied, and longer twenty-four-month

Lt P. G. de Zulueta with LCpl Evans 73, South Armagh 1979. *(Welsh Guards Archives)*

A patrol lies-up for the night, South Armagh 1979. *(Welsh Guards Archives)*

LSgt M. Lawson of Hawarden commands an APC driven by HRH The Duke of Edinbugh, Münster, 1971. *(Welsh Guards Archives)*

residential tours at permanent barracks such as Ballykinler and Ballykelly. All told, during Operation Banner 1st Battalion Welsh Guards completed one residential tour and seven roulement tours. Against that background the Battalion returned to Northern Ireland in 1972 under the command of Lieutenant Colonel M. R. Lee, this time to Belfast. The year 1972 was rapidly becoming the bloodiest one of the whole campaign, with more than 10,000 shooting incidents, compared to 1,756 in the previous year. In an effort to regain control of the situation, direct rule from Westminster had been introduced in March 1972. By then much of Belfast and Londonderry contained 'no-go' areas created by the Provisionals who sent out armed patrols and built barricades and roadblocks to 'protect' the communities they were allegedly defending. In essence, they controlled the Creggan and Bogside areas of Londonderry and several nationalist enclaves in Belfast, notably the Falls Road and Ardoyne. At the same time loyalist armed gangs had also started emerging, supposedly to defend the interests of the Protestant community across the Province and to target nationalist leaders. The most prominent of these were the Ulster Volunteer Force (UVF) and the Ulster Defence Association (UDA).

Such a situation could not be tolerated and 4,000 additional troops were sent to the Province. They were badly needed. On 21 July, a month after the Battalion arrived in Belfast, the Provisionals detonated twenty-two bombs in the city, killing nine civilians and two soldiers and injuring many more. Amongst the military casualties was Sergeant Philip Price of Trealaw in Glamorgan who was killed when a bomb exploded outside the Oxford Street bus station in Belfast city centre. That same night a Welsh Guards foot patrol was attacked but returned fire, hitting six PIRA gunmen. Later, it came to be known as 'Bloody Friday'. The response to the deteriorating situation was Operation Motorman which was mounted on 31 July and which succeeded in its objective of clearing the no-go areas in Londonderry and Belfast. At the same time smaller-scale operations were carried out in other places like Lurgan, Armagh, Coalisland and Newry. Some of the gloss was taken off the success when later in that day the Provisionals exploded three car bombs in the village of Claudy in County Londonderry, killing nine civilians, four of them Catholics.

At the end of October the 1st Battalion left the Province and returned to Münster. This was to be a regular pattern for the rest of the decade and into the 1980s – service with BAOR interspersed with roulement tours of Northern Ireland – and it attracted a mixed response within the Battalion. Older married men found it irksome to be separated from their families for up to six months at a time, but the younger newly joined Guardsmen tended to enjoy the experience because, after the intensive training at the Guards Depot, soldiering in BAOR or in London on public duties could seem tame by comparison. While in Northern Ireland the Battalion also came to understand that it was very much a junior NCOs' war, with most of the operations being conducted at section or sub-section level. Patrols became a fact of daily life, the most common formation being the 'brick', consisting of four men accompanied by one or two

The Escort to the Colour presents arms to the Regimental Colour, Queen's Birthday Parade, Berlin, 1979. *(Welsh Guards Archives)*

policemen. A typical day might see patrols being mounted from the company base at 06.30 with each brick under the command of a lance corporal, who had to know his 'patch' and had to react in the right way as the situation demanded. In most areas the Guardsmen had to endure episodes of physical and verbal abuse without responding and causing a confrontation with the local people. The rules of engagement were outlined in the 'yellow card rules', Army Code Number 70771, which provided specific guidance for the use of armed force.

A roulement tour could also be highly intensive, with little spare time and none of it usable locally because of the security situation. Accommodation in the urban centres was also basic, especially in the early days, and there was a good deal of bemusement at what was happening in the streets. The Battalion had faced terrorist activity in Aden but there was no comparison between confronting it in the *jebels* and desert wastes of southern Arabia and the streets and urban sprawl of Belfast, only an hour's flight from London. This was supposed to be a part of the United Kingdom, yet the Battalion was continually forced to witness behaviour that was only usually seen in the most bitter kind of counter-insurgency war. The conundrum was addressed by Captain Crispin Black, albeit from the perspective of the following decade when there was supposed to be movement back to 'normality'; even so some things in West Belfast remained very much the same

HRH The Prince of Wales watches training with the Commanding Officer, Lt Col C. R. L. Guthrie, MVO, Berlin, 1979. *(Welsh Guards Archives)*

throughout the thirty-odd years that the Battalion spent time in the Province:

> ... it's jolly difficult to police and control 100,000 people who profess to loathe your guts. We are blamed for everything in the Nationalist (IRA sympathizing) population in W[est] Belfast. If you have a hangover, if you live in a self-made slum, if one of your children is sick or even if you are generally browned off with the world, you can blame the Brits and especially the British Soldier. A funny custom really. Most of us, in even the smallest adversity, would blame ourselves or perhaps the Gods, but not the Republican parts of Northern Ireland. The fault is entirely with the British – even those most charming and engaging of fellow Celts, the Welsh.[4]

HRH The Duke of Edinburgh is towed out of barracks on his last day as Colonel of the Regiment, 1975. *(Welsh Guards Archives)*

The tour ended in October, allowing the Battalion to return to Münster once again before it moved to Chelsea Barracks, never the most popular of postings for young Guardsmen. The experience cannot have been all bad, though, because once again the Battalion won the Army Rugby Cup in 1973, for the third time in four years, beating 7 Signal Regiment 22–9 in the final.

Then it was back again to Northern Ireland for another four-month tour. This time it was to the rural area of South Armagh where the Battalion was based in Bessbrook Mill, the local linen mill which had been converted into a major military base complete with a busy helicopter landing area to supply surrounding outposts, travel by road being considered too dangerous due to the Provisionals' use of roadside bombs usually hidden in culverts. At that time South Armagh had not yet become the infamous 'bandit country' of the late 1970s and 1980s when the focus of terrorist activity had shifted from urban areas to the border with the Irish Republic, but it was still a dangerous place. Within a week of the Battalion's arrival, on 24 November 1973, the full seriousness of the deployment was brought home when Guardsman David Roberts of Holywell in Flintshire was killed in a radio-controlled landmine blast while taking part in a foot patrol in Carlingford Street, Crossmaglen. Thankfully this was the only casualty during the tour, but there were several other incidents in which patrols came under fire from Provisional 'active service units' (ASU), as the Provisonals styled their gangs. During the deployment in Bessbrook, two companies were again stationed in Belfast.

The tour ended in March 1974 when the Battalion returned to Chelsea Barracks and public duties. For Prince of Wales Company and No. 2 Company the end of the year brought separate tours to the Emirate of Sharjah, which had joined the United Arab Emirates (UAE) three years earlier, thus ending the period of the official British protectorate. Following a brief period of uncertainty, Sultan bin Mohamed Al-Qasimi, a noted author and historian, had come to power and was anxious that the UAE should keep open its friendly links with the UK. In December 1974 the Battalion moved to Caterham in Surrey which had become a station for London District following the removal of the Guards Depot to Pirbright in the previous decade. Situated on the edge of the North Downs, Caterham is within fifteen miles of London and it was there that the Troubles in Northern Ireland came back to haunt the Battalion.

Just after 21.30 on the evening of 27 August 1975, a bomb exploded in the public bar of the Caterham Arms in Coulsdon Road which was crowded with civilians and a large number of Welsh Guardsmen. The device had been planted beneath a bench seat by PIRA terrorists and no warning had been given. The explosion caused enormous damage and badly injured a number of people, including twenty Welsh Guardsmen; amongst the most severely hurt were Guardsman Paul Thomas, who lost both legs and an arm, Lance Sergeant Stephen Ollerhead, who lost a leg, and Guardsman Gareth Walters, who also lost a leg. First on the scene was a party of six off-duty policemen who had been playing bowls at St Lawrence's Hospital on the other side of the road from the pub. One of the police officers later remembered 'trying to clear the crowd that had gathered because at that time we were warned that the IRA was in the habit of planting secondary bombs to catch all those people who were trying to rescue those from the first explosion'.[5] It was the beginning of a determined IRA bombing campaign in London as the terrorists switched their efforts from Northern Ireland to England, killing thirty-five people and injuring many more over a period of fourteen months.

It was not until later in the year that the Provisionals' ASU responsible for the outrages was finally captured following a bizarre siege at 22b Balcombe Street in Marylebone in London which ended peacefully in December 1975 with the surrender of the PIRA gunmen responsible for the bombing – Martin O'Connell, Edward Butler, Harry Duggan and Hugh Doherty – who were eventually prosecuted and sentenced to life imprisonment. (After serving twenty-three years in UK jails, the four men were transferred to the high-security wing of Portlaoise Prison outside Dublin in early 1998 and were eventually released in 1999 as part of the Good Friday Agreement.) But it was by no means the end of the story. In the following year, on 30 July 1976, those Welsh Guardsmen who had been injured in the Caterham Arms bomb attack made a pilgrimage to the top of Snowdon. Led by their recently appointed Colonel, HRH The Prince of Wales, they travelled partly by royal helicopter and partly by the mountain railway to reach the summit. Courageously, Guardsman Thomas walked the last 200 yards unaided and it was generally agreed that the whole expedition had been a triumph and a sign of the Guardsmen's resilience. John Retallack's account noted that it was 'an event which must be unique in the history of any regiment in the British Army'.[6]

Belfast, 1972. (L to R) Maj J. I. C. Richardson, No. 2 Company, Maj C. R. L. Guthrie, Prince of Wales Company, and Maj (QM) Joyce, MM. (LSgt Joyce also features in Chapters 7 and 8). *(Welsh Guards Archives)*

During the stay at Caterham the Battalion was deployed on another tour to Cyprus, which lasted from October 1976 to May 1977. Since earlier

The Prime Minister, the Rt Hon. Margaret Thatcher, MP, visits Support Company in South Armagh, 1979. Standing behind her is Maj R. E. H. David. *(Welsh Guards Archives)*

Master Cook Parry (*left*) with his son Tony (*centre*) and the Master Cook's brother, Derek Parry 83, who later became WO2 RQMS at R, West Belfast, 1972. *(Welsh Guards Archives)*

deployments at Company and Battalion levels (1958 and 1968), the situation on the island had changed dramatically. Not only was part of the Battalion under the command of UNFICYP, having exchanged their khaki berets for light blue berets with the UN cap badge, but it was deployed in the peacekeeping role and could not open fire unless attacked or government property was under threat. This did not mean that tensions had receded, far from it. Two years earlier, in July 1974, the Greek military junta in Athens had carried out a coup d'état in Cyprus, to oust President Makarios and replace him with Nikos Sampson, the aim being to unite the island with Greece. This was followed almost immediately by a Turkish invasion on the pretext of restoring order. The Turkish air force began bombing Greek positions on Cyprus and airborne forces landed in the area between Nicosia and Kyrenia, where they were joined by armed Turkish-Cypriot militiamen. Turkish ground forces continued the invasion, landing from the sea and bringing with them tanks, trucks and armoured vehicles. A ceasefire was agreed at the end of the month, but by then Turkey had landed 30,000 troops on the island and captured Kyrenia, the corridor linking Kyrenia to Nicosia, and the Turkish Cypriot quarter of the capital, Nicosia. Although the junta in Athens fell at the end of July and constitutional order was restored in Cyprus with the return to power of President Makarios – thus removing the pretext for the invasion – Turkey consolidated its position. By the end of the summer, almost half of the island had been taken over by the Turks and 180,000 Greek Cypriots had been evicted from their homes in the north. At the same time, around 50,000 Turkish Cypriots had been moved to the areas under the control of the Turkish forces and settled in the properties of the displaced Greek Cypriots. With the island firmly divided, the UN operated a buffer zone, using its forces to keep the two sides apart.

When 1st Battalion Welsh Guards arrived in Cyprus in October 1976 the situation had been stabilized, but it still required tact and firm handling of the kind that the Guardsmen had perfected during the previous tours of Northern Ireland. On arrival, Battalion Headquarters and No. 2 Company and No. 3 Company settled in at Kitchener Camp, Polemidhia,

Field Marshal the Lord Guthrie of Craigiebank, GCB, LVO, OBE, DL

Charles Ronald Llewelyn Guthrie was born on 17 November 1938 and served as Chief of the General Staff (CGS), the professional head of the British Army, between 1994 and 1997 and then as Chief of the Defence Staff (CDS) between 1997 and 2001. He was appointed an Honorary Field Marshal in June 2012. Educated at Harrow and the Royal Military Academy, Sandhurst, he was commissioned in the Welsh Guards on 25 July 1959 and served as Adjutant of the 1st Battalion during the deployment to Aden in 1965. He served with 22 Special Air Service Regiment in many different places including the Middle East, the Far East and Africa. He returned to the 1st Battalion in Münster in 1970 and following one year as Military Assistant to the CGS, Field Marshal Lord Carver, in 1973, he became Commanding Officer of the 1st Battalion in 1977.

He then spent two years as colonel on the General Staff for Military Operations at the Ministry of Defence. Promoted to brigadier on 31 December 1981, he became Commander of 4 Armoured Brigade in 1982. In 1984 he was made Chief of Staff for I British Corps in Bielefeld. Following his appointment as General Officer Commanding North East District and Commander 2nd Infantry Division, based in York, on 18 January 1986, he was given the substantive rank of major general on 31 March 1986.

On 24 November 1987, Guthrie became Assistant Chief of the General Staff at the Ministry of Defence. On 2 October 1989 he was promoted to lieutenant general and appointed General Officer Commanding I British Corps, and, having been appointed KCB in 1990, he relinquished his command on 2 December 1991. He was appointed Commander of Northern Army Group and British Army of the Rhine on 7 January 1992 and, following promotion to general on 14 February 1992, became ADC to the Queen on 13 July 1993. He then became CGS on 15 March 1994, being responsible for providing strategic military advice to the British government on the deployment of troops for the Bosnian War. He went on to be CDS on 2 April 1997, in which role he advised the British government on the conduct of the conflicts in Kosovo and Sierra Leone, before retiring in 2001.

Field Marshal the Lord Guthrie of Craigiebank, GCB, LVO, OBE, DL, in a portrait painted by Anthony Oakshett in 1996 to mark his promotion to general and appointment as Chief of the General Staff. *(Anthony Oakshett)*

Lieutenant General Sir Christopher Drewry, KCB, CBE

Christopher Drewry was born 25 June 1947 and joined the Welsh Guards in 1969 after going to school at Malvern in Worcestershire and then reading Modern Languages at Oxford University. Graduate entrants to the Army at the time were much scarcer than today and, as an incentive, were excused any form of officer training, arriving directly in the Battalion with back-dated seniority. This was a less than satisfactory start to what would become a thirty-four-year career, but his first company commander (Charles Guthrie) put things right by sending him immediately on a three-month course where he joined thirty Guardsmen striving to become lance corporals. Sandhurst had nothing on that kind of initiation!

He was a platoon commander in Prince of Wales Company, first in Pirbright and then in Münster in the mechanized role, though the Battalion's tour in Germany was interspersed every year by four-month tours in Northern Ireland. A year training recruits in No. 8 Company at the Guards Depot Pirbright was followed by a return to the Battalion as Intelligence Officer and then a year as ADC to Major General Sir Philip Ward in Horse Guards.

By 1976 he had married and been appointed Adjutant of the Battalion serving in Caterham and Berlin before completing the Army Staff Course Division 3 in 1978–9. From Camberley he went to Aldershot as Chief of Staff 6 Field Force (UK Mobile Force) for his Grade 2 staff job. Arriving back in the Battalion in early 1982, he quickly found himself redeployed to the South Atlantic, where he commanded No. 2 Company in the Falklands War.

As a lieutenant colonel in 1984 he became Military Assistant to the Special Adviser (Peter Levene) to the Secretary of State in the MOD before taking command of the Battalion in the mechanized role at Hohne in 1985. High-readiness training in Germany and Canada and an operational tour in Belfast in 1986 completed his time as Commanding Officer.

Back in the MOD as Colonel, Army Plans, he became heavily involved in Arms Control negotiations at the end of the Cold War and in the 1989 Defence Review. As a brigadier he commanded 24 Airmobile Brigade in Catterick from 1990 to 1992 and was Director, Army Plans, in the MOD for three years thereafter.

A two-year tour as GOC United Kingdom Support Command (Germany) became his first posting as a major general, followed by another MOD tour as Assistant Chief of Defence Staff (Policy). He was Regimental Lieutenant Colonel Welsh Guards 1995–2000. Promoted to lieutenant general in 2000, his final job was to command the Allied Rapid Reaction Corps (ARRC) based in Rheindahlen, from where he went into retirement in early 2003.

Lt Gen Sir Christopher Drewry, KCB, CBE. *(Welsh Guards Archives)*

General Sir Redmond Watt, KCB, KCVO, CBE

Charles Redmond 'Reddy' Watt was born in 1950, the son of Lieutenant Colonel Richard Watt, who also served in the Regiment. After education at Eton and Christ Church, Oxford, he was commissioned in the Welsh Guards in 1972 and passed Staff College ten years later. In 1990 he was appointed Commanding Officer of the 1st Battalion and was promoted brigadier on 30 June 1993 and served in command of 3 Infantry Brigade in Northern Ireland in 1994 and 1995. He was appointed Director of Studies of the Army Staff College in 1996 and a year later became the Assistant Commandant (Land) at the Joint Services Command and Staff College. On 17 August 1998 he was promoted to major general and became General Officer Commanding the 1st (UK) Armoured Division which deployed to Bosnia as Headquarters Multi-National Division (South-West).

Watt became Major General commanding the Household Division and General Officer Commanding London District in 2000 in which capacity he had a significant role in the funeral of HM the Queen Mother in 2002. In 2003 he became Commander Field Army. In 2005 he was appointed General Officer Commanding, Northern Ireland, and from 2006 to 2008 he was Commander-in-Chief, Land Command. After retiring from the Army he became President of the charity Combat Stress and in 2011 was appointed Governor of the Royal Hospital, Chelsea.

A noted polo player who played off a two handicap while at Oxford, he subsequently became the highest rated Old Blue after the war when he peaked at five-goals. He was also a member of the Welsh Guards polo team (see separate feature) and was the highest goal player (five goals) since the Second World War.

near Limmasol, while Prince of Wales Company and Support Company moved to Alexander Barracks, Dhekelia, as part of the British Forces, Cyprus. On 1 November, as part of a UN redeployment, the UN contingent of the Battalion moved again to a new operational area to the west of Nicosia, with the result that the two halves of the Battalion were deployed along the so-called 'Attila Line' which divided Greek and Turkish forces. In both instances, the 'blue' and 'brown' elements were able to take advantage of the excellent weather conditions to engage in adventurous training including free-fall parachuting.

The deployment in Cyprus ended in May 1977 but it was only a prelude to a further posting to BAOR. This time the Battalion was based in West Berlin at Wavell Barracks. The only change from the previous posting was that by then former Nazi deputy leader Rudolf Hess was the only prisoner in nearby Spandau Prison, the other prisoners having been released between 1954 and 1966. During the two-year tour the Commanding Officer was Lieutenant Colonel Peter Williams, followed by Lieutenant Colonel Charles Guthrie, the Adjutant was Captain Christopher Drewry followed by Major Reddy Watt, and the Regimental Sergeant Major was Em Pridham. At the time, the Cold War was in one of its periods of relative thaw, with the Soviet Union showing an interest in détente largely due to its own internal economic problems and the impact of events elsewhere in the world. Earlier in the decade the US had concluded its involvement in the war in Vietnam in 1975 after North Vietnamese forces captured Saigon and North and

South Vietnam were united the following year. In the Middle East the inconclusive Yom Kippur War of 1973 had led to the disengagement of Egypt from hardline Arab politics, which helped to lay the foundations for a peace accord with Israel. Little noticed at the time, in April 1978 civil war broke out in Afghanistan following a Communist coup d'état; Soviet forces invaded in the following year.

The Battalion arrived in West Berlin at the beginning of 1977 to join the Berlin Field Force (as the Berlin Infantry Brigade became in April 1977) as one of the three resident infantry battalions which rotated every two years (the others being 1st Battalion Green Howards [followed by 2nd Battalion Royal Anglian Regiment] and 2nd Battalion Parachute Regiment). The main emphasis was on training not just in West Berlin itself but also at Sennelager and Soltau during the summer. Sport and adventure training also came into the equation with Guardsmen taking the opportunity to learn to ski at the Field Force Hut at Steibis in Bavaria, with the same centre being used in the summer months for canoeing, climbing, abseiling and hill walking. In the following year the strenuous 100-mile Nijmegen march provided training and adventure training in equal measure; although the Recce Platoon completed the course, the Commanding Officer was moved to admit that its members had to avoid 'such distractions as Dutch Army girls and crates of Amstel [beer]'.[7] But there was more to being stationed in West Berlin than having an agreeable time. Although the Cold War was relatively quiescent, West Berlin was always in a parlous position, being over a hundred miles inside East Germany, and, to remind the garrison of the realities of life, regular 'crash-out' exercises were held to test their reaction in the event of a Soviet attack. These were often held between 01.00 and 02.00 but it was always a point of honour which battalion was first out.

Operating in such a confined area, training exercises had to be geared to the topography and much use was made of the Grunewald forest, with the Havel being used for river-crossing training which would be of great use if the bridges were ever destroyed in time of war. During the Battalion's tour, the Chief of Staff to the Berlin Field Force was Major Mike Jackson of the Parachute Regiment (later General Sir Mike Jackson, Chief of the General Staff) and in his memoirs he provided an entertaining description of the perils and pleasures of mounting exercises on the Havel in the summer of 1978:

> In summer young Germans like to prance around naked in the sunshine and Berliners were no exception. On more than one of our exercises we burst out of the wood into a large clearing at the centre of the Grunewald to find bare Berliners disporting themselves. British soldiers in camouflage gear encountered German girls playing volley ball with everything on view – very good for morale![8]

Other highlights included public performances by the Band, the Corps of Drums and the Battalion Choir, all important contributions according to the Commanding Officer as West Berlin was 'considered in Germany as a centre of culture and art'. Ceremonial, too, played a part. Each May the Allied Forces Day Parade was held to underline the symbolic importance of the Allied contribution to the defence of West Berlin.

While the Battalion was in West Berlin it also contributed to a 'first' for one of the resident infantry battalions by creating a small pig farm which was used to raise funds for the Guardsmen's welfare. While this was an enterprising innovation, it had to be operated in a professional manner and the Adjutant put out a request for someone with previous experience of pig farming to take control. A volunteer was found in Lance Corporal Dickinson, a waiter in the officers' mess, who proved to be suitable and was promptly appointed to the honorary rank of Pig Major. The story quickly found its way to the British Forces Broadcasting Service in West Berlin, who equally promptly put in a request for an interview with the newly appointed Pig Major. After some nervous deliberation the Adjutant agreed.

> In the course of the interview, subsequently broadcast across the world, L/Cpl Dickinson was asked what difference he found between his current and previous jobs. Quick as a flash he replied that the officers insisted on being served from the left but the pigs didn't seem to mind.[9]

The tour ended in July 1979 and the Battalion returned to Pirbright, having left West Berlin with a number of sporting achievements under their belt. The hockey team won the Berlin Hockey League Cup and the BAOR Infantry Cup, beating 1st Battalion Cheshire Regiment in the final, while the soccer team managed the same feat by winning both the Berlin League and the BAOR Infantry Cup. Although the rugby team reached the final of the BAOR Cup, they were narrowly beaten by 1st Battalion Duke of Wellington's Regiment, the eventual Army Cup winners.

After post-tour leave the Battalion started training for another roulement tour of Northern Ireland, once again in South Armagh, which began at the end of October and finished in February 1980. Support Company was based in Newtownhamilton, No. 2 Company in Newry, Prince of Wales Company in Crossmaglen and No. 3 Company and Headquarter Company in Bessbrook. Before leaving for Ireland there was an intensive period of training on Salisbury Plain, then at the Lydd training area in Kent, which contained a model village that simulated the urban topography in Northern Ireland, and finally at the Stanford training area in Norfolk. Some idea of the physical character of the Battalion's tactical area of operations was given in the second issue of *The Leek*, a publication which is unique to the Regiment and which is published whenever the 1st Battalion is on operations to provide news, gossip, opinions and jokes. The description of South Armagh, though, was deadly serious. Comparing it to parts of Carmarthenshire or Breconshire, the author contrasted the natural beauty of the hog-back hills and rolling fields with the obdurate republicanism of the local population which made it a difficult place to police:

> Against this staunch republican attitude it is comparatively easy for the IRA to operate as it is safe country for them. They maintain this position by terror tactics on the local population. Conversely our position as the Security Forces is a difficult one, as intelligence is hard to get and we are literally looked on as an invading army. As the only police presence is in our forward bases at Crossmaglen and Forkhill, and that a very limited one, we solely represent the forces of law and order.[10]

On arrival, the Battalion came to the conclusion that nothing seemed to have changed since the previous tour in the area six years earlier, which was brought home to them on 13 November 1979, when eighteen-

HRH The Prince of Wales watches training with the Commanding Officer, Lt Col M. R. Lee, West Germany, 1972. *(Welsh Guards Archives)*

181 years service between them: *(L to R)* Capt (QM) B. D. Morgan, Maj (QM) I. A. James, MBE, Capt (QM) G. White, WO1 (RSM) E. L. Pridham, Capt (QM) D. J. L. Jones, Lt (QM) G. L. Evans, Capt (QM) W. E. Elcock. *(Welsh Guards Archives)*

year-old Guardsman Paul Fryer from Pontywaun in Gwent was killed while on patrol in Crossmaglen. A second Guardsman was wounded. The rest of the tour was dominated by the need to anticipate attacks of that kind and to attempt to forestall them. In the new year, Prince of Wales Company discovered and defused two landmines close to the border – which would certainly have caused casualties if detonated – while No. 2 Company did the same for a huge 600-pound radio-controlled bomb which had been placed in a lorry on the main Belfast–Dublin road. Fortunately the vehicle was recognized by Lance Sergeant Baker and after a complicated operation the site was made safe. Writing in *The Leek*, No. 2 Company made clear their detestation of this kind of attack: 'Many of us were sickened by this first-hand experience of the IRA's callous disregard for life. If this bomb had been detonated on one of Ireland's busiest main roads, many lives would have been lost.'[11] Luck was not always on the Battalion's side, however. In December 1979 two ration trucks were badly damaged and three Guardsmen wounded when a 300-pound bomb inside a cattle truck exploded just short of Bessbrook.

These were by no means the only incidents and not a week went by without further bombs being discovered in vehicles and, increasingly, in milk churns, usually hidden in culverts or concealed places beside the road. In January 1980, Prince of Wales Company came under fire from masked gunmen in a passing car but the patrol immediately returned fire. 'The car made a swift getaway. Another failed terrorist attack. The quick reaction of the patrol obviously gave the terrorists food for thought.'[12] To counter the various threats posed by the terrorists, the Battalion employed the following tactics: snap vehicle searches, house searches and patrolling by day and night. To prevent monotony, each platoon changed its task every forty-eight hours but, even in the midst of the difficulties and dangers of mounting patrols in a largely hostile community, there was always a chance of having a laugh, as No. 3 Platoon found when they interrogated a local man in Creggan early in the tour:

> Sergeant Harvey to man: 'Is this man your son?'
> The man: 'I think he is anyway.'
> Sergeant Harvey to man: 'How old is this son of yours?'
> The man: 'About twenty, but he hasn't really grown yet!'[13]

Welsh Guardsmen mounted in an FV432 APC firing a 7.62 mm GPMG, Münster, 1970. *(Charlie Carty)*

The tour in South Armagh was also enlivened by the mounting of longer mobile patrols through the surrounding countryside. With its rolling wooded hills, patchwork fields bounded by hedgerows and its narrow tracks and lanes, South Armagh was ideal territory for mounting ambushes and concealing terrorist activity. Intelligence-gathering was also an important element in any patrol's activities and to a great extent it was considered to be good infantry training for all those involved. By mid-tour it was also hectic: at Newtownhamilton, Sergeant Evans 80 of the Mortar Platoon wondered aloud if it might not be preferable going out 'dressed in a track suit and carrying a map and Silva compass'.[14] While patrolling in such an environment was always dangerous, it was more akin to 'real soldiering' and many Guardsmen preferred it to operating in urban centres such as Belfast and Londonderry. Bearing all that in mind, when the tour came to an end in February 1980 it was judged a success for the Battalion, a point made clear by the Commanding Officer, who was leaving for a new appointment in the Ministry of Defence:

> For many varied reasons the winter of 1979/80 will not be forgotten by any of us who were in South Armagh. We have known tragedy and notable success. The hours that have been spent on duty, patrolling, on guard and in ambush have been prodigious. Conditions have been uncomfortable, the weather foul, and the opportunities to relax few and far between. But, perhaps above all, it's the friends within the Battalion who will be remembered the most and the spirit which existed when times were not easy and often dangerous.[15]

During the tour the Battalion enjoyed visits from HRH The Prince of Wales, who arrived by helicopter under conditions of great secrecy, and also from Prime Minister Margaret Thatcher, who paid an informal visit to Newtownhamilton. Other highlights included the award of the MBE to Regimental Sergeant Major Pridham, who was later commissioned on posting to the Guards Depot as a platoon commander. His replacement was Regimental Sergeant Major Tony Davies 22, from Cardiff, on the Battalion's move to Pirbright the new Commanding Officer was Lieutenant Colonel J. F. Rickett, MBE.

After leave, it was back to Pirbright and public duties, with the first Queen's Guard for almost four years being found by Prince of Wales Company as 'the Battalion slipped into the routine of life in London District'.[16] But, of course, there was always more to Battalion life than a simple diet of public duties and spring drills. Summer saw Guardsmen involved in adventure training; while at the same time No. 3 Company moved to Earls Court to administer the Royal Tournament. In July, Support Company and the Corps of Drums took part in the Welsh Rugby Union's centenary celebrations in Cardiff where they preceded the Welsh National Squad during a parade through Cardiff. On the sporting front the hockey team reached the final of the Infantry Cup, the cricket team won the London District trophy, the polo team won the UKLF Inter-Regimental Cup, beating the 14th/20th Royal Hussars in the final, the first time an infantry regiment had won the competition since its inception in 1878. One member of the team, Guardsman E. Joseph, reckoned that he had 'played the best game of his life thanks to the stimulating Welsh music of the Regiment Band combined with the tremendous support of the many partisan supporters'.[17] It was also a successful year for the shooting team, who came eighth at Bisley, with Guardsman Jones 53 winning the prize for the Best Young Shot in the British Army. For Prince of Wales Company the highlight was undoubtedly the Review by HRH The Prince of Wales within the Quadrangle at Windsor Castle on 25 June 1980. The Review takes place every ten years and during his speech the Prince asked that the Company be renamed The Prince of Wales's Company, the title which will henceforth be used in this narrative.

At the end of 1981 the Commanding Officer reported that the year had been 'as busy, varied and successful as any in the recent history of the Regiment'.[18] He had good reason to feel so bullish about the performance of the 1st Battalion. In addition to finding the Windsor Guard 'on an almost permanent basis', the Battalion was engaged in security commitments at Heathrow Airport as part of Operation Trustee. This was a familiar role for all Foot Guards regiments in London District to supply back-up for the Metropolitan Police at Heathrow or Gatwick airports in the event of either facility being the target of a terrorist attack, a possibility that was taken very seriously by the government at that time. In January 1981 the Battalion took on the responsibility of being the Spearhead Battalion which was kept on readiness in mainland UK should a crisis develop in Northern Ireland that required instant reinforcement. Infantry battalions in this role usually had recent experience in the Province so that they arrived fully prepared and with all equipment packed. The requirement was to be ready to deploy headquarters, lead company and logistic elements within twenty-four hours and have the remaining companies and support assets on the ground within seventy-two hours.[19]

Spring brought a return to ceremonial duties with the square at Pirbright suddenly becoming exceptionally busy. At the St David's Day parade the leeks were presented by Major General J. M. Spencer-Smith and this was followed by the presentation of new Colours to the Battalion at Windsor on 14 May. At the Queen's Birthday Parade a month later, The Prince of Wales's Company and No. 2 Company found, respectively, the Escort and Number Two Guards. The Ensign was 2nd Lieutenant N. J. C. Drummond. The parade was almost marred when a man standing at the junction of The Mall and the Approach Road fired blanks from a starting pistol aimed at the Queen's carriage, fortunately without causing any damage or injury. He was apprehended almost immediately thanks to the quick thinking of a 'streetliner' from 2nd Battalion Scots Guards. During the summer there was adventurous training in Scotland and France while the shooting team, led by CSM Pritchard from Ross-on-Wye, covered itself in further glory by coming fifth at Bisley, firmly establishing itself 'as the top non-Gurkha shooting team in the Army'.[20]

At the end of July the Battalion returned from leave to participate in the wedding of their Colonel to Lady Diana Spencer, which was celebrated on Wednesday, 29 July. On the day itself the Battalion found the Royal Guard of Honour in Buckingham Palace, as well as three half companies of streetliners and a large party of ushers and other assistants in St Paul's Cathedral. This was in addition to mounting and dismounting Windsor Guard, the Queen's Guard and Tower Guard and finding the stand-by force for Heathrow Airport.[21] The summer ended with another appearance at the Cardiff Tattoo, while on the sporting front the rugby team made up for a disappointing loss in the semi-final of the Army Cup by winning the Army Seven-a-Side Cup and supplying three Army players in Lieutenant Jan Koops, Sergeant Davies 28 and Lance Sergeant Griffiths 28. At the same time the hockey team won both the London District Knock-Out and Six-a-Side Cups as well as providing the Army under-21 side with six players. The skiing team won the Infantry Cup at the UKLF Championships in Scotland and the polo team won the United Kingdom Inter-Regimental Cup

Regimental Polo

Although some Welsh Guards officers played polo before the Second World War, the Regiment's polo history really started with the founding of the Guards Polo Club in 1956 by HRH The Duke of Edinburgh, who was Colonel of the Welsh Guards at the time. Prince Philip played polo for the Welsh Guards with John Miller, Richard Watt, Bob Sale, Charles Guthrie and Mervyn Fox-Pitt, who went on to introduce polo in Scotland when he left the Army.

The Regiment won the Inter-Regimental Cup for the first time in 1976, beating 14/20th Hussars in the final. The Welsh Guards team was Jamie Collins, Sam Gaussen, Reddy Watt and Guardsman Eddie Joseph, who was the first private soldier to play in the winning team in the history of the tournament.

In 1977, Reddy Watt and Tony Ballard, who was an officer cadet at RMA Sandhurst, won the Inter-Regimental tournament playing for Sandhurst, and these two then joined Johnny Rickett and Simon Stephenson to win the tournament again for the Welsh Guards in 1980 and 1981. The Regiment did not play polo in 1982 because of the Falklands War, but on the Battalion's return won the Inter-Regimental tournament again in 1983 with Oliver Richardson replacing Johnny Rickett.

There was no polo in Hohne between 1984 and 1987, but the Regiment won in the UK in 1988 with the same team of Richardson, Ballard, Watt and Stephenson. After an operational tour in Belize in 1989, the Welsh Guards won again in 1990 and 1991, maintaining the Regiment's unbeaten record – remarkably the Welsh Guards were unbeaten in the Inter-Regimental tournament in the UK between 1976 and 1991, winning the tournament every time they entered it during that period.

After the end of the Cold War, as the Army reduced in size and was committed to operations in the Balkans, Iraq and Afghanistan, it was much more difficult to sustain a successful polo team, but many Welsh Guardsmen have played polo when time and operational commitments permitted, not least Rupert Thorneloe who was an experienced player when he joined the Regiment. In 2013 the Regiment entered the Inter-Regimental Cup again, with an Irish Guards officer in the team, and were beaten by the Royal Navy in the first round, but the foundations are being laid for more Welsh Guardsmen to play polo in the future.

Regimental Polo team, 1974: (*L to R*) 2Lt Plowden, Lt C. R. Watt, Lt B. J. D. Collins, Col S. C. C. Gaussen. *(Welsh Guards Archives)*

for the second time, as well as the United Services Cup, beating the 14/20th Hussars in the final. On the cultural front, the Regimental Choir took part in a performance of the Handel's *Music for the Royal Fireworks* in Hyde Park in front of the royal family on the evening before the royal wedding. Singing with the Morriston Orpheus Choir, the internationally known male-voice choir from Swansea, they were watched by a live audience of 400,000 and an estimated television audience of 500 million – 'some improvement on its previous concert audience of 79 in Woking!' With a six-week exercise in Kenya still to come in October/November, the Battalion could look back on an exciting year while anticipating the prospect of an equally stimulating year ahead.

The Battalion's year began in much the same way as the previous year, with Guards in London and at Windsor as well as providing a stand-by commitment to Operation Trustee at Heathrow. The most noticeable change was the Battalion's role in the trial of an experimental order of battle for a new 'Type B' infantry battalion. This involved three days' training in positional defence at the Battle Group Trainer at the Royal Armoured Corps centre at Bovington which enabled commanders to practise drills and tactics in an environment simulating battlefield procedures without wasting ammunition, fuel, vehicles and man-hours. This was followed in the middle of February by another Spearhead commitment, which also included specialized training for Northern Ireland at Hythe and Lydd, where the Rype village urban area provided realistic street scenes. It also gave an opportunity for many of the Guardsmen to play the part of rioters: according to the Commanding Officer, 'They seemed to take to it like ducks to water. Accent aside, one could well have been in Belfast and not Kent such were the hostility of the abuse and the accuracy of the missiles.'[22]

Sport, too, was progressing and there was success for the skiing team at the UKLF Championships at Aviemore where they won Open Team Championship, the Regular Army Inter-Unit Cup and the Infantry Championship. This was followed by more success for the rugby team when they beat 21 Engineer Regiment 12–6 in the Army Cup final, which was held that year at Sennelager. The shooting team, too, was hopeful of improving on the success it had enjoyed in the previous year, but events elsewhere were about to intrude on the Battalion's settled existence.

HM The Queen outside the gates of Buckingham Palace, taking the salute after the 1981 Queen's Birthday Parade. *(Reginald Davis/ Rex Features)*

CHAPTER ELEVEN

Falklands, 1982

JUST AS THE BATTALION was 'looking forward to the solid spring routine in London District', the first news arrived of an unusual event thousands of miles away in the South Atlantic. On 19 March a party of Argentine scrap-metal workers landed illegally at Leith Harbour on the island of South Georgia, which, together with the South Sandwich Islands, is a UK overseas territory.[1] Ostensibly the move was to make good a contract with the British company Christian Salvesen to demolish the old whaling station and other buildings, but this was a ruse by the Argentine military junta led by General Leopoldo Galtieri to set in train a series of moves that would lead to the invasion of South Georgia and the Falkland Islands, which had enjoyed British sovereignty since 1833. Over the years there had been periodic attempts by Argentina to reclaim the islands they call the Malvinas but they had come to nothing: *de facto* and *de jure* the Falklands were British, their 1,800 inhabitants considered themselves to be British and successive British governments had shown no inclination to quit the South Atlantic. All that changed when the Argentine workers raised their national flag in South Georgia and an awkward diplomatic stand-off quickly followed. Listening to the BBC News while on leave in Nelson near Caerphilly, Guardsman Simon Weston remembered Defence Secretary John Nott saying that no one should doubt British resolve to defend the country's interests in the South Atlantic.[2]

The background events which led to the outbreak of hostilities between Britain and Argentina on 2 April are now well enough known not to require any elaborate reiteration: the basic issue at stake was the sovereignty of the Falkland Islands and the determination of Prime Minister Margaret Thatcher not to give in to the illegal behaviour of the Argentine military

Battalion Headquarters, San Carlos Bay, Falkland Islands, 1982. *(Welsh Guards Archives)*

junta. Events moved rapidly. Following some cat-and-mouse military action on South Georgia, including the deployment of the Antarctic Survey Ship HMS *Endurance* with a detachment of twenty-two Royal Marines, the first Argentine military and naval units arrived in the Falklands and quickly seized control. By then the British government had been sufficiently alarmed to begin planning for an uncertain future. On 29 March, at a crisis meeting of the Cabinet, the Chief of the Naval Staff and First Sea Lord, Admiral Sir Henry Leach, advised that 'Britain could and should send a task force if the islands are invaded' and told Mrs Thatcher that, if ordered, a powerful naval task force could be assembled and made ready to sail in a matter of days.[3]

It was a courageous judgement and one which marked the spirit of the government's approach to the crisis. As Lawrence Freedman noted later in the official history of the campaign, while the Prime Minister did not ignore the Labour Opposition or fail to consult others, once that decision had been taken she 'did not look back'.[4] The composition of any force had still to be decided but it was apparent that a substantial ground element would be required in addition to Leach's naval task force, which consisted of 127 ships: 46 Royal Navy vessels, including three nuclear-powered submarines, 22 Royal Fleet Auxiliary ships and 62 merchant ships. By then, too, the main land force component was being decided and the chain of command fixed. Overall command was in the hands of the Commander-in-Chief Fleet, Admiral Sir John Fieldhouse, who would exercise command through Rear Admiral 'Sandy' Woodward, the Task Force Commander, flying his flag in the aircraft carrier HMS *Hermes*. The Land Force Commander was a Royal Marine, Major General Jeremy Moore, and command of the landing forces, initially 3 Commando Brigade, was given to Brigadier Julian Thompson. This expanded brigade group eventually consisted of three Royal Marine Commandos (40, 42 and 45), two Parachute Regiment battalions (2nd and 3rd), a reinforced artillery regiment, an air defence battery, two troops of the Blues and Royals, a reinforced engineer squadron, a reinforced light helicopter squadron, the Special Boat Service Squadron, and two squadrons of 22 SAS, as well as logistical and support units.[5]

With most of the task force under way, things began to move very quickly. On 17 April a council of war chaired by Admiral Fieldhouse was held on board HMS *Hermes* at Ascension Island, the mid-Atlantic

No. 4 Platoon, position overlooking Bluff Cove/Fitzroy Bay, Falkland Islands, 1982. *(Tracey Evans 58/Welsh Guards Reunited)*

Welsh Guardsmen prior to landing at San Carlos Bay, Falkland Islands, 1982: (*L to R*) LSgt Crowther, Gdsm Horner and Gdsm Evans 09. *(Gwyn Evans 09/Welsh Guards Reunited)*

Heavy Machine Gun Platoon provides overwatch during the landings at San Carlos Bay, Falkland Islands, 1982. *(MoD)*

Welsh Guardsmen dug-in in the Falkland Islands, June 1982. *(Welsh Guards Archives)*

Maj J. D. G. Sayers, Training Officer, and Lt R. L. Traherne, Anti-tank Platoon Commander, Falkland Islands, 1982. *(Welsh Guards Archives)*

LSgt Lane on the MV *Norland* which was used to repatriate 2,000 Argentine POW to Puerto Madryn in June 1982. Since there had been no formal cessation of hostilities, it was not clear what kind of reception the ship would get on arrival. As it was, the *Norland* was met without incident by an Argentine Type 42 destroyer, which the UK had sold to Argentina only a few years before. *(Charlie Carty)*

staging post which was so vital for the success of the operation. Amongst other matters discussed was the requirement for more ground troops as Thompson had already warned that 'the landing force as currently constituted cannot retake Port Stanley [the capital]'.[6] Basically, the plan was to land 3 Commando Brigade and to create a beachhead on the west side of East Falkland before building up forces for an assault on Port Stanley where the main elements of the Argentine forces were sited. For this plan to work, though, it was necessary to start planning for the insertion of what the Chief of the General Staff General (later Field Marshal) Sir Edwin Bramall envisaged as 'a reserve of brigade size'.[7]

There was considerable discussion about the composition of this force and what its remit might be. The designated Strategic Reserve for such tasks was 5 Infantry Brigade, but it was not fully established and had been denuded by the decision to remove 2nd Battalion Parachute Regiment and 3rd Battalion Parachute Regiment to reinforce 3 Commando Brigade. Its surviving infantry element was 1st Battalion 7th Duke of Edinburgh's Own Gurkha Rifles. There were also conflicting ideas about the rationale for such a force. The Ministry of Defence had already decided not to weaken BAOR by withdrawing an infantry brigade from West Germany and reducing Britain's NATO commitments. Consideration was given to using 1 Infantry Brigade, the United Kingdom's Mobile Force, but this was rejected on the grounds that it was a NATO-designated brigade. That left 5 Infantry Brigade as the only viable alternative and its deficiencies were made good by the addition of 2nd Battalion Scots Guards and 1st Battalion Welsh Guards, both 'first class battalions', according to Bramall. Both had recent experience of Northern Ireland, neither had any NATO commitment and, apart from public duties, both were available; it also helped that 1st Battalion Welsh Guards had only recently come off Spearhead and was fully fit and prepared. An announcement was made on 25 April.

Even at that stage it was not clear what role would be played by 5 Brigade, which was commanded by Brigadier M. J. A. (Tony) Wilson, late of the Light Infantry. However, if it was going to be used in a combat role, it needed a period of intensive battlefield training in ground conditions which replicated those found in the Falklands as far as possible. This was provided at Sennybridge in the Brecon Beacons between 20 April and 3 May during Exercise Welsh

Falcon, the largest field training exercise to have taken place in mainland Britain since the Second World War. The training was realistic and testing, being designed to accustom the men to becoming confident in the use of live ammunition, but there was later criticism in Army circles that by concentrating training on the platoon and company level Wilson had not done enough to refine the brigade's command structure.[8]

The preparations took on a new urgency when it was announced that the entire brigade would be transported to the South Atlantic on board the Cunard passenger liner RMS *Queen Elizabeth 2* (*QE2*), which was one of the many specially requisitioned civilian ships taken up from trade (STUFT). The voyage south would take three weeks, half the time taken by the rest of the Task Force, but no one really knew if the new Brigade would see any action. The uncertainty continued right up to, and beyond, departure, the mood being aided and abetted by the development and then failure of various diplomatic initiatives to find a peaceful solution and the escalation of hostilities between the opposing naval and air forces in the Falklands area.

Heavy Machine Gun Platoon, South Atlantic Campaign, 1982. Gdsm J. Yeo (*front centre*), LSgt B. Owen 52 (*centre rear*). *(Welsh Guards Archives)*

Before the Brecon exercise came to an end, the Battalion showed some of the versatility of the modern Guardsman by pressing ahead with the ceremony of granting of the Freedom of Carmarthen, wearing full home service uniforms. Bearskins and scarlet tunics took the place of combat clothing and the people of

Sgt T. Thorne handing out the post, Falkland Islands, 1982. *(Tim Reese 60/Welsh Guards Reunited)*

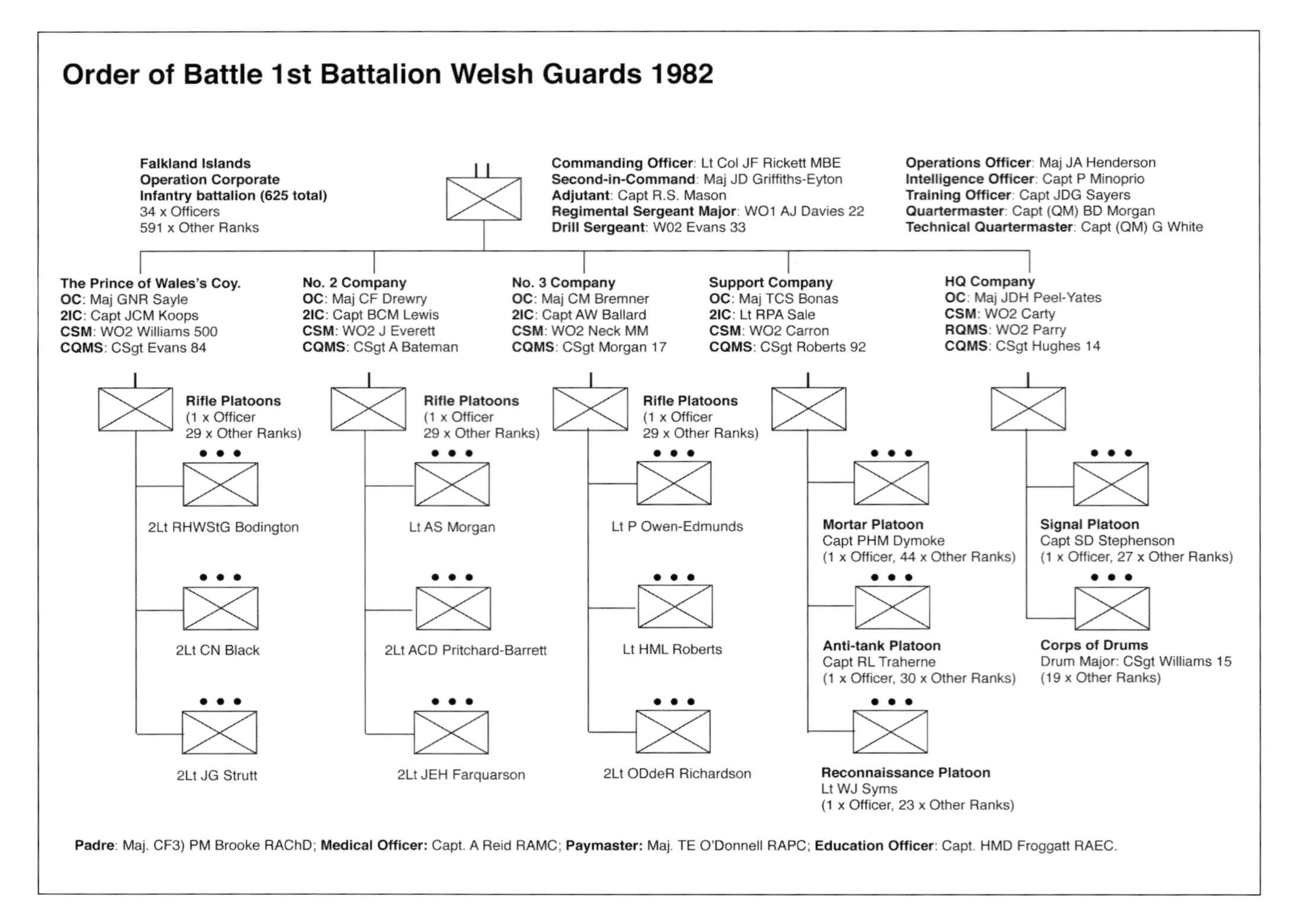

Carmarthen put on a display of warm affection for the Regiment when HRH The Prince of Wales, Colonel of the Regiment, accepted the Freedom on its behalf on 30 April. At the end of the exercise most of the Battalion then went on short embarkation leave before reassembling at Pirbright on the night of 11/12 May when, according to the Commanding Officer, Lieutenant Colonel Rickett, 'what might have seemed an extraordinary fantasy only a few short weeks before had become reality'.[9]

From Pirbright, fleets of buses took the Battalion to Southampton where a drizzly morning awaited them; the weather only started clearing in time for embarkation. In many respects the departure was quite surreal. From the outside, *QE2* looked just as it normally did, a smart Cunard ocean liner awaiting its passengers in its home port, but inside it was another story. Swimming pools had disappeared under ingeniously constructed flight decks and plywood hid the deep pile of carpeted corridors. In addition, military communications equipment had been brought on board as well as tons of military stores, including fuel and ammunition. Some supplies were stored on the open deck near the funnel and two strengthened helipads had been constructed. The liner left Southampton later that morning and, as recorded by the watching television cameras, it was one of the highly emotional leave-takings which helped to transform the Falklands war into a public event. As remembered by many of the Welsh Guardsmen themselves, there was a curious mixture of pride and solemnity; more than a few thanked the stiff eye-watering sea breeze which helped to disguise emotions, as, to their surprise, they were able to pick out the faces of loved ones in the crowds who thronged the dockyard area.

Once they were clear of the Solent, life on board the *QE2* was a curious blend of military discipline and cruise-ship comfort. There was a hefty daily training schedule, beginning with a dozen laps of the upper deck, each lap being equivalent to a quarter of a mile: in a final address to the Battalion before embarkation, the Commanding Officer made it clear that he expected everyone to remain match-fit. Stops were made at Freetown in Sierra Leone and Ascension Island; for the last stage, after 18 May, electronic

Preparing to disembark from HMS *Fearless*, San Carlos Bay, June 1982. *(Charlie Carty)*

silence was observed and the ship was blacked out during night-time hours. During this period the news from the Falklands was mixed. On 21 May a beachhead was established by 3 Commando Brigade at San Carlos, thereby removing any lingering doubts about whether or not 5 Brigade would see action – South Georgia had already been recaptured on Sunday, 25 April – but this story was balanced by news of incessant and successful aerial attacks on the Task Force. A number of valuable ships had been lost, including the SS *Atlantic Conveyor*, a Cunard

Guardsmen arriving ashore at Bluff Cove following the attack on the RFA *Sir Galahad*, 8 June 1982. *(Imperial War Museum)*

Gdsm R. Gill, Heavy Machine Gun Platoon, Falkland Islands, 1982. *(Welsh Guards Archives)*

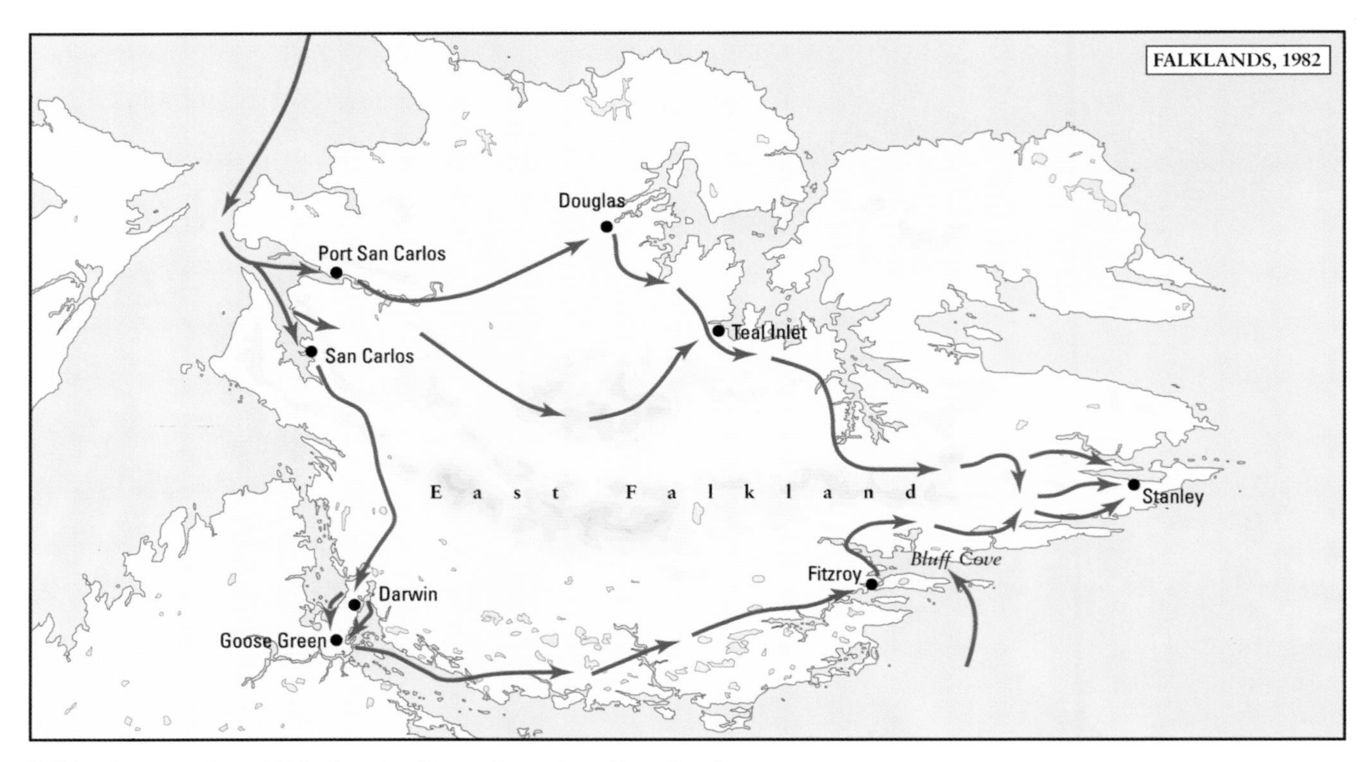

Falklands campaign, 1982, showing lines of attack on Port Stanley.

container ship which was carrying six Wessex and three Chinook helicopters. After being hit by two Exocet missiles on 25 May, it caught fire and the entire cargo was destroyed, a loss which would have a bearing on the subsequent land campaign.

From the very outset of *QE2*'s requisition it had been agreed that it would not be allowed to enter the immediate war zone, as its loss to Argentine air power would have been catastrophic to the British war effort. Accordingly, when 5 Brigade made landfall on 27 May, it was at South Georgia, 1,500 miles away, where they transferred from *QE2* to the P&O liner *Canberra* (also STUFT) in 'the haunting and brooding atmosphere of Grytviken Bay'. Ahead lay a three-day voyage on the smaller liner which started unloading the Brigade on the wet and misty morning of 2 June at San Carlos Settlement. There, according to Captain R. S. Mason, 'we crammed into Landing Craft and chugged ashore to be met by a BBC camera crew on the end of Port San Carlos jetty; this must mean that we are at war!'[10] Once the morning mist and drizzle had dispersed, revealing the rocky and boggy open ground around the settlement, the Welsh Guardsmen were immediately and sentimentally reminded of home. 'Just like Brecon,' said some, while others saw reminders of Pembrokeshire in the green-brown coastline with its sandy inlets ringed by small white-washed houses. Unfortunately, there the comparison ended: the terrain might have been picturesque but it was not ideal for digging in. Trenches quickly filled with water and the boggy ground with its thick clumps of heather – the never-to-be-forgotten 'diddle-dee' – made the going difficult. It was numbingly cold, there were fourteen hours of darkness and helicopters were at a premium.

Although the Guardsmen on the ground would not have appreciated it at the time, the British forces were gradually gaining the upper hand. The *QE2* had also brought a divisional headquarters led by General Moore and the land forces were rearranged to take account of the new tactical situation. On 28 May, 2nd Battalion Parachute Regiment had fought the first successful land battle at Goose Green and Brigadier Thompson had begun moving the bulk of his brigade across the north of East Falkland for the eventual investment of Port Stanley. For 45 Commando and 3rd Battalion Parachute Regiment, this entailed a hard cross-country march across boggy, heather-tufted land towards Mount Kent. Thompson then moved his headquarters to Teal Inlet while Wilson established 5 Brigade headquarters at Darwin, which had been repossessed after Goose Green. With all his ground forces now firmly established in the northern sector, the way was open for Moore to exploit the southern flank for a pincer attack on Port Stanley.

On landing on East Falkland, Moore was not

The A Echelon (Battalion resupply group); Falkland Islands 1982. *(Welsh Guards Archives)*

5 Brigade establishes a bridgehead, San Carlos Bay, Falkland Islands, 2 June 1982. *(Imperial War Museum)*

unhappy with the situation he found: a Brigade Administrative Area had been established at San Carlos, psychological superiority had been achieved at Goose Green and ground forces were advancing on Mount Kent. Intelligence about the enemy's positions and intentions was still proving to be what Thompson later described as 'an educated guess', although an Argentine map captured on Mount Kent suggested that Brigadier General Mario Menendez in Port Stanley was expecting an attack from the south, thus reinforcing Moore's decision to deploy 5 Brigade along the south of East Falkland from Fitzroy and Bluff Cove which had been seized by 2nd Battalion Parachute Regiment in a daring forward move on 2 June. This would provide Wilson with a jumping-off point for 2nd Battalion Scots Guards and 1st Battalion Welsh Guards, but they still had to get there over difficult terrain and take their heavy equipment with them. Without helicopters they either had to march the thirty-five miles or they could be carried by sea. The Welsh Guards tried both methods. On 3 June the Commanding Officer drew up a plan for a night march to move the Battalion over the Sussex Mountains to a position called High Hill some three miles north of Darwin. To assist them, two sections of the Recce Platoon were flown out at last light to secure the area.

HRH The Prince of Wales talks to Gdsm S. Weston at the South Atlantic Campaign Medal Parade, Buckingham Palace. Sgt Fisher is on Gdsm Weston's right. *(Welsh Guards Archives)*

An Argentine IAI Dagger flies over RFA *Sir Galahad*, 24 May 1982. *Sir Galahad* was damaged in attacks that day. *(Imperial War Museum)*

A Rapier anti-aircraft missile system overlooking Fitzroy Bay. This failed to work due to salt water corroding the electronics. *(Imperial War Museum)*

What followed next was deeply dispiriting. At 02.00 the Battalion moved out of its Assembly Area, taking as much kit with them as they possibly could – in most cases this amounted to 100 pounds per man packed into their Bergen backpacks. One small and totally inadequate civilian tractor and trailer accompanied them to help with the mortars and the .50 Browning machine guns and there was the promise of assistance from up to ten tracked Snowcats that were returning to Goose Green. However, these were running short of fuel and were unable to help; according to the Battalion's Battle Account, this proved the last straw, despite the other vehicle's best efforts:

> Undaunted, the Battalion continued up into the mountains but before too long the one tractor and trailer proved unable to cope with the appallingly muddy conditions and only a herculean effort by the men of The Prince of Wales's Company who managed to bodily lift the vehicle out of the mud allowed the return of the Battalion with all its kit to the Assy [Assembly] Area.[11]

On their return, the Battalion was then informed that shipping was to be made available to take them to Bluff Cove and they should be ready to embark from the Brigade Assembly Area by 12.45 on 5 June. The first plan was to embark The Prince of Wales's Company and No. 2 Company with all of 2nd Battalion Scots Guards on board HMS *Fearless*, one of the Navy's two assault ships for amphibious operations, while the remainder of 1st Battalion Welsh Guards embarked on the Landing Ship Logistic (LSL) *Sir Tristram*, but this was countermanded at 14.30. Some idea of the confusion can be gained from the Battalion Battle Account: 'Order and counter order were rife. Troops who had been waiting all day in the cold and wind in readiness for embarkation had to be told that embarkation had been delayed yet again.' That evening the Commanding Officer and the Operations Officer flew to Moore's headquarters on board HMS *Fearless* for an orders group to determine the next move. This proved to be equally complicated and led to further delays and difficulties, largely due to political decisions taken in London. Fieldhouse had told Moore that he was unwilling to risk using either HMS *Fearless* or her sister ship HMS *Intrepid* beyond a point to the north of Lively Island because of the threat posed by an identified land-based Exocet missile system in the Port Stanley area.

Under this new plan, and taking into account Fieldhouse's warnings about the use of the larger ships, the two Foot Guards battalions would be transported separately. On the night of 5/6 June, HMS *Intrepid* would embark 2nd Battalion Scots Guards and take them to a position south of Lively Island where they would be trans-shipped into four landing craft utility (LCU) for the final stage to Bluff Cove. On the following night, 1st Battalion Welsh Guards would follow on board HMS *Fearless*, which would rendezvous with two LCUs off Direction Island and use these, plus its own two LCUs, to get to Bluff Cove. At first the men were happy to be on the move and were able to dry out once on board in the 'sanctuary' of the ship's vast tank deck. At 0100 the Battalion prepared to trans-ship to the LCUs, but only two were available, the second pair having failed inexplicably to reach the rendezvous point. Keen to retain the momentum, Moore decided that the Battalion should go in two waves. Half would travel immediately (No. 2 Company, Battalion Headquarters, Machine-gun Platoon, Anti-tank Platoon, Recce Platoon, Defence Platoon and half of A Echelon (Battalion resupply group), while the rest (The Prince of Wales's

RFA *Sir Galahad* on fire, Bluff Cove, 8 June 1982. *(Imperial War Museum)*

Company, No. 3 Company, Mortar Platoon and the other half of A Echelon) would carry out a similar move the following night. 'That agreed, the two LCUs gently nudged their way out of the stern of the LPD [*Fearless*] and in a ghostly grey light headed for Bluff Cove.' Colonel Rickett went with the leading half of the Battalion: he was later to remember his parting words to Moore: 'General, you must promise me that you'll look after my men, because I can't do so.'[12]

After four hours at sea, half of the Battalion landed at Yellow Beach 1 with four Land-Rovers and, 'with minimum of fuss', moved to an assembly area in a quarry two miles to the south of Mount Challenger. As dawn broke on 8 June, the day was bright and clear and the morning and early afternoon were spent moving stores, preparing a recce patrol and liaising with 2nd Battalion Scots Guards and 42 Commando, which formed the left flank. The main concern was the exact whereabouts of the rest of the Battaliòn with whom contact had been lost – the Commander's Diary Narrative stated that 'comms to Bde were terrible'.[13] Not even Brigade Headquarters knew what had happened to the group which was under the command of Major Guy Sayle, The Prince of Wales's Company. In fact, they had returned to San Carlos where the original intention had been to use *Fearless* again that night to transport them to Bluff Cove so that 'the whole battalion would be operational as soon as possible'. However, at 14.00 (7 June) it was decided that the men and their equipment would board the LSL RFA *Sir Galahad* for onward transport to Fitzroy, where they would trans-ship to LCUs for the three-hour trip to Bluff Cove. Accompanying them would be a Rapier air defence unit and 16 Field Ambulance, RAMC, and all its attendant stores and vehicles, which were duly loaded by 02.00 (8 June). The original plan was to offload the Welsh Guards at Bluff Cove to allow *Sir Galahad* to continue with the rest of its load to Fitzroy.

By that late stage it was clear that *Sir Galahad* could not possibly complete its voyage by first light, thus increasing the threat of Argentine air attack. Sayle was summoned to the bridge by the master, Captain Philip Roberts, who wanted to delay the journey to the following night, 'because it seemed sensible not to arrive in daylight'. A signal to that effect was sent to *Fearless*. In response, Moore's staff ordered Roberts to sail immediately but only to travel as far as Fitzroy where the Rapier unit and the Field Ambulance would be offloaded. As the Welsh Guardsmen had settled down for

Landing craft from HMS *Fearless* en route to San Carlos Bay. Sea King helicopters can be seen on the *Fearless*'s deck. *(Imperial War Museum)*

the night, Roberts thought nothing would be gained by informing them of the change of plan and that 'this new information could be passed to them in the morning when they were called'.[14] It was only when *Sir Galahad* arrived at first light at Port Pleasant near Fitzroy that Sayle realized that they were in the wrong place and that, 'In the first hour it wasn't a question of who got off first – no one could have because there were no means of getting off, there weren't any landing craft available.' He informed Captain Roberts that he was not prepared to act until he had received 'orders what to do either from my battalion or from brigade'.[15] Also at Port Pleasant was *Sir Galahad*'s sister ship *Sir Tristram*, which was busy offloading ammunition; her presence, plus the arrival of *Sir Galahad*, must have been noted by Argentine spotters on Mount Harriet ten miles to the north.

Unfortunately, at this point matters became heated as a disagreement broke out between Sayle and the two landing officers working for Commodore Amphibious Warfare (Commodore Michael Clapp, RN). Major Ewan Southby-Tailyour, Royal Marines, and Major Tony Todd, Royal Corps of Transport, both wanted the Welsh Guards to vacate the ship as quickly as possible to avoid the danger of Argentine air attack. 'Nobody is going to Bluff Cove,' warned Southby-Tailyour, 'unless they walk.'[16] Sayle and Major Charles Bremner, commanding No. 3 Company, disagreed with this and made it clear that they would not move their men from *Sir Galahad* as their existing orders were to get to Bluff Cove by the afternoon of 9 June with all due speed and carrying all heavy equipment and ammunition in preparation for the advance on Port Stanley. There was the added complication that their Commanding Officer and Brigade Headquarters still did not have any clear understanding about their present whereabouts; so confused was the situation that one staff officer believed that the entire 1st Battalion was already at Bluff Cove.[17] To the company commanders on the spot, it seemed 'to be unwise to go wandering off into a battle area with apparently no means of communicating and without a clear picture of where the other units were and what they were doing and on a journey the majority of which would be completed in the dark'.[18] Furthermore the estimated distance of the march was fifteen miles due to doubts over the condition of a vital bridge at Fitzroy.

While this was happening, a Sea King helicopter which had accompanied *Sir Galahad* had begun unloading the Rapier detachment and this was followed by the Field Ambulance. But it all took time. The next solution was to take the Welsh Guards round to Bluff Cove using the solitary LCU working at Fitzroy unloading ammunition. Once that task had been completed, it would take the men and their ammunition in shifts while other heavy equipment would be carried by local tractors with trailers. This was accepted but then there was another hitch when it was discovered that the LCU's ramp had been damaged and could not link up with *Sir Galahad*'s stern ramp. A repair was attempted but to no avail, and a further change of plan was ordered to bring the LCU alongside *Sir Galahad* so that the LSL's crane could transfer the heavy equipment while the men would disembark by scrambling nets. This would be a lengthy and tedious operation and one of the officers in No. 3 Company, Lieutenant Hilarian Roberts, warned his men that it would not be over until nightfall: 'The men of both Companies were either helping with the winching, filing out of the tank deck to scramble down into the boat or simply waiting to get off. Some were still up in the main superstructure in the canteen and television rooms.'[19] They had been there for five hours and the Argentine forces had used the time to good effect.

Unknown to anyone in the area, the Argentine observers on Mount Harriet had seen the two ships at Port Pleasant and had passed the information back to the mainland where a sizeable air operation was ordered from the Rio Gallegos base, 450 nautical miles away. The attack would be made by eight McDonnell Douglas A4 Skyhawks and six IAI Daggers of the 5th and 6th Fighter Groups, while four Dassault Mirages of the 8th Fighter Group would make a decoy attack to the north in order to lure away the Sea Harrier combat air patrol that had been ordered to cover Fitzroy. Having left their base in the late morning, the aircraft arrived in Falklands' airspace at 13.30 when their presence became known to the Brigade Assembly Area at San Carlos and an air raid warning was issued – by then three Skyhawks and one Dagger had been forced back due to technical problems. The faster Daggers were the first to arrive and would have attacked *Sir Galahad* and *Sir Tristram* from the west had they not sighted the frigate HMS *Plymouth* alone in Falklands Sound. The British warship was attacked and badly damaged, but replied with heavy defensive fire and the Daggers, running short of fuel, were forced to return to the mainland, pursued by the Sea Harrier combat air patrol.

No such problems awaited the five Skyhawks, although, as their commander, First Lieutenant Carlos Cachon, was to remember later, they too had to rely on luck. While heading towards Fitzroy they flew through a rain squall and, finding the bay deserted, turned right to begin the flight home. Below them they could see formations of soldiers – men of 2nd Battalion Scots Guards – who put up a screen of small arms fire. It was while they were passing over this area that

Elements of the 1st Battalion Welsh Guards move into position around Mount Harriet, Falkland Islands, 12 June 1982. *(Imperial War Museum)*

Cachon's wingman noticed the LSLs in Port Pleasant. The Skyhawks straightened up, banked to the left and pressed home their attack, scoring direct hits on both ships with three 500-pound bombs and strafing the superstructure with 20 mm cannon fire. Three Skyhawks attacked *Sir Galahad*. One bomb exploded in the engine room, another in the empty officers' quarters, but the one which did the most damage hurtled through the hold into the open tank deck where the bulk of the men were waiting to trans-ship. A fourth bomb bounced off the deck harmlessly into the sea. The effect of the attack was immediate and devastating for the men on the tank deck where the bomb set off a chain reaction of ignited fuel and ammunition. In the enclosed space, terrible damage was done to those who took the full force of the explosion, men like Guardsman Wayne Trigg of Holyhead who was serving in the Mortar Platoon:

> Then it was one big bang. There were bodies, screaming, shouting. The ship was one big fireball. I covered my face first as the blast came from the centre of the ship to the back end. Because the back and the front ends of the ship were both closed there was no way for the flames to escape. They rebounded and came back and caught most of my back, my legs. I actually caught fire.[20]

Hilarian Roberts felt the fireball as it rushed past him and recalled his hands 'turn as if to grey rubber gloves', and he remembered the Company Quartermaster Sergeant telling everyone to 'keep calm' and he repeated the words himself as he and a dozen others were led by a narrow hatch up to the deck and safety. Later he told friends that he knew something was wrong when one of his men appeared in front him, saluted and said, 'I think we ought to get off this ship, sir.' Guardsman Simon Weston, who was waiting with the rest of the Mortar Platoon near the stern, also managed to escape. Although horribly burned, he fought his way to a stairwell where CSM Brian Neck, No. 3 Company, was calmly guiding the survivors to safety amidst the fire and the explosions. The moment was later recalled by Major Bremner as being in the best traditions of the regiment:

> There was no panic. The CSM, Neck, bawled instructions and instinctively they were obeyed. He was doing everything as he did in barracks, making everything appear normal; he organized everything in a typical Foot Guards way. We found a way up and out of the tank deck; a queue was formed and men filed out. One man got half-way up a companionway when the CSM called him back, against the flow of traffic: 'Did you fill in your ADAT [Army Dependents' Assurance Trust] Form?' 'Yes Sir,' he cried. 'Good. Now you see the bloody point of it. Go on, get a move on, you're holding everyone up.' It was a brilliant move that relieved the tension.[21]

No. 2 Company moves towards Port Stanley, Falkland Islands, 1982. *(Imperial War Museum)*

Gallantry Awards, Falklands Campaign

Officer of the Most Excellent Order of the British Empire (OBE)
Lieutenant Colonel J. F. Rickett, MBE

Member of the Most Excellent Order of the British Empire (MBE)
23877373 WO1 (Superintending Clerk) L. Ellson
23876522 RSM A. J. Davies

Military Cross (MC)
Captain A. J. G. Wight

Military Medal (MM)
23929678 CSM B. Neck
24498796 LCpl D. Loveridge
24599314 Gdsm S. Chapman

Mention in Despatches
Captain J. D. G. Sayers
Lieutenant W. J. Syms
24008055 CSM G. Hough
24386530 LSgt D. Graham

Commander-in-Chief Fleet Commendation
Major J. D. Griffiths-Eyton
Captain (QM) B. D. Morgan
24405437 LSgt D. Shaw

Commander-in-Chief UKLF Commendation
Major M. R. Senior

Roll of Honour, Falklands

Soldiers serving with 1st Battalion Welsh Guards
24364397 Lance Corporal B. C. Bullers, Army Catering Corps
24422490 Lance Corporal A. Burke
24332979 Guardsman J. R. Carlyle
24579323 Private A. M. Connett, Army Catering Corps
24513849 Guardsman I. A. Dale
24263842 Guardsman M. J. Dunphy
24584832 Guardsman P. Edwards
24185183 Sergeant C. N. Elley
24598683 Guardsman M. Gibby
24511408 Guardsman G. C. Grace
24520370 Guardsman P. Green
23929722 Guardsman G. M. Griffiths
24555311 Guardsman D. N. Hughes
24400658 Guardsman G. Hughes
24339321 Guardsman B. Jasper
24398540 Private M. A. Jones, Army Catering Corps
24562019 Guardsman A. Keeble
24125031 Lance Sergeant K. Keoghane
24578382 Guardsman M. J. Marks
24442460 Private R. J. Middlewick, Army Catering Corps
24608405 Guardsman C. Mordecai
24220163 Lance Corporal S. J. Newbury
24498671 Guardsman G. D. Nicholson
24513947 Guardsman C. C. Parsons
24503713 Guardsman E. J. Phillips
24562309 Guardsman G. W. Poole
24565127 Craftsman M. W. Rollins, Royal Electrical and Mechanical Engineers
24495304 Guardsman N. A. Rowberry
24463538 Lance Corporal A. R. Streatfield, Royal Electrical and Mechanical Engineers
24433054 Lance Corporal P. A. Sweet
24454603 Lance Corporal C. C. Thomas
24497060 Guardsman G. K. Thomas
24436475 Lance Corporal N. D. M. Thomas
24446382 Guardsman R. G. Thomas
24508985 Guardsman A. Walker
24433056 Lance Corporal F. C. Ward
24516114 Guardsman J. F. Weaver
24090540 Sergeant M. Wigley
24472259 Guardsman D. R. Williams

Welsh Guards serving with other units:
G Squadron, 22 Special Air Service Regiment
24221177 Sergeant J. L. Arthy
24076141 Sergeant W. J. Hughes
24184150 Sergeant P. Jones

A Sea King Mark 2A helicopter of 825 Naval Air Squadron, flown by Lt Cdr H. Clarke, RN, winches up a survivor from a liferaft to the rear of the burning RFA *Sir Galahad. (Imperial War Museum)*

Others who helped in the initial rescue were Lance Corporal Dale Loveridge and Guardsman Stephen Chapman. Loveridge, Chapman and Neck were each awarded the Military Medal. While leaving the ship by the nets, CSM Neck slipped and dislocated his shoulder, an old rugby injury, but by then the rescue was in full swing. Those on shore started pulling lifeboats ashore while Royal Navy helicopters flew mission after mission to pick up survivors or used their rotors to fan the life-rafts to safety. The rescue was captured by a visiting television news team who had been invited by Wilson to film the 5 Brigade build-up and who ended up taking some of the most dramatic footage of the war. This was seen by a worldwide audience and one of the most enduring images was the sight of *Sir Galahad* burning and the frantic rescue attempts, including the harrowing vision of Guardsman David Grimshaw from Llandudno being carried by stretcher, supporting his shattered left leg in the air.

In the midst of the mayhem, courage and discipline prevailed amongst the survivors and played a key role in preventing the tragedy from escalating. Many Guardsmen in The Prince of Wales's Company remembered with gratitude the sight of the 'tall figure of our company second in command [Captain Jan Koops] moving around the deck checking that no one had been left behind in the debris and chaos'. His bravery and energy were an example to everyone that day: he was instrumental in preventing panic and imposing order on a disastrous situation (it helped that he was the captain of the Battalion rugby team). The last people to leave *Sir Galahad* were a small group of Welsh Guardsmen who courageously made a final search for survivors before leaving the stricken vessel, which was in real danger of blowing up as a result of the raging fires and the exploding ammunition.[22]

For the Commanding Officer and the rest of the Battalion, the first inkling of the mishap had arrived when they saw the Skyhawks flash over their position at Bluff Cove and expended 10,000 rounds of small arms and machine-gun fire at the aircraft. 'I didn't know they were there, nobody at Brigade HQ knew they were on that ship,' said Rickett later. 'Nobody had any information, any news about it at all.' Like wildfire the news spread from Fitzroy; the Commanding Officer heard while visiting No. 2 Company and quickly set off by motor-bicycle with Regimental Sergeant Major Tony Davies. When they arrived at Port Pleasant, the scene that greeted them was one of organized chaos and one that Rickett has never been able to forget:

> The survivors were struggling back, all the casualties without their equipment. They were in a state of shock – and we were under [another] Argentine air attack at the time – they were like a lot of sheep wanting help, wanting guidance. And the cry was 'Sir, what shall we do?' And I said, 'Stick to me because I'm lucky, I haven't been hit by anything yet.' Somehow that steadied them and it was all right.[23]

In fact it would take at least a fortnight for the two companies to be properly kitted out again. No. 3 Company was evacuated that night to San Carlos while The Prince of Wales's Company spent the night in a sheep shed at Fitzroy where they were taken under the wing of 2nd Battalion Parachute Regiment. The following day they were taken back to *Fearless* by helicopter. Bereft of all their kit and equipment, their war was over and they spent the rest of the campaign guarding the Force Maintenance Area at San Carlos. Even at that stage it was not possible to gauge the losses; it was not until after the fall of Port Stanley, which brought the war to an end on 14 June, that the full extent was known.

All told, forty-eight men were lost on *Sir Galahad*, thirty-two of whom were Welsh Guardsmen, with many more wounded, most of them suffering from burns and broken limbs. By then, three other Welsh Guardsmen serving with the SAS had lost their lives on 19 May when a Sea King helicopter crashed into the sea, killing all twenty passengers, during a sea transfer prior to the San Carlos landings. They were Sergeant John Leslie Arthy, Sergeant William John Hughes and Sergeant P. Jones. Before the campaign ended, Lance Corporal Thomas 03 was killed on Mount Harriet during the last phase of the operation.

Inevitably the attack on *Sir Galahad* had an effect on the Battalion that went beyond the loss of casualties, terrible though these had been. It deprived 5 Brigade of half a battalion of well-trained soldiers, plus a good deal of ammunition, transport and supplies. It also meant that the surviving 1st Battalion could only play a reduced role in the attack on Port Stanley, which began almost immediately as General Moore was keen to keep up the momentum of the advance. For Captain R. S. Mason and others, 'A feeling of despair and terrible loss gave way to grim and purposeful resolve as two companies of marines joined us and we marched towards Stanley.'[24] On the night of 10 June the Battalion supported 42 Commando's attack on Mount Harriet, securing their start line, destroying an enemy machine-gun post in the mouth of a cave with Milan missile fire and acting as the brigade reserve. Two nights later, on 12 June, the Battalion was in reserve as the Scots Guards attacked Mount Tumbledown from the west, while the Gurkha Rifles engaged Mount William from the north. If these assaults were successful, it would be the task of the Welsh Guards to exploit forward to Sapper Hill to the east, the last obstacle before reaching Port Stanley. This would be the final phase of the brigade attack; because of the danger posed by indiscriminate Argentine mining, the men would be lifted by helicopter to the start line. No less dangerous were the stone-flows, which were described as 'slippery, jagged rocks piled haphazardly in our path … too long to by-pass, too dangerous to risk quick movement and broken ankles, and broad enough to delay progress significantly'.[25]

In spite of the dangers and difficulties, this should have been the Battalion's compensation for all the hardships they had endured, but no sooner were the first units in position for the attack than a message came over the net announcing an imminent ceasefire. The war was over: 1st Battalion Welsh Guards had been in the Falklands for less than two weeks, although for most of the men it must have felt like a lifetime. Once settled in Port Stanley, the men began the massive task of clearing up; according to the Commanding Officer, it was an unglamorous, 'difficult, dangerous and time-consuming job with the ever present risk of booby traps and unexploded ordnance'. It was not just enemy ordnance that caused problems: on 13 July, while clearing the runway at Port Stanley, a party of men, including six Welsh Guardsmen, were seriously injured when two Sidewinder missiles were accidentally discharged from a Harrier aircraft.

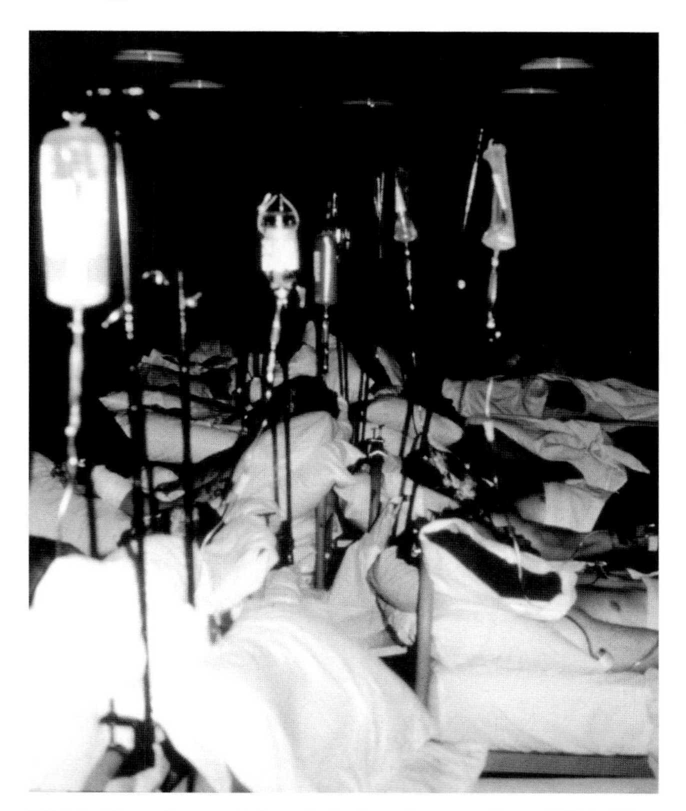

Welsh Guardsmen injured during the attack on RFA *Sir Galahad* recover on board the SS *Canberra. (Imperial War Museum)*

Garrison Sergeant Major W. D. G. Mott, OBE, MVO

Billy Mott was the first Welsh Guardsman to be appointed Garrison Sergeant Major of London District. This ceremonial post is the most senior non-commissioned officer appointment in the Guards Division. A native of Ellesmere Port, Billy joined the Army in 1979. As a lance sergeant, he served as an instructor at the Guards Depot and at the Royal Military Academy Sandhurst, he was a colour sergeant, company sergeant major and regimental sergeant major. Prior to being appointed Garrison Sergeant Major, Headquarters London District, he was the GSM at Headquarters Northern Ireland.

Billy Mott is one of three brothers who have served with distinction in the Welsh Guards. John Mott finished as the RSM of the Royal Military School of Music, Kneller Hall, while Nick Mott served as a major in the role of Quartermaster of the 1st Battalion and was promoted lieutenant colonel in 2014.

GSM Mott has taken part in more Trooping the Colour parades than any other serving Welsh Guardsman. He was involved in the ceremonial arrangements for Queen Elizabeth, The Queen Mother's funeral and the Golden Jubilee State Procession. He was also a pivotal figure in the repatriation ceremonies for those killed in action during operations in Iraq and Afghanistan. For the wedding of Prince William to Catherine Middleton in April 2011, GSM Mott was presented with a new rank insignia, reviving the original one made for sergeant majors appointed to the court of King William IV in the early nineteenth century. This incorporates the royal coat of arms worn by selected warrant officers class 1 of the Household Division, placed over four chevrons sewn in gold thread, the traditional badge of a sergeant major.

The Mott Brothers: (*L to R*) John, Billy and Nick. *(Welsh Guards Archives)*

The Argentine prisoners-of-war also had to be dealt with and The Prince of Wales's Company provided the guard for thousands of captured or surrendered Argentine soldiers who were repatriated to Puerto Madryn on board SS *Canberra*. This could have been a disconcerting task, guarding the representatives of a power which had wreaked so much havoc on their friends, but 'the Guardsmen carried out their duties with faultless humanity, sense and good humour'.[26] A second consignment was handled by No. 2 Company (Major C. F. Drewry) which provided the guard for 593 prisoners-of-war, including General Menendez, on board the MV *St Edmund*, a Sealink car ferry. All were high-category personnel who had been kept behind until mid-July against an outbreak of further hostilities.[27] They included Argentine Special Forces personnel who were found to be constantly plotting to attack the guard force and hijack the ship.

On the return to Port Stanley, the MV *St Edmund* embarked the rest of the Battalion for the nine-day voyage back to Ascension Island. Before leaving the islands, No. 3 Company constructed a simple but effective memorial in the shape of a Welsh Guards leek; later, in November 1983, it was joined by a ten-foot-high Celtic cross which stands above the eastern promontory at Port Pleasant and looks across the waters to where the Welsh Guards had to face the horror and confusion of war. Its inscription reads, '*Yn angof ni chant fod*' ('We will remember them'). At the same time *Sir Galahad* was declared a war grave and towed out to be sunk in the waters of San Carlos Deep.

On 29 July the Battalion arrived at RAF Brize Norton on board four VC10 transport aircraft and immediately went on six weeks' leave. On returning to Pirbright on 8 September, there was a period of intensive training for public duties before the Battalion embarked on a tour of Wales, ending with the laying up of the Colours in Llandaff Cathedral on 16 September. A detachment was also found for the City of London Falklands Parade on 12 October. Just under a year later, on 12 July 1983, a Service of Memorial and Thanksgiving was held at Llandaff Cathedral in the presence of the Colonel, HRH The Prince of Wales. It was conducted by the Dean of Llandaff, with the Address being given by the Reverend D. M. T. Walters, a former chaplain to the 1st Battalion Welsh Guards.

Welcome though these commemorations were, there was an understandable desire within the Battalion to put the past behind them. It was easier said than done. The *Sir Galahad* incident was the biggest single loss of

The Regimental Memorial, Bluff Cove, Falkland Islands. with detail on right *(Imperial War Museum)*

life during the conflict and it was inevitable that there would be a certain amount of introspection. One of the survivors was Guardsman (later Lt Col) Nicky Mott 84 of The Prince of Wales's Company, who was on the vessel with his brother Guardsman Bill Mott 88. He is on record as saying that 'the whispers went on for years, it certainly affected the Regiment'.[28] Later, when the war was being analysed, the biggest single complaint within the rest of the Army was that the Guardsmen were not fit for fighting over such rough terrain, but this was a lazy generalization. While there may have been doubts about their ability to adapt to amphibious operations, the two Foot Guards battalions were fully capable of coping with the physical conditions on land – as 2nd Battalion Scots Guards showed during their successful engagement with the Argentine forces on Mount Tumbledown. And it should not be forgotten that earlier in the year 1st Battalion Welsh Guards had undergone an intensive period of tactical, fitness and shooting training in order to be ready for its Spearhead duties which surpassed standards reached by other battalions at the time.

As for the bombing of *Sir Galahad,* it too became a subject for discussion and debate which took some years to die down. A Naval Board of Inquiry was convened on 12 July 1982 and, while its report highlighted shortcomings such as the failure of the Rapier air defence system at Fitzroy and the contribution made by the collapse of the communications system, which meant that 5 Brigade Advance Headquarters knew nothing of the whereabouts of the Welsh Guards, the main comment under the heading 'Landing Beach and Assets at Fitzroy' was sympathetic to the decisions made by the two company commanders: 'It is considered that no one individual was to blame for these events; each one, small in its way, added up to a very difficult situation.'[29] On one level that is a fair assessment. In common with most setbacks, whether or not in war, the bombing of *Sir Galahad* was due to a steady accumulation of events which were either not adequately addressed or were wrongly assessed. Plain bad luck also played a role, as did poor communications and an absence of reliable information. For example, the report underlined the fact that no one knew if the bridge at Fitzroy was passable to infantrymen: given that lack of information, the decision not to proceed on foot was 'reasonable'. The other deciding factor was the lack of helicopters which meant that 'there was a clear necessity to move the WG by sea' at a time when the advance on Port Stanley was a military priority.

The Naval Board of Inquiry was an attempt to piece together the events at Fitzroy that day but it did not tackle important issues which affected the transport of the Welsh Guards. These included the failure of the two LCUs to rendezvous with *Fearless* off Direction Island on the night of 6 June and the issuing of the Sail Order to the Master of *Sir Galahad* the following day, even though it was known that the ship would arrive at its destination in broad daylight and in full view of the enemy. The absence of air defences at Fitzroy has also never been addressed satisfactorily and some of the problems surrounding communication still require robust analysis. In the overall direction of the war, 5 Brigade's southward advance into the Port Pleasant area was a bold move to complete the investment of Port Stanley, but it was always in danger of being compromised by the Argentine air threat. In warfare, choices have to be made and chances taken. Ultimately that is the price of command, but in this case the bill was picked up at Fitzroy, not by those who gave orders or failed to give them, but by the Guardsmen at the sharp end, on board the *Sir Galahad.*

CHAPTER TWELVE

An Ever-Changing World, 1982–1996

UK, Germany, Belize, Northern Ireland

On the Battalion's return to London District, it had a new Commanding Officer in Lieutenant Colonel R. F. Powell. In January 1984 1st Battalion Welsh Guards became the first Foot Guards battalion to serve in BAOR in a station other than Münster when it moved into Campbell Barracks in Hohne, a sizeable garrison town to the south-west of the Lüneberger Heide, roughly equidistant between Hamburg and Hanover. Not only was this station situated in a military landscape dominated by BAOR's presence in northern Germany but it is also surrounded by flat monotonous heath and rolling moorland. Only after arriving did the Battalion discover that Hohne is also near to Bergen-Belsen, the site of the infamous Nazi concentration camp. As the Commanding Officer pointed out, it was a punctuation mark of sorts with the beginning of the year being spent 'Janus style, with the battalion looking back with affection on four years in Pirbright and forward with anticipation to four years in Hohne'. The new role was as a mechanized infantry battalion in 22 Armoured Brigade and the Battalion soon found itself on the BAOR 'roller-coaster', the Commanding Officer noting in his report for the first half of 1984 that the training cycle offered very little time to draw breath: 'The rest of April offered no respite but only diversification. The Prince of Wales's Company went straight from Soltau onto Site Guard to be relieved a week later by No. 2 Company. Officers and NCOs of 3 Company meanwhile completed two weeks' training in the new voice procedure with 5th Royal Inniskilling Dragoon Guards Battle Group, in preparation for their visit to Canada on Exercise Medicine Man 4.'[1]

A change from the training cycle came at the end

Gdsm Vancorler and Gdsm Thomas 74 of The Prince of Wales's Company on patrol, Ballykelly, 1994. *(Welsh Guards Archives)*

Signpost at Airport Camp, Belize, 1989. *(Welsh Guards Archives)*

of the summer when Guardsmen appeared on the square in tunic. These were The Prince of Wales's Company and No. 2 Company which formed a marching detachment from the Welsh Guards for the fortieth anniversary celebrations of the liberation of Belgium. The highlight was the mounting of two parades in the Grand Place in Brussels on 3 September. In the morning the marching detachment led 130 members of the Association, all veterans from the liberation, and in the evening at beating retreat the order was reversed. On both occasions the detachment was under the command of Lieutenant Hilarian Roberts whose father Brigadier J. M. H. Roberts was also present and had served in the Welsh Guards in 1944. In the week that followed there were two further beating retreats, wreath-laying and visits to the battlefields at Wavre and Hechtel, with 'Old Comrades taking groups of Guardsmen round the village explaining the battle as they saw it'. The commemoration concluded with a lunch given by the Bürgermeister of Hechtel.

There were also moments of unexpected catharsis. As happens so often when fighting takes place in a built-up area, there will be civilian casualties and Hechtel had been no different: thirty-five civilians were killed. Of these, eleven had been executed by a German firing squad on suspicion of being collaborators and 121 houses had been destroyed by artillery or mortar fire. But one memory transformed that tragedy. At the height of the fighting in the town, a Welsh Guardsman had been ordered to keep watch on a row of windows and to open fire if he saw any suspicious movement. Believing German soldiers to be inside one house, when he saw a curtain moving he shot at the window; then a woman ran into the street shouting that a young boy had been shot in the chest. The shocked Guardsman ran into the building to tend to the boy who was rushed to hospital in Brussels for emergency treatment. Forty years later, during the reunion ceremonies, the story was told again and 54-year-old Peter Jensen stepped forward to come face-to-face with his erstwhile assailant, Guardsman Don Walker, and 'a friendship was started between these two men who had met in such strange circumstances'. That was not all. During the lunch the names of the casualties, British and German, were read out and,

The Commanding Officer inspects the Corps of Drums, Elizabeth Barracks, Pirbright, 1988. (*L to R*) Capt R. H. Bodington, CSgt (Master Tailor) Parry, Lt Col D. P. Belcher, WO1 Evans 84, RSM. *(Welsh Guards Archives)*

while this was happening, Meirion Ellis, who had served in the 2nd Battalion, noticed that his friend Llewellyn 15 was strangely moved. He asked for the reason for his tears and back came the reply: 'It was because I was thinking of the young German lad I shot in the face with my Sten gun. I can see him now, and I wondered if he was amongst those whose names were read.'[2]

During the winter the Battalion participated in Exercise Stag Rat, with the US Army providing the enemy in a battle across Lower Saxony, and that set the tone for the year that lay ahead with training in Bavaria, Münsterlager and at Putlos near Oldenburg in Schleswig-Holstein, where one of the highlights was a visit to a nudist beach: Guardsmen were promptly told that they would have to leave unless they too removed their clothing. The only comment from one of the Guardsmen was 'That's nothing compared to Miners' Week in Barry!' In preparation for a deployment to BATUS in Canada, a Welsh Guards battle group trained at Soltau with 1st Royal Tank Regiment. It was a busy period for the whole Battalion – 'hectic but also varied and interesting', according to the Commanding Officer – with the men being pushed to their limits. Whether skiing in Bavaria, taking part in field firing at Sennelager, adventure training almost anywhere in West Germany or participating in larger-scale exercises with other NATO troops, the Battalion had rarely had its hands so full.

In the middle of the tour, between March and July 1986, the Battalion returned to Northern Ireland as the Belfast roulement battalion. Two years earlier the PIRA had mounted their most audacious attack on the British mainland by bombing the Grand Hotel in Brighton during the Conservative Party conference,

Lt Col D. P. Belcher, CO 1st Battalion Welsh Guards, inspects WO2 (DSgt) Downes and WO2 (DSgt) Roberts 15, accompanied by WO1 (RSM) Evans 84, Elizabeth Barracks, Pirbright 1988. *(Welsh Guards Archives)*

Maj Gen Sir I. Mackay-Dick, General Officer Commanding London District and Major General Commanding the Household Division, inspects the Corps of Drums, Wellington Barracks, London, 1996. Drum Major Scattergood is to the left with Drummer Hughes 79 immediately to the front. *(Welsh Guards Archives)*

No. 3 Company, anti-tank (Milan) detachment in action at British Army Training Unit Suffield, Canada, 1984. *(Welsh Guards Archives)*

LSgt N. Mott gives directions to LCpl Davies 04, Jungle School North, Belize, 1989. *(Welsh Guards Archives)*

killing two people and wounding many others but failing to harm their intended target, Prime Minister Thatcher. This gave added impetus to the need to find a political solution and within a year the Anglo-Irish Agreement was signed in London and Dublin. Predictably, perhaps, it was rejected by the unionist parties in Northern Ireland. By the time the Battalion arrived in Belfast the strategic situation was that the PIRA had settled down to fight a 'long war' in the knowledge that they could not achieve an immediate military victory. At the same time the British Army conceded that there was a limit to what could be done purely by military means and that the only realizable aim was containment of the violence by retaining ground and holding the ring.

Against this background there had been a change in tactics. Dividing platoons into smaller units known as 'bricks' and 'multiples' helped them to be more effective on the ground, while a closer working relationship with the RUC provided essential local knowledge as well as building stronger relationships with local communities. Avoiding the same routes and common patterns of behaviour proved to be a hard nut for the terrorist to crack'.[3] Other improvements from the previous tours included provision of better accommodation and the Battalion found itself housed in four bases: Battalion Headquarters in Springfield Road; The Prince of Wales's Company in Fort Whiterock, close to 'the two hardest Republican areas of all: the Ballymurphy Estate and the Turf Lodge';[4] No. 2 Company was in North Howard Street Mill; and No. 3 Company in Musgrave Park Hospital where it was co-located with the A Echelon (Battalion resupply group). Support Company, was split into two platoons which went to The Prince of Wales's Company and No. 2 Company was strengthened by a platoon from 50 Missile Regiment, Royal Artillery, and a platoon from 1st Battalion King's Own Scottish Borderers. Facilities for recreation had also improved which made life more tolerable for the Guardsmen at a time when it was important to get the balance right between maintaining vigilance and the need to keep up a good public face.

HRH The Prince of Wales reviews The Prince of Wales's Company in the Inner Quadrangle, Windsor Castle, 1990. (*L to R*) Lt M. Kitto, Capt M. Rudd (temporary equerry to HRH), Maj R. L. Traherne, Brig J. F. Rickett, CBE (Regimental Lieutenant Colonel), HRH The Prince of Wales. *(Welsh Guards Archives)*

Even though patrolling yielded results with the arrest of suspected terrorists, tensions remained high throughout the tour and there were times when the Battalion's patience was stretched to the limits. In one incident in Ballymacarrett, when a rioter managed to make good his escape, a local resident started shouting insults at the Welsh Guards patrol, insisting that he was a veteran of Arnhem and had lost a leg fighting for the Crown. The alleged response from the patrol sergeant was that he 'ought to 'ave more f—— sense, then'. A few minutes later, after Dr Ian Paisley, leader of the Democratic Unionist Party, had unexpectedly arrived on the scene, this was greatly embellished by the complainant, who alleged that he had been told by the same sergeant that he might lose the second leg in Belfast. The next day the incident was reported in Stormont by Dr Paisley, a noted opponent of the Anglo-Irish Agreement, who criticized the behaviour of the patrol and the attitude of Welsh Guardsmen. Fortunately for posterity and the good name of the Regiment, the journalist Kevin Myers had also witnessed the incident and had noted the sudden improvisation which had made it seem that 'an Ulster war hero [was] being abused by the IRA-loving Welsh Guards'.[5] Small wonder that the Commanding Officer had insisted that public relations should play 'a vital part in the success of the tour' – winning hearts and minds is always a vital ingredient in any counter-insurgency war. It was a delicate balance and maintaining it could cause tensions amongst members of the security services; one unnamed Welsh Guardsman interviewed by the historian Max Arthur understood the paradox that one man's terrorist could be another's freedom fighter:

> At times, I can see the IRA's point of view. If I'd been born in the Falls Road I know I'd be in the IRA. They want a free Ireland. They see the British Army as oppressors, and back home if I was a Welsh Nationalist and Wales wasn't part of Britain and English troops came in, I'd throw bricks, stones and petrol bombs. Most times out there I was dedicated to my job, stopping these terrorists, and that's why I enjoyed it so much. But other days I could see those IRA guys are really dedicated to their cause, prepared to die for it, and I respect them for that.[6]

In August, following leave, the Battalion returned to Hohne and a further cycle of training began with field firing at Sennelager. Training cadres continued over the winter, interspersed with inter-company competitions, including the 300 Williams Rugby Cup which was won by No. 3 Company, and the Leuchars Cup Patrolling Competition, which was held on the Soltau ranges.

Ronald Reagan accompanied by Major T. C. S. Bonas inspecting a Guard of Honour found by The Prince of Wales's Company during the President's State Visit to London, June 1988. *(Welsh Guards Archives)*

300 Williams Cup

The 300 Cup has been the premier sporting competition of the 1st Battalion since the cup was presented by WO2 John Williams 300 in the 1970s. 300 had foresight and imagination and, as a rugby fanatic, he had a vision whereby each company would field a rugby team and compete annually for his trophy. His vision endures and has always, and will continue to be, an opportunity for rugby players of all abilities to display their skills with the hope of being selected for the Battalion rugby squad.

The competition takes place over the course of one day, usually at the start of the season, with the five companies competing in a league system, culminating in a final. The 300 Cup is not for the rugby purist. Passion and company pride can often drive players to compete fanatically and often the desire to win comes at the cost of displays of rugby genius.

300 has often travelled to wherever the 1st Battalion may be serving to watch the 300 Cup and present his trophy along with awards for veteran, young player and player of the tournament. In 2009 the 300 Cup was played shortly after the Battalion had returned from Afghanistan and 300 was present for the day. No. 3 Company secured the coveted 300 Cup with young player of the tournament presented to Guardsman Craze, who went on to secure a place in the first XV. The player of the tournament was awarded to Guardsman Divavesi, who also secured a place in the first XV. The biggest cheer for the day, however, was saved for Colour Sergeant Andrew Morgan Brown 16, a veteran of forty-four years of age. 16 had left the Battalion at the end of twenty-two years' service in 2006 but, along with several others, had returned as a reservist for Operation Herrick 10. 16's final act was a fantastic performance in the 300 Cup before he returned to civilian life in Llanelli.

The 300 Cup continues to feature prominently in the Battalion programme of events thanks to the foresight of WO2 John Williams 300.

Leuchars Cup

The Leuchars Cup is an inter-company patrolling competition set up by Major General Peter Leuchars when he was the Commanding Officer in the early sixties as a means of preparing for its operational tour of Aden. The first competition was held at Sennybridge during pre-tour training and was won by Vyvyan Harmsworth. Although not an annual event because of varying Battalion commitments, the Leuchars Cup is always fiercely contested and is an excellent vehicle for testing all soldiering skills, teamwork, endurance and perhaps, above all, company and platoon spirit. Major General Leuchars often presented the Cup to the winning platoon and victory is always regarded as a prestigious achievement. The winners receive pewter mugs.

Major General P. R. Leuchars, CBE, presents the Leuchars Cup to No. 2 Platoon, The Prince of Wales's Company, 1990s. *(Welsh Guards Archives)*

Bearer party for the coffin of the Duchess of Windsor, commanded by Capt P. Owen-Edmunds, leaves the West Door of St George's Chapel, Windsor, watched by HM The Queen and HRH The Duke of Edinburgh, Windsor, April 1986. *(Welsh Guards Archives)*

In November 1988 the tour came to an end and the Battalion moved to Elizabeth Barracks at Pirbright and a return to public duties with a new Commanding Officer, Lieutenant Colonel Paul Belcher (Regimental Sergeant Major M. Hughes was also a new appointment). In the following year the Battalion deployed to Belize for a six-month tour of duty as the resident infantry element in the garrison of what had formerly been known as British Honduras. Although Belize had become independent in 1981, a British military presence remained in the country largely to deter the threat posed by neighbouring Guatemala, which had made several invasion threats between 1948 and 1975. At the time of the Battalion's deployment in April 1989 the garrison numbered 1,600 personnel, with its headquarters at Airport Camp near Belize City and three other camps to accommodate company groups – Holdfast in the west and Rideau and Salamanca in the east.

Until 2010 the country also provided the facilities for the British Army's jungle warfare training. Throughout the deployment in Belize patrols were sent regularly into the jungle, a standing requirement for the British garrison in the country. The jungle, the patrol's habitat for five days, proved to be a dirty place, overgrown, desperately hot and humid. Although most of the patrolling was conducted along recognized tracks, the need to keep on a bearing meant that the Guardsmen had to use machetes to hack their way through the deep and unyielding undergrowth. At night, if they were lucky, they reached a village and might be invited to sleep there; if not, it was a case of setting up camp and settling into the pitch-black night that always falls about 18.00. One Guardsman's account said it all:

A patrol deploying by Lynx helicopter, Ballykelly, 1993. *(Welsh Guards Archives)*

> The easiest way to describe it is like it's an overgrown allotment. You fight your way through the undergrowth and the thing is you have to walk through it even if there's a fallen tree in the way because you've got to keep on a bearing. It's pretty hard going. You can be walking about forty minutes to cover a kilometre. You always camp up before last light. What you do is reach your objective at least by half past four and radio in. Then you set up your hammock, your mossie [mosquito] net and your poncho because you never know when it's going to rain. One of the things I don't like about the jungle is that it could be sunny one minute and the next it could be tipping it down.[7]

FV432 Armoured Personnel Carrier, Hohne, 1986. *(Welsh Guards Archives)*

If the terrain had been all flat, as it looked from the air, progress during the day might have been simpler, but the Belizean jungle has layered itself over hills and into river valleys, up creeks and through ravines. The first thing that struck the Guardsmen on their patrols was its scale. The attap trees, often rising forty feet up into the sky, formed an airy canopy, while down below in the damp earth their roots snaked out into the vegetation, thus providing traps for the unwary. Added to those difficulties were the dangers posed by the denizens who regarded the jungle as their home and resented man's intrusion – the scorpions, tarantulas and lethal reptiles such as the coral snake and the small but deadly fer de lance. One sergeant was bitten by a pit viper on his first day in the jungle – luckily only one of its fangs struck but his mates were shaken by the incident. 'Basically, everything wants to bite or sting you,' said one. 'That's Belize all over.'

Not every Guardsman was called upon to do jungle patrols – no more than eighty in any company, according to one company commander – while many carried out other duties such as manning observation posts along the Guatemalan border. At all times the patrols and the OP parties were accompanied by one or more soldiers of the Belize Defence Force, the country's standing armed service. The eventual aim was that they would take over responsibility for their country's defence and internal security. This did happen. In 1994, following Guatemala's recognition of Belizean independence, the main British force withdrew from Belize, leaving a training and support unit which remains in place to this day to assist British Army training in Belize, including that undertaken by the Welsh Guards, who have returned a number of times since and did so again in 2014.

HM The Queen, Colonel-in-Chief, receives a bouquet of flowers, St David's Day, 1990, Elizabeth Barracks, Pirbright. Brigadier J. F. Rickett, OBE, Regimental Lieutenant Colonel, is on the left of the photograph. *(Welsh Guards Archives)*

The tour of Belize came to an end in October 1989 and the Battalion returned to Pirbright. It could hardly have been anticipated at the time, but ahead lay a sea-change in international politics which would have an impact on the commitments and purpose of the British Army.

Even as the Battalion was preparing to leave Belize and return home, momentous events were taking place in Eastern Europe which would reshape the certainties of the previous forty-five years by bringing the Cold War to an end. In the middle of August 1989 the Hungarian government started the process by effectively opening its border with Austria and allowing East German citizens to use their country in transit to West Germany. This triggered a similar move in Czechoslovakia and was followed by mass demonstrations in East Germany itself. On 18 October, President Erich Honecker of East Germany resigned and his regime began to implode. Three weeks later, on 9 November, the border in Berlin was breached as people on both sides of the Berlin Wall began to pull it down and stream across to the other side. It was a historic occasion: the fall of the Berlin Wall, the most significant icon of the Cold War, turned out to be the first step towards German reunification, which was formally concluded on 3 October 1990. In the aftermath, other Eastern European countries began to slough off dependence on the Soviet Union and to reject the Communist political philosophy which they had shared during the Cold War as members of the Warsaw Pact.

That was not all that happened in 1990. While these momentous events were taking place in Europe, an equally significant incident had shaken the Middle East and would soon have wider implications for the rest of the world. On the night of 2 August, Iraqi forces invaded neighbouring Kuwait in pursuit of a territorial claim engineered by President Saddam Hussein in an attempt to improve Iraq's access to the waters of the Gulf. Within twenty-four hours, most resistance had ended and the Kuwaiti royal family had fled to Saudi Arabia, leaving their country under Saddam's control as the nineteenth province of Iraq. The scene was then set for a lengthy diplomatic stand-off and confrontation between Iraq and the West: on 29 November 1990, the UN Security Council passed Resolution 678 which gave Iraq until 15 January 1991 to withdraw from Kuwait and empowered member states to use 'all necessary means' to force Iraq out of Kuwait after the deadline had passed. In pursuit of that aim, the US launched Operation Desert Shield on 8 August 1990 and started deploying forces in the region. The UK and France followed suit on 14 September.

Other than supplying personnel – twenty-three served in various capacities, including with the Army Air Corps – the Battalion was not directly involved in the ground operations to oust Iraqi forces from Kuwait in January and February 1991. For a short period it was thought that there might be a role for the Battalion as it had been deployed to Kenya and was considered to be 'acclimatized' for service in the Middle East. As the conflict unfurled, and because the Guardsmen were considered to be ready for deployment, the Battalion was effectively on standby to move straight to the Gulf from Kenya, if required, as battlefield casualty replacements (BCRs), following on from the Scots Guards (who did deploy as BCRs). In the event it became evident that they would not be needed.

However, the conflict and the collapse of

Assault river crossing, Belize, 1989. *(Welsh Guards Archives)*

Brigadier J. F. Rickett, Regimental Lieutenant Colonel, visiting No. 2 Company, Northern Ireland, 1993. (*L to R*) Major W. J. Syms, Brig Rickett, Lt Col T. C. R. B. Purdon. *(Welsh Guards Archives)*

Communism in eastern Europe did have a bearing on the way in which the Battalion would operate in the future, a point that was underlined by the Regimental Lieutenant Colonel in his foreword to the *Welsh Guards Magazine*: 'It has been a most significant year, starting with universal concern before the overwhelming success of the Gulf war, followed by immensely important developments in Germany and the Soviet Union, and capped by the Government Defence Review, disarmingly titled "Options for Change".'[8] It was perhaps noteworthy that the changes were taking place at the very time that the Welsh Guards celebrated the Regiment's seventy-fifth anniversary. New Colours were presented at a ceremony at Buckingham Palace on 30 May 1990 and the 1st Battalion Trooped the Colour at the Queen's Birthday Parade on 16 June.

The *Options for Change* review was introduced in 1992, its aim being to cut the size of the armed forces in response to the changed global situation. For the Army this meant a reduction in personnel to 104,000 and cutbacks to the infantry by way of a number of amalgamations and disbandments – in the Foot Guards, the Grenadier Guards, Coldstream Guards and Scots Guards all lost their 2nd Battalions, which were placed in suspended animation in 1993. While the Welsh Guards escaped most of the turbulence, they were affected in other ways, notably in the ceremonial commitment. For example, as a result of the review, the Queen's Birthday Parade was cut back to six Guards, one sentry post was removed at each of the three royal palaces, the Tower Guard was reduced and there were adjustments in the spacing for streetliners

HM The Queen presents leeks, St David's Day, 1990, the 75th Anniversary of the Raising of the Regiment, Elizabeth Barracks, Pirbright. (*L to R*) Brigadier J. F. Rickett, CBE, HM the Queen, Major (QM) A. Davies, MBE. *(Welsh Guards Archives)*

WO2 (CSM) Covell having just dug-in his FV432 APC, Hohne, 1986. *(Welsh Guards Archives)*

Members of the Anti-tank Platoon with a Milan missile launcher, Exercise Trumpet Dance, Fort Lewis, USA, 1998. *(Welsh Guards Archives)*

Guardsmen of No. 3 Company experiencing a taste of armoured warfare during Exercise Medicine Man 5 in Canada, 1999. *(Welsh Guards Archives)*

HM The Queen inspects the Battalion prior to the Presentation of New Colours, Buckingham Palace, 1990. *(Welsh Guards Archives)*

on all state occasions. At the same time, the Guards Depot at Pirbright was closed to make way for an Army Training Regiment, with a Guards Company to train recruits for the seven regiments of the Household Division. After passing out, the recruits moved to an Infantry Training Battalion at Catterick, which included a Guards Training Company responsible for Phase 2 training. At Regimental Headquarters there were reductions and changes in staffing levels, the most significant being the post of Regimental Lieutenant Colonel, which ceased to be full-time, although it would be held by a serving officer, with the day-to-day management instead being left in the hands of the Regimental Adjutant from 1989. The last officer to hold this post full-time was Lieutenant Colonel Charles Dawnay. Many found these changes disconcerting, but they were part of a new drive to make economies and to stretch the available budget. As Brigadier J. F. Rickett, the next Regimental Lieutenant Colonel, put it, 'everything costs money and it is a tightly controlled resource'. In the Battalion there was a new commanding officer in Lieutenant Colonel T. C. R. B. Purdon, who transferred from the Irish Guards to succeed Lieutenant Colonel C. R. Watt, Commanding Officer since 1990. Secondly, I. P. Dyas took over as Regimental Sergeant Major from N. Harvey, who was commissioned before moving to Cardiff to run the Liaison Team at Maindy Barracks.

Throughout 1991 the Battalion had been kept busy, beginning the year training in Kenya on Exercise Grand Prix 2, which combined live firing at Mpala Farm and adventurous training in such exotic locations as Mount Kenya and the Masai Mara. At the end of the summer, having been reviewed by the Prince of Wales at Windsor on 22 May, The Prince of Wales's Company (Major R. L. Traherne) left for the South Atlantic to train in the Falklands, the first visit since the war nine years earlier. In October, home service dress was put away in preparation for a two-and-a-half year deployment to Shackleton Barracks, Ballykelly, in Northern Ireland. Originally a wartime RAF base, this had remained an RAF station until 1971, but by the time of the Battalion's deployment it was one of the fourteen British Army barracks in the Province. The Battalion's role was Brigade and Province Reserve, supplying companies or platoons to

Humber Pig APCs from No. 3 Company in West Belfast, 1986. *(Welsh Guards Archives)*

all of Northern Ireland's six counties. As a result, No. 2 Company (Major P. H. M. Dymoke, CSM A. Bennett) was deployed as a third operational company to Belfast, while No. 3 Company (Major Hugh Bodington, CSM F. K. Oultram) moved location nine times in one month as various brigades and battalions vied for its services. In practice this meant that the Battalion had four rifle companies with clearly defined roles: two were designated as operational, one would be on leave and the fourth would be engaged in training. (With no requirement for heavy weapons, Support Company – Major Martin Syms, CSM M. Frost – became in effect a fourth rifle company.) A routine was established involving deployments to Rosslea in County Fermanagh on the border with the Irish Republic and to Kinawley, also in County Fermanagh, where Support Company narrowly missed catching the terrorists who had just blown up a television mast.

It was a strange period in the Troubles. Behind the scenes, attempts were being made to find a political solution, but low-level violence continued to be perpetrated by both nationalist and loyalist terrorist groups as they struggled to maintain their positions and gain some advantage. In March 1993, the second year of the deployment, the PIRA exploded two bombs in Warrington, Cheshire, killing two children and wounding fifty-six civilians, while a few days later the Ulster Defence Association, using the name Ulster Freedom Fighters, shot dead four Catholic civilians and a PIRA volunteer at a building site in Castlerock, County Londonderry.

The Battalion began its first accompanied tour of the Province at a time when peace was in the offing, but there was still a long way to go before any ceasefire was called and vigilance was still required. By the end of the year the Commanding Officer was able to report that 'the initiation of the Guardsmen into hardened grizzled Northern Ireland veterans is almost complete'. Much of the work was unglamorous and frequently boring but it was all necessary for operational effectiveness at a time when it was essential to maintain an air of normality. While based in Rosslea, No. 3 Company described their deployment as a continuous loop – as in the movie *Groundhog Day* in which a day is repeated again and again. The only let-up arrived when Lance Sergeant Williams 25 observed figures advancing towards him while using a thermal imaging device and alerted the post in readiness for a possible PIRA attack. When nothing happened he had to admit that the figures were in fact rabbits. By way of contrast, The Prince of

The Battalion Warrant Officers, Elizabeth Barracks, Pirbright, 1991. (*L to R*) WO2 (CSM) Owen 52, WO2 (Drill Sgt) Roberts 15, WO1 (RSM) Harvey, WO2 (CSM) Evans 13, WO2 (CSM) Harford. *(Welsh Guards Archives)*

Lt R. H. Talbot Rice instructing on the lance corporals' promotion course, Hohne, 1986. *(Welsh Guards Archives)*

WO2 (DSgt) Roberts 15 instructs a young officers' course, Elizabeth Barracks, Pirbright, 1990. (*Front row L to R*) Capt M. W. B. G. Dyer, Lt E. N. Tate, Capt R. J. A. Stanford, Lt G. Bartle-Jones. *(Andrew Chittock)*

Wales's Company reported that 'An IRA gunman decided to attack a patrol from No. 2 Company but was forced to disappear rather swiftly into the darkness and across the border hotly pursued by a cascade of 5.56 mm tracer.'

Sport and adventure training played their usual role in keeping the Battalion busy and, although the Commanding Officer admitted that the former had had to take a back seat, the rugby team, trained by Major Ray Evans, the Quartermaster, won the Northern Ireland Rugby Cup. Adventure training was not neglected either. In addition to groups of Guardsmen departing to Cyprus for parachuting, to France for canoeing and to Scotland for climbing, training and selection took place for Captain J. H. B. Warburton-Lee's ambitious Roof of America Expedition. This followed on from the equally successful Roof of Africa Expedition which had taken place in 1990 as part of the regiment's seventy-fifth anniversary celebrations and whose objective had been to climb seven African mountains ranging from 10,000 to over 19,000 feet high – from Jebel Toubkal in Morocco to Mount Mulanje in Malawi. This time the scope would be even larger, covering similar peaks in both North and South America and including 'dog-sledding and snowmobiling in Arctic Alaska, mountaineering in Alaska and Argentina, and kayaking, rafting, trekking and sailing in Chilean Patagonia en route for Cape Horn'.[9] The main difference was that candidates were also selected from other regiments in the Household Division but, as the Commanding Officer wryly remarked, 'Numerous members of the Battalion attended these [selection] camps; some have recovered.'

Building close ties with the local community was also high on the agenda in Northern Ireland. The Regimental Band visited in December 1992 and marched through Limavady and Coleraine town centres and the finale of their tour was a concert with the Regimental Choir in one of the former aircraft hangars. This proved to be a great success not least because it was open to the local population, many of whom had never set foot inside Shackleton Barracks. Other successful initiatives saw the dog section working with local young people from the St John Ambulance to help them overcome their fear of dogs, and the cookhouse managed a number of local events ranging from providing cakes at local fetes to catering for 1,500 hungry children. The Master Chef, WO2 Mason, estimated that he and his staff had supplied 26 tons of chips, 224,000 eggs and 52,000 pints of milk. There was also hospitality for a group of local Catholic priests, many from Republican neighbourhoods, who toured the Barracks, met families and were entertained to lunch in the officers' mess. Similar functions were held for local farmers and local councillors as the Battalion played its part in being part of the community – an essential 'hearts and minds' ingredient in winning this kind of protracted insurgency.[10] Inevitably external factors impinged on the tour as the effects of *Options for Change* began to be felt, with a number of officers and NCOs taking redundancy and clerks transferring to the newly formed Adjutant General's Corps.[11]

LSgt Nicholls and LSgt Roberts 16 of The Prince of Wales's Company, United Nations Verification Mission, Angola, 1995. *(MoD)*

Gdsm R. Price 64 of The Prince of Wales's Company, United Nations Verification Mission, Angola, 1995. *(MoD)*

In March 1994, Battalion Headquarters was given the task of commanding Operation Rectify, whose aim was to ensure that the security base at Crossmaglen was rebuilt in safety. By then the PIRA had developed a new kind of mortar which was capable of damaging

HRH The Prince of Wales talking to LCpl Perry and LCpl Jones 95 of the Corps of Drums and their partners, St David's Day, 1995, Tern Hill, Shropshire. *(Welsh Guards Archives)*

existing defences; during the reconstruction work, roads into the area had to be secured against the possibility of attack, the Battalion acting, as one officer put it, as 'hired hands'. On three occasions Battalion Headquarters took up to seven rifle companies under command and the Battle Group deployed to South Armagh, itself under command of 3 Infantry Brigade (Brigadier C. R. Watt, Welsh Guards). Logistically, the operation was a sizeable undertaking and the Quartermaster and Technical Quartermaster had the difficult task of resupplying over 800 men by helicopter. Every department of Headquarter Company was involved and all contributed to the support given to the duty companies in the field. The operation was a notable success in preventing any terrorist attack and the Battle Group earned the praise of the General Officer Commanding, Lieutenant General Sir Roger Wheeler, and the Commander Land Forces, General Sir John Wilsey, both of whom visited the troops on the ground. It was also a reasonably predictable existence with a pattern to company duties – operations, training and leave – which the Guardsmen liked. Once the families had accepted the reduced security levels in Northern Ireland, they too enjoyed the experience and were sorry to leave once the tour ended in 1994.[12]

With a nice sense of timing, the Battalion's tour at Ballykelly ended just as the first IRA ceasefire was being announced. It came into being on 31 August 1994, with the loyalist paramilitaries, temporarily united in the 'Combined Loyalist Military Command', reciprocating six weeks later. Although these ceasefires failed in the short term and broke down in February 1996, they marked an effective end to large-scale political violence and paved the way for the final ceasefire in July 1997. On completing this first accompanied tour of Northern Ireland, and buoyed up by the prospect of 'some political progress in this unfortunate province', the Battalion moved back to England in the early summer of 1994, this time to Clive Barracks at Tern Hill in Shropshire as part of 143 (West Midlands) Brigade. Again, this was another former RAF base which had been transformed into Army accommodation. There was also a new Commanding Officer in Lieutenant Colonel R. L. Traherne, while Regimental Sergeant Major J. W. Harford had taken over from I. P. Dyas in December 1993. In April 1995 the Battalion provided the infantry element for a UN force based on 9 Supply Regiment, Royal Logistic Corps, which deployed to Angola on Operation Chantress to supply protection for civilian contractors in the south-west of the country. This was supplied by The Prince of Wales's Company under the command of Major Hugh Bodington; the force was based on the south-west Angolan coast near Catumbela.[13] Between September 1995 and January 1996 the Battalion provided assistance to Liverpool city authorities during a strike by firefighters in protest at cutbacks to their service. This presented Headquarter Company with a considerable logistical challenge, with Guardsmen being deployed in the city at various TA centres during periods of industrial action. The Light Aid Detachment also had the formidable task of maintaining the 'unreliable and unstable' Green

Sergeant (Drum Major) Johns 02 leads the Regimental Band of the Welsh Guards through the centre of Coleraine, County Londonderry, March 1993. *(Welsh Guards Archives)*

Roof of Africa and Roof of the Americas

In 1990, to celebrate the Regiment's seventy-fifth anniversary, the Battalion mounted the Roof of Africa Expedition, an ambitious adventure which aimed to go to any country in Africa with a peak over 10,000 feet and climb the highest in each country, joining up each mountain objective in a 39,000-mile overland expedition.

Starting in Morocco, with an ascent of Jebel Toubkal (13,671 ft) in the High Atlas Mountains, the first of five teams made their way deep into the central Sahara, through Senegal, Mali and Niger. The second team climbed Mt Cameroon (12,799 ft) before continuing the vehicle journey across the central African jungle to the Mountains of the Moon in Eastern Zaire. They climbed Margarita Peak (16,763 ft) in the Ruwenzoris and Karisimbi (14,800 ft) in the Virungas of Rwanda before heading on to Kenya. The third team climbed Mounts Kenya (17,050 ft), Kilimanjaro (19,340 ft) and Mulanje (10,000 ft) in Malawi before heading for Zimbabwe. A fourth team kayaked the entire 450-mile length of the River Zambezi in Zimbabwe before the fifth team completed the expedition with a major mobility challenge through Namibia, Botswana and Zambia, culminating in an extremely demanding crossing of the jungle in Zaire to finish in Cameroon.

The eleven-month expedition was packed with adventures: paragliding over the Sahara, camping with pygmies, surviving several hippo attacks and digging, winching and nursing the four-vehicle convoy through desert and jungle.

Following on from the success of Roof of Africa, in 1994, the Welsh Guards again took the lead on behalf of the Household Division in planning an even more elaborate expedition aimed at taking relays of Guardsmen, junior NCOs and young officers to face a series of demanding challenges. The Roof of the Americas Expedition set out to reach the geographical and physical extremes of North and South America.

Over fifteen months, the six teams dog-sledded and snowmobiled over 2,000 miles through Alaska's Arctic winter to reach the easternmost and northernmost extremities of North America; climbed Mt McKinley (20,237 ft), the highest mountain in North America; kayaked and rafted the entire 225-mile length of the Colorado River as it passes through the Grand Canyon; made the first source-to-mouth descent of the Mazaruni River through the jungles of Guyana; climbed Mt Aconcagua (22,837 ft), the highest mountain in South America; trekked, rafted and explored the remote parts of Chile, before taking a boat down the Beagle Channel to Cape Horn. Again, the expedition's major objectives were linked in a vast overland journey full of adventure and challenge for young soldiers.

Roof of America Expedition, Phase 1 Team. View from the back of a dog sled during their 2,000-mile journey through Arctic Alaska to reach Cape Prince of Wales and Point Barrow, the westernmost and northernmost points of the Americas, February 1994. *(John Warburton Lee)*

Roof of America Expedition, Phase 3 Team on their 225-mile journey by raft and kayak through the Grand Canyon, Arizona, June 1994. *(John Warburton Lee)*

Capt Sir Richard Williams-Bulkeley, Bt, when he was High Sheriff of Gwynedd in 1993, with his son, also Richard, when he was ADC to Gen Sir Charles Guthrie, C-in-C BAOR. Sir Richard is the grandson of Major R. G. W. Williams-Bulkeley, MC, who transferred from the Grenadier Guards to join the Welsh Guards in 1915 (*see* Chapter 1). *(Welsh Guards Archives)*

Goddess fire engines and the Guardsmen had to summon up reserves of patience and good humour especially over the weekend of 5 November when 'the youth of Liverpool proceeded to set alight all available combustible material which included just about every waste paper bin in the city'.[14]

For the first time, the Welsh Guards commemorated St David's Day in 1996 by holding the parade in Wales where The Prince of Wales's Company (Major Robert Talbot Rice) represented the Regiment at Cwrt-y-Gollen Camp in Crickhowell in Powys. The leeks were presented by Lady de Lisle, widow of Lord Glanusk, who had commanded the Training Battalion in 1941. In the following month the Battalion moved to Wellington Barracks and public duties in London, leaving the Commanding Officer to remark:

> Once in London, Battalion Headquarters found itself with the difficult task of steering the Battalion through the Household Division 'silly season'. With a full programme of public duties, Guard Mounting from Horse Guards, the Queen's Birthday Parade, Garter Service and State visits of the Heads of State of France and South Africa, whilst also preparing for our pending exercise in Canada and tour of South Armagh; Battalion Headquarters soon became an area that any young officer with a brain (a rare breed indeed!) learned to stay away from.[15]

The return to London District could have been a signal for the Battalion to throttle back during the 'silly season' of public duties, Guard Mounting and the Queen's Birthday Parade, but the life of the modern Guardsman is not just about certainties. It is mainly about adaptability and variety and, as ever, those virtues were not lacking in 1996. During the summer there were KAPE tours in Wales, an opportunity for Keeping the Army in the Public Eye, which took No. 3 Company to Penally, an attractive coastal village in Pembrokeshire which already had long-standing links to the Army, being home to an old training camp, a modern firing range and also to historic training trenches from the First World War. More to the point, in the local school Guardsmen were able to offer pupils insights into modern soldiering, including the somewhat alarming information – for the teacher at least – that guard duty at Buckingham Palace could be enlivened by 'the good-looking girls outside'. At the same time Support Company went to Rhyl in North Wales where the Mortar Platoon spent five days with young teenagers, all 'interested potential recruits', whose curiosity alone testified to the success of the scheme. Faced by the need to maintain recruiting levels at a time when the size of the Army was shrinking, the Regiment understood the need to keep a high profile in Wales; the KAPE tours were very much part of the process. This was a matter of increasing concern: at the Regimental Association's previous Annual General Meeting, the Swansea and West Glamorgan Branch had suggested that much more could be done to project the Welsh Guards in the media and in Job Centres in South Wales. A sign of the times was the first arrival of recruits from Commonwealth countries, especially from the Pacific islands.

At the end of the summer, the Battalion deployed to Canada for the now familiar Exercise Pond Jump West, travelling to Wainwright for exercises in September and October. Supported by 8 Alma Battery, Royal Artillery, and a troop from 9 Parachute Squadron, 21 Royal Engineers, the Battalion carried out some worthwhile training and realistic live-firing exercises, all under the expert eye of the Small Arms School Corps. Hard training was complemented by R&R and adventure-training packages which included riding, mountain biking, white-water rafting and freefall parachuting, all in the heart of the Rockies. The exercise ended with an all-arms battle group live-firing attack and an inter-company ice hockey match. As was noted at the time, 'It is still being debated which event was the more violent.'

Battalion Shooting Team

It has been said that Welshmen wearing the leek in their cap have been associated with marksmanship since before the Battle of Agincourt in 1415. Certainly shooting has always been taken seriously in the Welsh Guards, with notable successes in the 1980s. The bedrock to that success was laid down during the Battalion's time in Berlin by CSM Ernie Pritchard; coming 8th in 1980 and 5th in 1982 in the Major Units' Championships at Bisley.

The Battalion's posting to Hohne in 1984 provided the opportunity to capitalize on this experience and, under the watchful eye of CSM Frank Ward, the team won the Major Units' Trophy at Bisley for three consecutive years, 1983–5. Individuals from the team also shot as members of the British Army Rifle Team, competing internationally in Canada in 1984, and with the Army VIII; many of the team also qualified as members of The Army 100, the best 100 rifle shots in the Army.

The recipe for success was that the team was written off from duties for the best part of three months every summer, travelling back to stay in D Lines at Pirbright; and was run with a singleminded focus on marksmanship and fitness, supported by a dedicated 'butt party' who ran the ranges. Ammunition never seemed to be in short supply, with the team allegedly taking the British Army of the Rhine's unused allocation and firing over 200,000 rounds a year. The SLR rifle, GPMG, SMG and Browning pistol were each shot in separate competitions, with all scores contributing to the Major Units' Championship.

Another key to success was continuity amongst the most experienced shots, including Sgt Rowlands, Sgt Evans 13, Sgt Owen 52, LSgt Salmon, LSgt Brinkworth, LSgt Edwards 28, LCpl Taylor 95, Gdsm Jones 53 and Gdsm Crocker. As Army Rifle Association rules changed, young soldiers and an officer also had to be included: LCpl Gibbon, LCpl Teague, Gdsm Pritchard 92, Gdsm Evans 80, Gdsm Spillane, Gdsm Dawson, Gdsm Pridmore and Lt Talbot Rice were amongst those who also shot in the team.

Members of the Battalion shooting team practising on Pirbright Ranges, 1985. (*L to R*) LSgt Salmon, Gdsm Crocker, Gdsm Jones 53, Sgt Evans, Sgt Owen 52, LSgt Brinkworth. *(Welsh Guards Archives)*

The shooting team seniors with the Major Units' Trophy, 1985: (*L to R*) Sgt Owen, LSgt Brinkworth, CSM Ward, Sgt Evans, Sgt Rowlands. *(Welsh Guards Archives)*

The shooting team, LDRA, 1973: *Back row* (*L to R*): LCpl Carlyon, Gdms Watts, Gdsm Davies 97, Gdsm Dyas, Gdsm Williams 98, Gdsm Greenway, Gdsm Williams 33, Gdsm Tucker 08, Gdsm Lewis 73. *Centre row*: LSgt Richardson, Gdsm Liversage, LCpl Williams, Gdsm Hall, Gdsm Wilson, LCpl Thomas, Gdsm Lanham, Gdsm Minett, Gdsm Keenan. *Front row*: CSM Tubb, CQMS Gatrell, CSM I. M. Skyrme, Capt H. R. Oliver-Bellasis, Capt C. F. Drewry, Sgt Evans, CSM Davies. *(Welsh Guards Archives)*

CHAPTER THIRTEEN

Peacekeeping, 1997–2005

London, Northern Ireland, Wales, Bosnia

THE RETURN TO LONDON was a prelude to a further deployment to Armagh in Northern Ireland in the spring of 1997, with the Battalion under the command of Lieutenant Colonel A. J. E. 'Sandy' Malcolm and with WO1 F. K. Oultram as Regimental Sergeant Major. It was still an unsettling time in the Troubles. In the previous year the IRA had revoked their two-year-old ceasefire and announced the decision with a bomb attack in London's Docklands area, killing two people and causing £85 million of damage to the capital's financial centre. This was followed by other assaults on the British mainland, culminating in a massive bomb attack on the city centre of Manchester on 15 June 1996, causing damage which was valued at £411 million, making it the largest and most expensive bombing in the UK since the Second World War. Shortly after the Battalion arrived in South Armagh, the IRA signed a second ceasefire in July 1997 as part of the peace negotiations which came to be enshrined in the 'Good Friday' agreement of 10 April 1998. Following this, the focus turned again to a political solution and it is fair to say that the moves on the political front were gradually being matched by changes within Northern Ireland itself – for example, the housing estates in Crossmaglen, once run-down and neglected, had been

Peacekeeping in Bosnia 2002: (L to R) Capt S. E. Birchall, Signals Officer, Lt Col R. H. Talbot Rice, Commanding Officer, Sgt Williams 75 and CSgt Weekes. *(Welsh Guards Archives)*

smartened up with BMWs and Mercedes now being seen parked outside neat houses and a general air of prosperity within the town and the surrounding area. One example could stand for many: Crossmaglen's once dilapidated cattle market had been transformed into a pedestrian square and, on the surface at least, people were more welcoming to the security forces. For the older Guardsmen, who remembered the enmity shown by the locals during previous tours, this was a welcome improvement and a sign that things were being normalized as far as everyday life was concerned. That was certainly the feeling when The Prince of Wales's Company Group (Major Talbot Rice, CSM Davies 86) returned to its old stomping ground in Crossmaglen for its third tour of duty in thirty years:

> Although there remains a large section of the community which will not even acknowledge the presence of a soldier – or if they do it is only to offer abuse – the majority will at least nod a hello; and a few can be quite friendly; especially when they are outside the town and don't think that their reactions are being watched.[1]

Even so, there was still a need to maintain the observation towers outside the town and these provided one of the reasons for the Company Group's presence. Even at that stage of the conflict, there were still ten surveillance towers across South Armagh, all on sites leased or requisitioned from civilians, civilian organizations or various government agencies. During the course of the Battalion's tour, the question of the cost of maintaining these surveillance towers was raised in Parliament; the Secretary for Defence responded that, in South Armagh at least, they were still part of the landscape and had a military purpose: 'The security situation is kept under constant review and the configuration of the military infrastructure adjusted accordingly. However, there are no plans to withdraw any of the towers from use over the next six months.'[2] This meant that they still had to be maintained and, towards the end of the tour, Major Talbot Rice was able to report that 'the tower commanders had become quite house proud: Sergeant Thomas 35 has turned G10 into a static version of the USS Enterprise and Colour Sergeant Harman and Sergeant Jones 59's multiples had demonstrated excellent DIY skills at G20'. For the duration of the tour, The Prince of Wales's Company Group was over 150 strong and included the Milan platoon and the

Lt Col C. F. B. Stephens, Regimental Adjutant, with HM Queen Elizabeth, the Queen Mother, at the Field of Remembrance, Westminster Abbey, November 1991. Lt Col Charles Stephens was Regimental Adjutant 1976–1978 and 1990–2007 and is one of the longest serving officers in the Regiment. *(Welsh Guards Archives)*

Corps of Drums, with other personnel from Headquarter and Support companies filling key posts. Other changes in the Battalion's order of battle included the creation of Battalion Tactical Headquarters, divided into Operations, Intelligence and Signals, which worked on a twenty-four-hour cycle divided into day and night shifts, with the result that everyone had 'that unhealthy white glow of people who have not seen the sun for half a year' and there were continual cries of 'Keep the noise down, I'm trying to sleep.'[3]

A word needs to be said about the composition .of the company groups. Two multiples were the equivalent of a platoon with one multiple usually commanded by the platoon commander and the other by the platoon sergeant. This structure evolved through the need to develop improved patrol tactics which would allow the ground to be dominated more effectively without establishing set patrol patterns. This was important because predictable movements were often observed and noted by 'dicking' (scouting, observation and early warning techniques).[4]

The configuration of the Battalion into company

2Lt M. B. D. Jenkins, Mortar Platoon Commander with Guardsman Lentle on patrol in Belize, August 1989. *(Welsh Guards Archives)*

groups was not without military significance because, for all that 1997 marked another stage in the peace process, Northern Ireland was still an unquiet place. An early indication of the potential for violence came ten days into the tour when a multiple from No. 3 Company Group (Major Bathurst, CSM Jones 73) came under fire from a PIRA gunman firing a Barrett 0.50 calibre sniper rifle in the vicinity of Forkhill. During the attack, on 29 March, an RUC constable was badly wounded but the multiple commander, Sergeant C. D. B. Price, and one of the team commanders, Lance Sergeant Thomas 53 (later awarded a Mention in Despatches in the Operational Honours List), showed great presence of mind by organizing medical assistance under fire and recovering evidence which led to the arrest of members of the sniper team, including Michael Caraher, Bernard McGinn and Martin Mines.[5] This was a significant moment as the so-called 'South Armagh Sniper' had become an iconic figure for the Republican movement and the generic name covered the identity of a number of gunmen who caused the deaths of nine British service personnel between 1990 and 1997. However, by the time the Battalion deployed in South Armagh, the greatest threat was assessed to be that of dissident Republicans who refused to accept the ceasefire. As a result, new 'splinter' organizations such as the Real IRA (RIRA) and Continuity IRA (CIRA) emerged. Loyalist marches caused additional trouble, most notably at Drumcree, where the Battalion was heavily involved in operations to contain the resulting outbreak of Republican violence.

Corps of Drums (Machine Gun Platoon) on Exercise Trumpet Dance, 1998. (*L to R*) Dmr Coburn, LSgt Hughes and Dmr Barrington. *(Welsh Guards Archives)*

WO2 Davies 86 CSM, The Prince of Wales's Company, and LSgt Evans 62 with a major find – not weapons but armour plates used to protect IRA snipers firing the .50 Barrett rifle from a vehicle. Crossmaglen, South Armagh, 1997. *(Welsh Guards Archives)*

As happened in previous tours, each company group had its own Tactical Area of Responsibility (TAOR) and each had been reinforced for the deployment – for example, No. 3 Company Group had a platoon (two multiples) from No. 2 Company Group as well as the Mortar Platoon for the duration of its stay at Forkhill. No. 2 Company Group (Major Treadgold, CSM Badham) was based in Newtownhamilton (NTH), which had changed very little and which the Guardsmen likened to an early 1970s episode of the television soap *Coronation Street*, although one more generous soul said it was 'just like Merthyr'. In the early stages the threat level remained high, although it was noted that 'the inhabitants of NTH were generally amiable and multiples could proceed around town relatively unhindered'.

Around this time, computers were becoming essential battlefield tools to aid intelligence gathering and analysis. Guardsmen were sent to the Interactive Learning Centre at Headquarters in Lisburn to develop IT skills. With advances in technologies and improvements in software, the Battalion became much more effective at monitoring suspicious activity. Systems now in common use by police forces everywhere were used to great effect. This had a huge impact on the Army's ability to prevent or mitigate terrorist attacks.

Communications also improved and, as they did so, Guardsmen had to become conversant with the new systems. As No. 3 Company Group found at Forkhill, 'every man had to be capable of monitoring up to five separate nets as well as being able to talk confidently over the radio'. In comparison to earlier tours, company commanders felt that this was the biggest single challenge for their men. It required long hours of training and it was all a far cry from the early days in Northern Ireland when Combat Net Radio (CNR) – initially of the Larkspur and then Clansman series – did not always work in urban areas where almost all of the first operations took place. This was a major problem (but not a new one for the Army) as CNR was not optimized for urban areas and, as Ministry of Defence scientists pointed out, the laws of physics imposed strict limits on the capabilities of communications equipment. It was not until the 1980s that the deployment of Cougarnet across Northern Ireland gave most units a robust, reliable, deployable and secure means of tactical communication.

Generally speaking, though, the tour was considered a success. The only discordant note involving communal violence came on 12 July, perhaps predictably, when the PIRA decided to mortar Newtownhamilton using a timer on the device, which detonated just after midnight, 'scattering mud and debris over a wide area', and giving Guardsman O'Leary and Guardsman Jakeman 'both on Sangar duty the best seats in house'. In that respect the tour of South Armagh had not been dissimilar to earlier operational tours undertaken by the Battalion. Brief spurts of adrenaline had been balanced by longer spells of routine duty with little spare time to indulge in sporting or recreational activities. Each company base had its own multi-gym which was much in demand as Guardsmen worked hard to stay fit in the otherwise cramped quarters where 'Sky TV (what did the Army do before this invention?) was pumped into every room and a gaming machine, sun bed and highly stocked NAAFI gave the Gdsm some respite from the pressures of everyday life.' Visitors also offered a welcome change and brought fresh challenges which allowed Guardsmen to reveal hidden talents as greeters and briefers, No. 3 Company Group noting that,

Lt Col Colonel R. H. Talbot Rice presents a prize to Lt H. Llewelyn Usher, Battalion Boxing Competition, 2004. *(Welsh Guards Archives)*

> Lance Sergeants Humphreys and McKeon and Guardsmen Thomas 49, Hemmings and Parry 34 found themselves regularly at the centre of attention from the new Secretary of State for Defence [George Robertson], the Chief of the Defence Staff [General Sir Charles Guthrie], the Chief of the General Staff [General Sir Roger Wheeler], the General Officer Commanding Northern Ireland [General Sir Rupert Smith], his counterpart in London District [Major General Sir Evelyn Webb-Carter] and endless Major Generals and Brigadiers from Procurement, Military Operations and every department one can think of.

As Major Ben Bathurst pointed out, it was an ideal way in which to showcase to senior officers the manifold abilities of the modern Welsh Guardsman.[6]

Towards the end of the tour, Princess Diana was fatally injured on 31 August 1997 in a car crash in the Pont de l'Alma road tunnel in Paris. Although she was by then divorced from the Regimental Colonel, HRH The Prince of Wales, she was given a ceremonial funeral in Westminster Abbey on 6 September which was watched by a worldwide television audience. Pall bearers for the coffin were chosen from The Prince of Wales's Company under the command of the Adjutant, Captain Richard Williams, MC, and the party was made up of the following Guardsmen: John Jones (Caerphilly), Llyr Owen (Talysarn, Gwynedd), Christopher Treharne (Llanelli), Phillip Bartlett (Brecon), Patrick Dewaine (Haverfordwest), Gareth Thomas (Bridgend), Ken Sweetman (Rhyl) and Carl John (Cardiff). Second-in-command of the party was WO2 (Drill Sergeant) Paul Cunliffe (Llanrwyst). Because of the Battalion's deployment, the duty of pall bearing was initially allocated to the Scots Guards; but when the Commanding Officer insisted on 1 September that he could provide ten men to be at Chelsea Barracks for rehearsals the following day, the Major General agreed that the honour should fall to the Welsh Guards. Having been plucked from their duties in Crossmaglen after a selection process at midnight, timed so that those on a late patrol could be considered, the party was flown by helicopter to

HRH The Prince of Wales presents a leek to Major R. S. M. Thorneloe on St David's Day. To the right of HRH is Lt Col R. H. Talbot Rice and to the left is Captain M. Browne, BEM, QM. *(Welsh Guards Archives)*

London on 2 September and immediately started practising in Chelsea Barracks for their role.

Three years earlier, in April 1994, while serving in Cambodia with the United Nations during the transition to democratic elections, Captain Williams (son of Brigadier Peter Williams, who had commanded the Battalion 1974–7) had been awarded the Military Cross, one of only three awarded to Welsh Guardsmen since the end of the Second World War. (The other two were awarded to Captain [later Brigadier] Aldwin Wight during the Falklands War and Major Jimmy Stenner while serving in Iraq.) Held prisoner by the Khmer Rouge, Williams single-handedly freed UN personnel held by Pol Pot's troops and, on another occasion, unarmed and trapped in an ambush by guerrillas, he inspired seven Indonesian companions to push the twenty-five-strong Khmer Rouge killers back into the jungle.

Once again, the end of an operational tour saw the Battalion return to London and public duties. They arrived to find that recruiting was proving to be a challenge, as it was across the rest of the Army – a result of numerous factors including the buoyant economy and the availability of more training courses and additional opportunities for tertiary education for young people. For the Battalion, it did not help that Wellington Barracks was considered to have detrimental effect on manning. If so, that simply provided a spur and, after KAPE tours in Wales in 1998, the Battalion was the second-best manned in the Household Division. Indeed, for the first time on record, the Battalion's recruiting figure for the year was over 200 and the target for the next year was immediately set at 300. The key seemed to lie in providing potential recruits with evidence of the variety of experiences to be had while soldiering as a Welsh Guardsman. In April 1998, No. 3 Company deployed to Belize for a six-week tour on Exercise Native Trail, taking with them a platoon of trainee Guardsmen, commanded by Captain Matt Collins, Irish Guards (later killed in Afghanistan), who had been released six weeks early from their initial training at the Infantry Training Centre, Catterick. While this was happening, the rest of the Battalion started training for the ceremonial season, which in 1998 included the Queen's Birthday Parade at which the Regiment's Colours were trooped. Major Robert Talbot Rice commanded the Escort to the Colour, 2nd Lieutenant Henry Bettinson carried the Colour, and CSM Davies 86 and CQMS Harman were respectively right and left guide.

Almost inevitably perhaps, June was followed by a brisk rise in tempo when the Battalion was put on forty-eight-hour standby for deployment back to Northern Ireland, after an upsurge in violence at Drumcree during the traditional marching season. Although the situation was brought under control and the Battalion was stood down, it was immediately warned to provide personnel to cover emergency services during the firefighters' strike in Essex under the terms of Operation Fresco. With good reason, at the end of the year the Commanding Officer was able to produce the following assessment of what had been an action-packed period:

> As this article is written, Battalion Headquarters is just about the only part of the Battalion in Wellington Barracks. Elements of The Prince of Wales's Company are in Macedonia, on a NATO Peace Support Operations Exercise, in Poland with the SAS, and at Fremington with the Guards Adventure Training Wing. 2 Company have decamped en masse to the South of France for what they have assured everyone is a highly arduous adventure training exercise. 3 Company are honing their fighting skills on Company training, and Support Company are on some well-earned leave. At the same time we have half a Company with 1SG [1st Battalion Scots Guards] in Northern Ireland, and a section of men on a rugby tour in the USA. This snapshot provides a very clear view of how busy and how varied the life of a modern day Guardsman really is.[7]

A weapons find in Bosnia by No. 7 Platoon, No. 3 Company, 2004: Maj G. Bartle Jones, WO2 Jones 59, Lt H. Llewelyn Usher, Pl Sgt: Sgt Matthews. *(Welsh Guards Archives)*

The following year saw no let-up as the Battalion entered a period of preparation for Exercise Trumpet Dance, a planned deployment to the USA which took place between 20 February and 6 April 1999 when the Battalion was based at Fort Lewis, nine miles south-west of Tacoma in Washington State. The training facility embraced the vastness of the Yakima Training Center, a high-desert wilderness area encompassing 320,000 acres and covered with sagebrush, volcanic formations, dry gulches and large rock outcroppings. The area's vast flat valleys and intervening ridges made it entirely suitable for large-scale training, albeit carefully coordinated around the mating habits of the sage grouse, a protected species in the area. The Battalion took advantage of the opportunities offered at Fort Lewis, with the live firing in the final exercise 'making so much noise that we had a great deal of difficulty in convincing the neighbouring American Corps headquarters that we were a Battalion and not a Brigade'. Continuous rain failed to dampen enthusiasm but it left some Guardsmen admitting that they had 'begun to harbour doubts about whether Wales really is the wettest country in the world'. Also fully involved in the exercise was the Corps of Drums which allowed its members to fulfil their role as machine-gunners, providing fire support from their GPMGs and winning praise from their US hosts, not just for their endurance and stamina but also for 'the high quality of singing that emanated from the back of the 4-ton trucks as Lance Corporal O'Brien led the Corps' singing efforts, maintaining morale however wet or tired they were'.[8] The exercise also included two days' skiing at Snoqualmie and the usual R&R package which provided Guardsmen with the opportunity to travel widely in North America, some heading over the border into Canada while others ventured south towards the delights of California.

In September there was a change of Commanding Officer when Lieutenant Colonel Andrew Ford took over from Lieutenant Colonel Sandy Malcolm, who had had, at one point during the earlier deployment in Northern Ireland, 1,200 troops under the command of 1st Battalion Welsh Guards. Commissioned into the Grenadier Guards in 1977 and a veteran of the fighting in Kuwait in 1991, Ford transferred to the Welsh Guards in August 1999 before taking command. He became the first Grenadier to command the Battalion since Murray-Threipland in the early days of the Regiment's history. Under his leadership the Battalion prepared for its next tasks, and moved to Bruneval Barracks in Aldershot to become part of 12 Mechanised Brigade, the third deployable brigade in 3rd (UK) Division. This came about as a result of the Strategic Defence Review of 1998 (SDR), the first attempt by the newly elected Labour government to address defence issues in a rapidly changing world. With its catch-line, 'Making the World a Safer Place', the review's aim for the Army was to focus on mobility, precision firepower and force protection by producing six deployable Brigades which would provide 'the ability to undertake two brigade-size operations (one of which could be sustained indefinitely) simultaneously at short notice'. One of these formations would be 12 Mechanised Brigade, which would evolve from 5 Airborne Brigade, an existing light brigade, 'by transferring in an armoured infantry battalion and two mechanised infantry Battalions, and the AS90 self-propelled gun, whose awesome firepower has been demonstrated in Bosnia as essential in subduing the warring factions'.[9]

It was said at the time that the Army was the principal beneficiary of the SDR, largely in recognition of the fact that of all the services it had suffered the most from overstretch in the 1990s. Coupled with perennial personnel shortages, the combination of heavy and continuing commitments in Northern Ireland and Bosnia had placed it under pressure throughout the decade and the strain had fallen mainly on the infantry and on combat service support units. As a result of these changes the new Regimental Lieutenant Colonel, Major General Sir Redmond Watt, was able to put the SDR into context in his foreword to the last edition of the regimental magazine of the twentieth century:

Bearer party for the 42-stone, lead-lined coffin of Diana, Princess of Wales, 1997. *(Welsh Guards Archives)*

> The Battalion is now in the part of the Army which follows the formation readiness cycle. This means that they will now take part in high quality all arms training so that they can be held on high readiness as part of 12 Mechanised Brigade for operations overseas. This is part of an exciting new role and will give us access to some of the best training and operational experience that the Army has to offer.[10]

Before the close of the century, the British Army found itself committed to further operations overseas. With the death of President Tito in 1980, the multinational, multi-ethnic federation of Yugoslavia began to crumble, the process reaching a climax in 1991 when the republics of Croatia and Slovenia declared their independence. The federal government, dominated by Serbia, rejected these declarations and war broke out in the Balkans that same year. After European Union monitoring had failed to halt the conflict, the UN intervened, first through the implementation of sanctions in July 1992 and then through the authorization of a number of peacekeeping operations. A UN Protection Force (UNPROFOR) was deployed in Croatia and Bosnia between 1992 and 1995 to deliver humanitarian assistance and to provide protection for so-called 'safe areas'. However, the weakness of the UN mandate was highlighted following the Srebrenica massacre of 1995, when UN peacekeepers were left helpless under the limited terms of their mandate to halt the slaughter of Muslims. Exhausted by economic sanctions, however, and under the threat of further action from NATO, the combatants agreed to bring the war to an end under the Dayton Peace Accords of 1995.

British military forces subsequently formed part of the post-conflict mission which had been mandated under UN Security Council Resolution 1031, initially known as the Implementation Force (IFOR) and subsequently as the Stabilisation Force (SFOR). The latter was provided with an ambitious and broad mandate, ranging from traditional conflict-prevention to the modern roles of institution-building and reconstruction tasks. However, during 1998, war began to break out in the Serbian province of Kosovo, an area dominated by ethnic Albanians that had been left out of the Dayton settlement. International diplomacy in the shape of the Rambouillet Accords failed to resolve the issue; brutal repression and widespread displacement of ethnic Albanians by Serbia followed. The UN proved unable to act, and NATO threats against Serbia, led by President Slobodan Milošević, were ignored. In March 1999 the Alliance began a controversial air campaign against Serbia and introduced plans for the deployment of ground forces in Kosovo (KFOR).

The earlier of these events also lay behind the evolution and philosophy of the SDR. Coinciding with this change in British defence policy, the Household Division set out a revised plan promulgated by Major General Sir Evelyn Webb-Carter, which identified key elements in the Division's approach to its unique operational and ceremonial roles, namely Excellence, Professionalism, Discipline and Leadership. Of course, none of this was particularly new as those ingredients had been central to the ethos of the

Commonwealth Soldiers

The first Commonwealth soldiers started to arrive in the Battalion in the late 1990s. There are currently forty-three serving Commonwealth soldiers, the highest-ranking being a sergeant. There are ten Commonwealth nations represented in the Battalion. The majority come from African nations but the largest single group is from Fiji. Welshmen and Fijians share a common passion – rugby. The Battalion rugby squad is a natural home for Fijian Guardsmen and the team has been strengthened by their speed and robustness. African soldiers also assimilate well into the Battalion. The mixture of cultures has proved invaluable on recent operational tours to Afghanistan where cultural empathy is a battle-winning asset.

Soldiers from the Commonwealth serving in the Regiment, 2012. (*L to R*) Gdsm Agu from Nigeria, Gdsm Saho from the Gambia, Gdsm Nyekanga from Ghana (wounded in action), Gdsm Sapak from Ghana and Gdsm Jawarra from the Gambia. *(MoD).*

WO1 (RSM) K. Oultram receives the Queen's Colour. Queen's Birthday Parade, 1998. *(MoD)*

Capt R. S. M. Thorneloe, Adjutant in Brigade Waiting, Queen's Birthday Parade, 1998. *(MoD)*

Regiment since its foundation but, as an old century was gradually coming to a close, the mantra was worth repeating. During this period the Battalion Second-in-Command and Adjutant also changed, with Major Ben Bathurst and Captain Rupert Thorneloe handing over respectively to Major M. C. J. Hutchings, Grenadier Guards, and Captain Dino Bossi. The Battalion also received a new Regimental Sergeant Major in M. Cooling, who took over from Captain Martyn Miles on his being commissioned and posted as Regimental Recruiting Officer, based in Maindy Barracks, Cardiff.

Substantial changes took place during the period from September 1999 to March 2000 as the Battalion prepared to assume the title of a battle group. Standing operating procedures for a battle group headquarters were developed and tested under simulated conditions at the Combined Arms Staff Trainer (Warminster); new appointments included a Battle Group Warfare Officer (Captain Richard Williams-Bulkeley) and an Intelligence Surveillance Target Acquisition Officer (Captain Jimmy Stenner); effects rated planning was introduced and the empowering concept of mission command was developed and practised by all ranks. All this preparatory work would pay huge dividends in the training year of 2001 and during operations later on.

On settling into Aldershot in the spring of 2000, the Battalion's first task was to come to terms with the Saxon armoured personnel carrier, their 'battle taxi', which they used while operating in the mechanized role. Developed by GKN Sankey in the early 1980s as a lightly armoured 4 x 4 wheeled vehicle capable of carrying a section of infantrymen quickly into battle, it had proved itself serving with BAOR, but it had limited off-road mobility and was fast becoming obsolescent. When the Battalion received its vehicles, some were found to be in a parlous state of repair but they were put to rights and were soon reported to be in good order. The new role and change of station also signalled a change of pace, with a greater emphasis on operational training. In October 1999 (while still based in London), The Prince of Wales's Company (Major Mark Jenkins, CSM Evans 24) set off for Bavaria on Exercise Gothic Dragon for training in difficult terrain, and the following July deployed to Belize for Exercise Native Trail, where training at the jungle school included contact drills, obstacle crossing, navigation and a tactical river crossing. At the same time No. 3 Company (Major Lloyd, CSM Tutt) was sent to South Armagh as a reinforcement company for 1st Battalion Devonshire and Dorset Regiment and found that, although the ceasefire remained in place, sporadic violence was never far away, necessitating a 'a fairly heavy patrolling schedule'. The Regimental Band also saw changes. In February 1999, Joanna Williams (flute) from Builth Wells and Lucy Ellis (French horn) from Tywyn, Gwynedd, made history when they became the first female recruits to join the Band after a one-year musical training course at the Army School of Music, Kneller Hall, prior to joining the Regiment.

For Support Company (Major Guy Bartle-Jones, CSM Lewis 80), the new year and new century brought a change of title to Manoeuvre Support Company, while the Corps of Drums changed from machine gunners to assault pioneers. Other innovations included the Signals Platoon becoming the Communications Information Systems (CIS) Platoon. Perhaps the greatest asset that the Battalion possessed during this period was the very high quality of its staff at Battle Group Headquarters who proved to be a highly motivated group of young officers, keen to learn, keen to experiment and keen to succeed in the complex environment of the all-arms arena.

The basic training of Guardsmen was also put under the microscope and some reforms were introduced. The former Infantry Training Battalion at Catterick had expanded threefold with the creation of three training battalions necessitating a name change to the Infantry Training Centre. The Guards Training Company, responsible for Phase 2 training, formed part of the 1st Battalion with Light Infantry training companies; the 2nd Battalion trained the line infantry regiments and the 3rd Battalion was composed of training companies for the Parachute Regiment and the Gurkhas. In 2000, trials took place to rationalize and reform this system to produce a new structure aimed at reducing wastage and producing a better product by centring all basic infantry training at Catterick. The result was the Combat Infantryman's Course (Standard Entry) which began in January 2002 and became an immediate success. Lasting twenty-six weeks, it was conducted entirely at Catterick and covered all aspects of infantry training and handling all platoon weapons, including the 94 mm anti-tank weapon. Instructors with the Guards Training Company, whose course was two weeks longer to take account of the additional requirement for drill, announced themselves satisfied with the outcome:

> The course is a significant improvement on what has gone before. It has been designed by us, for us, and is intended to produce not just an Infantryman, but a Guardsman ready to take his place either on operations or public duties.[11]

Queen's Birthday Parade, 1998. Lt Col A. J. E. Malcolm, Field Officer in Brigade Waiting, leads the Parade. *(MoD)*

Commonwealth Exercise Iron Anvil, BATUS, Canada, 2001. (*L to R*) Major B. J. Bathurst, Capt G. R. Harris, WOI (RSM) M. Cooling. *(Welsh Guards Archives)*

CVR(T) Scimitar with 30 mm Rarden cannon belonging to the Reconnaissance Platoon during exercise Iron Anvil, BATUS, Canada, 2001. Capts B. Ramsay, J. D. Stenner and R. L. Gallimore to the front of the vehicle. *(Welsh Guards Archives)*

In time the Centre produced 80 to 100 Welsh Guardsmen a year and all were members of Number VIII Company which consisted of instructors and recruits alike, the idea being to inculcate the ethos of the Regiment at an early stage in the training, coming together for important occasions such as St David's Day.

In the following year the tempo of training quickened again, 2001 being the brigade-training year within the force-readiness cycle. This included a tactical engagement simulation exercise (TESEX) held on Salisbury Plain in July 2001 when lasers were fitted to all weapons and the exercising troops wore sensor vests to replicate the effect of live fire. This was followed by Exercise Iron Anvil, a return to Canada, in August and September, which involved Battle Group and Brigade TESEXs in an all-arms environment. The Battle Group rose to the challenge magnificently: 'after action reviews' consistently showed Guardsmen 'winning' battle after battle. Officers and men were fit, disciplined and motivated to succeed. On one occasion, the Battle Group dismounted its Saxon vehicles and conducted a night march over twenty-three miles with full loads of kit, weapons and ammunition, infiltrated the enemy armoured battle group, attacked it from the rear and comprehensively defeated it. The 'enemy' Commanding Officer was most surprised to have been captured by extremely proud, if exhausted, Welsh Guardsmen! It can be fairly said that this training under Lieutenant Colonel Ford's tenure laid the basis for the succession of operational tours which followed.

All this was undertaken ahead of the Battalion's first operational deployment outside Northern Ireland since the Falklands War twenty years earlier. In March 2002, as part of Operation Palatine, the Battalion deployed to Bosnia-Herzegovina as the United Kingdom Battle Group within the Multinational Division (South West) for a tour that the new Commanding Officer (Lieutenant Colonel Robert Talbot Rice) forecast would be a 'busy summer in the Balkans'. The Battalion Battle Group was based in Banja Luka, the second largest city in Bosnia-Herzegovina, and the main centre of influence in the Republika Srpska which came into being in the early 1990s and was predominantly Serb in its ethnic make-up. During the tour the Battalion Battle Group had under its operational command C Squadron Household Cavalry Regiment, and B Battery, 1 Royal Horse Artillery, with the rifle company groups at the following locations: The Prince of Wales's Company (Major Alex Lewis, CSM M. O'Driscoll) at the Mrkonjic Grad Shoe Factory with three rifle platoons and a composite support weapons platoon (eight multiples), equipped with Saxon and Land-Rover snatch vehicles; No. 2 Company Group (Major Alex Macintosh, CSM M. D. Evans) at the Banja Luka Metal Factory with three rifle platoons and a composite weapons platoon (seven multiples) and a patrol base at Prijedor; No. 3 Company Group (Major Guy Bartle-Jones, CSM P. R. Jones) at Mrkonjic Grad Bus Depot with two rifle platoons and the Assault Pioneer Platoon (six multiples) and patrol bases at Kotor Varos and Knesevo ('think Rorke's Drift but

Lying in State of HM Queen Elizabeth, the Queen Mother

The planning of the military commitment for royal funerals is an ongoing requirement which has the codename 'Bridge' prefixed by a river or town – thus Tay Bridge was the lying in state and the funeral of HM Queen Elizabeth, the Queen Mother.

HM Queen Elizabeth, the Queen Mother, died on Easter Saturday, 30 March 2002, at the great age of 101. Her long life had been one of dedication and service both to her husband, King George VI, and to her country. Her background was Scottish and she had family connections with both the Grenadier and Coldstream Guards.

It was decided that the lying in state should last, without interruption, from midday on Friday, 5 April, for three days, in Westminster Hall, the great Norman meeting place at the east end of the Houses of Parliament. All Household Division regiments were to take part in the watches, each watch lasting six hours. Only officers below the rank of lieutenant colonel were to be on duty. The watch was split into twenty-minute vigils. The Regiment was allocated the watch from 06.00 to 12.00 on Saturday, 6 April. Rehearsals took place the previous day in the gymnasium of Chelsea Barracks. The coffin was draped with the Queen Mother's personal standard. It rested on top of the catafalque, which was in the centre of the Hall, with steps going up to a room where the officers could relax. At each corner of the catafalque stood an officer in home service clothing, head bowed and sword point resting between his feet. It is a considerable strain to stand in this position without moving, seeing only the feet of people moving past; thus two reserve officers stood at the top of the stairs should one of those on duty need relieving. At exactly twenty-minute intervals the officer in charge of the watch would stand at the top of the stairs and, with a double tap of his sword on the ground, signal for the relief to take place, with the officers marching in silence in slow time. Even in the early hours of Saturday morning, the queues were enormous with thousands of people waiting patiently to file past the catafalque and pay their last respects to a great Queen and Queen Mother.

After three days of lying in state, the funeral of Her Majesty took place in Westminster Abbey on 9 April 2002. It was a great privilege for the Regiment to be involved and an unforgettable experience for those officers taking part.

Four Welsh Guards officers keep vigil around the catafalque, during the lying in state of HM Queen Elizabeth, the Queen Mother, in Westminster Hall, April 2002. *(Welsh Guards Archives)*

without the Zulus', was one comment). During the tour, the mission was to 'maintain a safe and secure environment in order to help the International Community to attain its strategic objective of a peaceful and secure Bosnia-Herzegovina'.

Throughout the tour the situation remained seemingly safe and secure, but recent history underlined the fact that this was still a volatile region. During the 1992 conflict it had witnessed ethnic cleansing on a horrific scale, Prijedor being second only to Srebrenica in the enormity of the killing, in this case mainly of 5,200 Croats and Bosnian Muslims also known as Bosniaks. If any reminder were needed of the history of violence in the Balkans it was only necessary to recall an incident in 1851 when an Ottoman warlord, Omer Pasha Latas, bloodily suppressed the Prijedor region and then warned his rival chieftains not to eat fish from the River Sava, 'for they have been feeding on Bosniak flesh which I drove into the river Bosnia at Doboj'.[12] The rough, though majestic, mountainous topography was also a challenge and the mainly rural population subsisted in

Map of Bosnia during 1st Battalion's deployment 2002.

Queen's Birthday Parade, 1998. Field Officer in Brigade Waiting: Lt Col A. J. E. Malcolm, OBE; Adjutant in Brigade Waiting: Capt R. S. M. Thorneloe; Capt of the Escort to the Colour: Maj R. H. Talbot Rice, OC The Prince of Wales's Company; Subaltern of the Escort: Lt G. C. G. R. Stone; Ensign of the Escort: 2Lt H. G. C. Bettinson; Regimental Sergeant Major: WO1 K. Oultram. *(Welsh Guards Archives)*

a basic economy with a lifestyle that seemed to have changed little since the Middle Ages. The local infrastructure was rudimentary and the whole operational area was awash with weapons and ammunition. Coupled to an existing bellicosity which went back through several generations to the years of the Ottoman Empire, the region around Banja Luka (850,000 square kilometres) had most of the ingredients required for reigniting conflict or at the very least destabilizing the surface harmony. That much became clear when the Battle Group conducted a search for weapons in the community during the second half of the tour. Despite the fact that there had been many similar operations in the past, this search produced over 1,000 military rifles and pistols as well as 165,000 rounds of ammunition and twenty-two large-calibre mortars. A team led by Guardsman

The entire Welsh Guards Battle Group mounted in AT105 Saxon 4x4 APCs training in Suffield, Canada. This picture was taken a few days before 9/11, in September 2001, a date that dramatically changed the course of the Regiment's history. *(Welsh Guards Archives)*

Bowen of No. 2 Company found two recoil-less anti-tank weapons, though a vintage 'bomb' discovered by the same company turned out to be a weapons canister dropped into the area during the Second World War to arm local partisans. Again, this was another pertinent reminder of the area's bellicose past.

A major part of the Battle Group's life was taken up with operations in support of the civilian population as part of the normalization process. That kind of outreach work is essential in any post-conflict environment, both to restore confidence and to rebuild the infrastructure. With a budget of around £325,000 to spend on nine major projects, there was enough to keep the Guardsmen busy as 8 Platoon discovered when they helped the Royal Engineers to rebuild the so-called Ross Bridge which linked Republika Srpska with the neighbouring Federation, which had been partially destroyed during the recent fighting. Further down the scale but no less welcome was the refurbishment and re-equipment of a school outside Banja Luka with a generous donation of £3,000 given to the Commanding Officer by Eton College. Work of that kind was not only a practical gesture which benefited the local population but it also showed them that SFOR was a friend and not an army of occupation. An ability to speak a few words of Serbo-Croat helped and one or two Guardsmen claimed to have become quite proficient after one or two glasses of 'slivo'. Anti-smuggling operations were also part of the routine and The Prince of Wales's Company spent three weeks in Operation Andrea 4, which exposed some of the limitations of using SFOR in that unfamiliar policing role.

Perhaps the most memorable ingredient in winning local hearts and minds was the arrival of the Regimental Band in July to stage a sell-out open-air concert within the grounds of historic Banja Luka Castle, to an audience of around 4,000 Bosnians. Many more were able to watch it live on national television. During the concert, the Welsh Guards Choir gave a stirring version of 'Cwm Rhondda' and 'God Bless the Prince of Wales'. The visit was not only a roaring success, which pleased the locals, but according to the regimental magazine, it also had 'the unexpected effect of educating the Battalion. They were able to see at first hand the positive effect the Band can have, and how skilled the Bandsmen are.'[13]

Training, too, was a priority and the whole Battle Group made good use of terrain which provided numerous opportunities for live firing and adventurous training. R&R was not neglected either and, in his quarterly report to HM The Queen, the Commanding Officer was able to offer reassurance that all the

The Prince of Wales's Company in Belize, 1998. 2Lt H. G. C. Bettinson leads his platoon through the Intermediate Jungle School. *(Welsh Guards Archives)*

Guardsmen in the Battle Group had taken full advantage of facilities available on the coast, particularly on the island of Brac, where everyone, including the visiting Regimental Lieutenant Colonel, was able to try scuba diving, windsurfing and other fun activities, including riding 'on a giant inflatable banana towed behind a speedboat'.[14] On a more serious note, the tour ended in September on a high with everyone feeling that they had accomplished a thoroughly worthwhile task in a challenging yet attractive part of Europe. The only discordant note was the announcement on returning to the UK that tour leave had been postponed to allow the Battalion to train once again for a role in Operation Fresco to cover for striking fire-fighters. This could have caused some ill-feeling, but the Guardsmen reacted with considerable stoicism and settled into the intensive training for an operation which eventually involved 30,000 service personnel. The Battalion was given responsibility for covering Warwickshire as well as Coventry and parts of Birmingham. The first period, a two-day strike, began on Wednesday, 13 November, and ended on Friday, 15 November, but when another round of reconciliation failed, further strike periods were planned.

Once deployed, and having mastered the elderly and cumbersome 'Green Goddess' fire engines, the Guardsmen's main complaint was that the worst aspect of the work was the 'hanging around'. Not that they were left with nothing to do. Support Company (Major R. G. B. Pim, CSM R. J. Brace) tackled a huge fire at the National Agricultural Centre at Stoneleigh, as well as an equally large conflagration at a carpet showroom in Rugby. The Prince of Wales's Company was equally tested when it fought and contained a large fire at a factory complex in West Bromwich. As the Commanding Officer pointed out, 'the soldiers acted splendidly and the fires were successfully contained with no loss of life'.[15] There was a certain amount of relief when the strike was called off, albeit temporarily, in December, as it allowed a welcome three weeks of leave. Not only did this make up for the disappointment of the cancelled leave earlier in the year, but it allowed thoughts to turn to the Battalion's next move, to RAF St Athan in south Wales in May 2003, when it would operate as a rear-based Northern Ireland battalion (except for the Close Observation Platoon which was permanently forward in Ballykelly).

This was something of a red-letter day in the

Capt D. Bevan, Adjutant and WO1 (RSM) Davies 86, in Bosnia, 2002. *(Welsh Guards Archives)*

history of the Regiment. Although it had an administrative presence at Maindy Barracks in Cardiff, the Battalion had never been stationed in its entirety in Wales. Unlike the Scots Guards, which are frequently based in Scotland, the Welsh Guards have never been based in their country of origin for any appreciable period of time. All that was about to change and for the Battalion it was both an opportunity and a challenge. For a start, it would entail spending part of the summer in Northern Ireland to provide support to the recently formed Police Service of Northern Ireland. The Battalion would be based in Ebrington Barracks in Londonderry, built in the 1840s and shortly to be handed back to the city in response to the lowered terrorist threat. The Battalion started arriving ahead of the marching season, which passed without incident. Although Northern Ireland was found to be a more secure environment than it had been in the past, a small number of PIRA terrorists remained active in the area. On 15 June a large 1,200-pound car bomb was found abandoned on the edge of the city which 'tested every level' of No. 3 Company in dealing with it. Apart from occasional excitements of that kind, which were themselves reminders of the recent past, the three-month tour was relatively incident-free and the majority of the Battalion began returning to RAF St Athan in July with 'everyone looking forward to taking full advantage of the opportunities presented of being so close to home'.[16] For some, though, there was no rest. At the end of August the rugby team embarked on its first ever tour of South Africa, the plans having been first mooted in Bosnia. Although it

No. 2 Company training exercise Native Trail, Belize, 2000. OC No. 2 Company, Maj H. G. C. Bettinson, CSM No. 2 Company, WO2 Topps. *(Welsh Guards Archives)*

WO1 (RSM) Davies 86 and Gdsm Hughes 57 fighting a serious fire at a carpet warehouse in Rugby during Operation Fresco, November 2002. The Battalion had been given responsibility for the whole of Warwickshire and part of the West Midlands. *(Welsh Guards Archives)*

Training in Bosnia, 2007. *(Welsh Guards Archives)*

Queen's Birthday Parade, 1998. *(Welsh Guards Archives)*

ended in three defeats against superior opposition, the general feeling was that 'It was a great way to pick up the Rugby side after almost two years of not being able to play, due to various operational commitments.'[17]

As part of the Northern Ireland operational cycle, the Battalion continued to send companies to Londonderry on a routine basis but the commitment ended on St David's Day 2004 when The Prince of Wales's Company flew back to RAF St Athan, arriving by C-130 Hercules and being greeted by HRH The Prince of Wales. The 'Jam Boys' were then presented with their leeks while still wearing their combat clothing. The gesture seemed to mark the conclusion of operational duties in Northern Ireland and the beginning of a new phase which marked the Regiment's arrival in Wales, a moment characterized by the Commanding Officer as a chance to regroup: 'After a long period in which the needs of the Army had been put first, the needs of the individual took preference for a change.'[18] The emphasis within the Battalion then switched to reviving sporting interests, adventurous training, preparing soldiers for promotion and granting much-needed leave.

Being situated in the Vale of Glamorgan in south Wales, RAF St Athan was well placed to allow the Battalion to make connections with the community and the opportunity was taken to participate in local charitable and sporting events and to encourage recruitment. Some soldiers lived out and were able to spend time with their families and friends, always a bonus, and the consensus within the Battalion was that the choice of station had been good for Welsh Guardsmen and their families. Only the North Walians could find any fault, as they discovered the truth of the nineteenth-century traveller George Borrow's observation that north and south were well and truly divided by the high heart of mid-Wales, the 'empty grandeur of Plynlimon'.[19] In May there was a two-week exercise in Brecon to perfect conventional soldiering skills. By then rumours of a forthcoming deployment to Iraq had been realized and that had become the principal focus of training. The familiar backdrop of Lydd, used for training for Northern Ireland and Bosnia only a few months previously, took on a new cultural flavour during the summer, with instructors from Salisbury's Babylon School of Languages on hand to teach basic Arabic phrases and to explain the religious and cultural context in which the Battalion would be operating later in the year. For everyone this would provide the biggest change in emphasis as they prepared to meet the demands of a very different type of warfare.

CHAPTER FOURTEEN

The Global War on Terror, 2005–2009

Iraq, Bosnia, UK

In the morning of 11 September 2001 the world changed for ever when two hijacked passenger airliners – American Airlines Flight 11 and United Airlines Flight 175 – slammed into the twin towers of the Word Trade Center in New York, causing them to collapse two hours later and creating immense damage in Lower Manhattan. It soon became clear that this was not a random act or a tragic accident but part of a well-planned operation against the United States of America. On the same day, a third plane, American Airlines Flight 77, crashed into the Pentagon (the headquarters of the United States Department of Defense), leading to a partial collapse in its western side. A fourth plane, United Airlines Flight 93, had been targeted to hit the United States Capitol in Washington, DC, but crashed into a field near Shanksville, Pennsylvania, after its passengers tried to overpower the hijackers. Almost 3,000 people died in the attacks, including all 227 passengers and 19 hijackers aboard the four aircraft. Suspicion as to those responsible quickly fell on a group known as al-Qaeda, which had come into being in 1979 to resist the Soviet presence in Afghanistan and had then evolved into an anti-American terrorist group that had already attacked US targets in east Africa and the Middle East. Its leader was Osama bin Laden, a

Battle Group Headquarters and No. 3 Company, Al-Amarah, Iraq, 2005. Commanding Officer: Lt Col B. J. Bathurst, OBE; Operations Officer: Capt R. W. Gallimore; RSM: WO1 T. Harman. *(Welsh Guards Archives)*

wealthy Saudi Arabian Islamic fundamentalist who had been given shelter by the extremist Taliban regime in Afghanistan. Three days later, a joint resolution, 'Authorization for Use of Military Force against Terrorists', was passed by the US Congress, permitting US Presidents to fight terrorists and any nations protecting them. This was followed on 7 October by a bombing campaign against Taliban targets and al-Qaeda camps in Afghanistan and the subsequent invasion of the country by a US-led coalition which included UK forces.

The news of the attack on the USA arrived when the 1st Battalion was engaged in training operations in Canada (see previous chapter) and was greeted by everyone with a fair degree of astonishment. Little could anyone have known, though, that the 9/11 outrages would have a huge impact on the Welsh Guards and the rest of the country's armed forces in the years that lay ahead. Once the 'War on Terror' (a

HRH The Prince of Wales inspects The Prince of Wales's Company, Clarence House, 2006. The Colonel is seen here talking to the CQMS, CSgt Griffiths 50. To the extreme left is the Regimental Lieutenant Colonel, Col A. J. E. Malcolm, OBE. *(Welsh Guards Archives)*

Patrol from No. 3 Company in Basra, Iraq 2005. Note Snatch Land-Rover in background. *(Andrew Chittock)*

Warrior from the Scots Guards provides support for HQ Company, Basra, Iraq 2005. *(Andrew Chittock)*

Guardsman from No 3 Company returns fire from behind a burnt out Iraqi T-54 tank, Basra, Iraq 2005. *(Andrew Chittock)*

The Commanding Officer, Lieutenant Colonel B. J. Bathhurst, with the Chief of Police of Al Amarah, speaking to Iraqi News in Camp Abu Naji, Al Amarah, Iraq, 2005. *(Welsh Guards Archives)*

Guardsman McNally of The Prince of Wales's Company on patrol outside Basra Police Station, Iraq, 2005. *(Andrew Chittock)*

description coined by President George W. Bush on 20 September) had been launched, it soon became clear that it would have a global reach. Following the invasion of Afghanistan and the ousting of the Taliban regime, the attention of the US and its allies switched to Iraq which had been listed as a State Sponsor of Terrorism in 1990 at the time of the invasion of Kuwait and which had emerged as a possible possessor of weapons of mass destruction (WMD). In pursuit of that claim, in November 2002, the UN passed Resolution 1441 which called on Iraq 'to comply with its disarmament obligations' or face 'serious consequences'. As a result, UN weapons inspectors were allowed into the country but failed to find any significant WMD stockpiles and soon admitted that their work would take many more months to be completed.

While this was happening, Congress authorized President Bush to use force against Iraq and the US began a diplomatic effort to create a 'Coalition of the Willing', which included the UK but not France or Germany. The war against Iraq began on 19 March

Major J. D. Stenner, MC. This portrait, by Diana Blakeney, was commissioned following Major Stenner's death while on active service in Iraq 2004.

Queen's Birthday Parade, 2008. (*L to R*) LCpl Sangin, CSgt Hoosan and LCpl Whatling. *(MoD)*

2003 with an air campaign, followed almost immediately by a ground invasion which quickly defeated Iraq's forces. Baghdad fell the following month after a mere six weeks of fighting and Saddam Hussein's regime quickly crumbled, allowing President Bush to announce on 1 May that major combat operations had been concluded. It was an overly optimistic statement because the fighting in Iraq quickly descended into a series of insurgency conflicts against the US-led coalition and the newly developing Iraqi forces and the post-Saddam government. The most serious fighting in January 2004 centred on the city of Fallujah and culminated in November with a coalition success which cost the lives of ninety-five US personnel and around 1,350 insurgents; it quickly spread into the southern city of Basra and the surrounding countryside. Major Jimmy Stenner, who had been awarded a Military Cross for his part in the early stages of the invasion of Iraq, was tragically killed in Baghdad in December 2003, depriving the Welsh Guards and the Army of one of the most talented officers of that generation. It was against that background that the 1st Battalion deployed to southern Iraq in September 2004 on Operation Telic 5 to be part of 4 Armoured Brigade, which itself was part of the British-commanded Multinational Division (South-East).

During the tour, 1st Battalion Welsh Guards was divided into a number of parts. The Reconnaissance Platoon, commanded by Captain Tom Smith, assisted by CSM P. R. Jones 40, was first into action with 1st Battalion Black Watch on Operation Bracken, based at Camp Dogwood, in support of the United States Army and Marine Corps. Thereafter they reverted to 4 Armoured Brigade. The Prince of Wales's Company,

Warrant officers' convention, Tower of London, May 2010: (*L to R*) WO2 (RSWO) Smith 63, WO2 (CSM) Nelson (Irish Guards), WO2 (CSM) Dunn, WO2 (DSgt) Brown 89, WO2 (RQMS) Scholes, WO1 (RSM) Roberts 99, WO2 Cunliffe, WO2 (CSM) Ryan, WO2 (CSM) Taylor, WO2 (CSM) Jenkinson, CSgt (Master Tailor) Wilson. *(Welsh Guards Archives)*

commanded by Major Dino Bossi, assisted by CSM P. Robinson, was selected for the most independent location, Old State Building in the middle of Basra City. There, they came under the command of 1st Battalion, Duke of Wellington's Regiment Battle Group, and were subjected to repeated attack by insurgents using mortars, rocket-propelled grenades and small arms fire. No. 2 Company, commanded by Acting Major Giles Harris, assisted by CSM Darren Pridmore, was originally earmarked to take over the security and escort role in Basra Palace, but at the last minute their task changed to that of Security Sector Reform, working out of the Shatt-Al-Arab Hotel in Basra under the command of 1st Battalion Duke of Wellington's Regiment Battle Group. This entailed training and mentoring the Iraqi police forces.

The remainder of 1st Battalion Welsh Guards first formed up as a Battle Group on the day of the road move 185 miles to the north of Basra. As the Commanding Officer, Lieutenant Colonel Ben Bathurst, pointed out in his overview of the operation: 'We arrived in Camp Abu Naji (the name means "Father of the Nation" which is what the British were called in Iraq in the 1920s), just outside the city of Al Amarah in Maysan Province. To give a sense of proportion, the distance and direction from Basra City to Al Amarah is about the same as London is to Newport but there the similarities end!'[1] The Welsh Guards core of the Battle Group consisted of Battalion Headquarters, with the Commanding Officer, the Second-in-Command, Major Harry Lloyd, and Regimental Sergeant Major Terry Harman, along with Headquarter Company, under Major Keith Oultram, including the Intelligence, Operations and Signals' staff and all the Quartermaster and Transport elements. Heavier firepower was provided by A Squadron, Royal Dragoon Guards, under Major Dan Rawlins, with Challenger 2 main battle tanks, along with Burma Company, Duke of Wellington's Regiment, under Major Phil Wilson, with Warrior armoured fighting vehicles. For about half of the tour this force was supplemented by another company of Warriors, provided by Left Flank Scots Guards, commanded by Major Andrew Speed, and for a shorter period towards the end by Right Flank, Scots Guards, commanded by Major Lincoln Jopp, MC. Reconnaissance along the border with Iran was provided by B Squadron, Queen's Dragoon Guards, under Captain Arthur Chamberlain, with Combat Vehicles Reconnaissance (Tracked). Patrolling in Snatch (armoured) and Wolf (unarmoured) Land-Rovers was No. 3 Company under Major Charlie Antelme, assisted by CSM Mike Monaghan. Also fully involved was Support Company which had inherited some FV432 armoured vehicles; while these provided added mobility, they also involved 'some frantic thumbing of pamphlets' before they could be used.

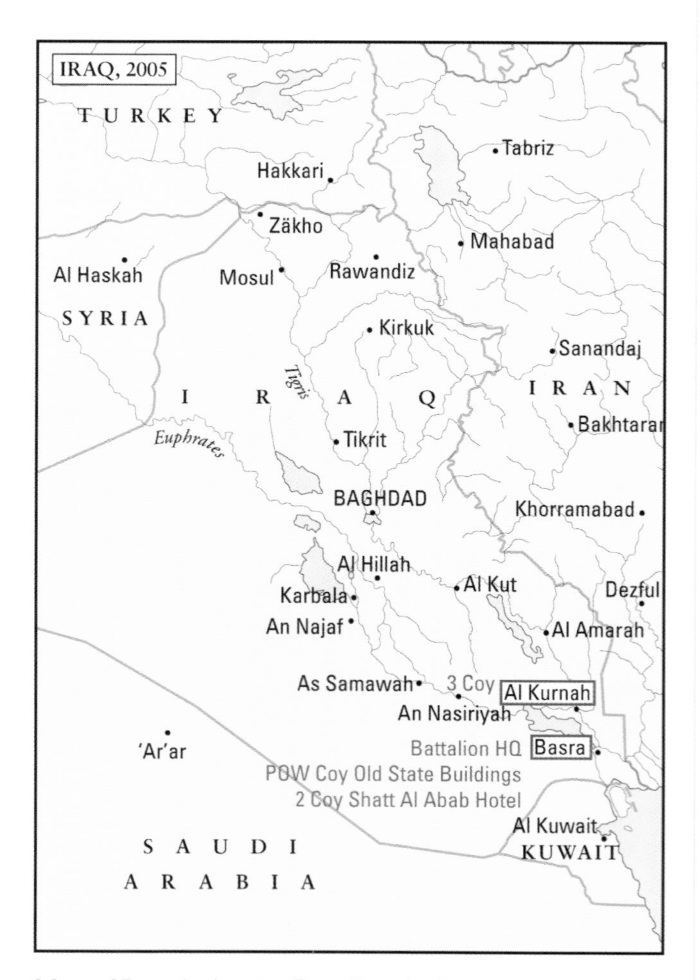

Map of Iraq during 1st Battalion deployment 2005.

On arrival it was soon clear that all of the component parts faced a daunting situation. Unlike the bulk of the rest of the country, Maysan province had not been liberated by coalition forces and this had allowed the Iraqi Army to melt away, leaving a power vacuum which had been filled by local tribal militias. During the long reign of Saddam Hussein, the people of Maysan province had been largely anti-government but this did not imply that they welcomed the presence of coalition forces. Weapons were in abundance, as were young, disaffected men willing to use them. Basra, too, was imploding. The third-largest city in Iraq, it was the first to fall to coalition forces and initially it was reasonably calm, reflecting perhaps its earlier cosmopolitan social make-up and past history of religious tolerance. By late 2004, though, violence in

Basra was steadily increasing. In the wake of an uprising by the Jaysh al-Mahdi militia in Najaf and Karbala earlier in the year, many of the fighters had moved south to Basra and fuelled the violence, as the fight between the Shia factions for control of the city's lucrative resources intensified, particularly during the period before the elections which were scheduled to take place in January 2005. Insurgent groups (and their militias) such as the Sadrist Trend (Jaysh al-Mahdi), Islamic Supreme Council of Iraq (Badr Corps) and Fadhila all vied for control of the oil infrastructure and smuggling network, the security forces and the provision of public services and state resources.[2] A British military briefing conducted before the arrival of the Battle Group spoke of 'a complex insurgent environment with many disparate elements working on parallel tracks without being truly united at the top'.[3] Targeted assassinations, kidnapping, sectarian violence, gunfights, and widespread criminality characterized this phase of the struggle, which persisted throughout the period of the tour.

For the previous battle group – formed by 1st Battalion, The Princess of Wales's Royal Regiment (1 PWRR) – this situation meant that their tour had involved almost continual warfare and the brutal nature of the fighting was reflected in the casualties (two killed and forty-eight wounded), as well as in the number of gallantry awards made to the regiment, including one Victoria Cross awarded to Private

No. 2 Company skydiving from 12,000 feet above Perris Valley, California, 2003. (*L to R*) Gdsm Page (in red), Sgt Goodman, 2Lt Richards, Sgt Williams 88, Lt Bishop. *(Welsh Guards Archives)*

Queen's Birthday Parade, 2008: (*L to R*) Lt Col R. J. A. Stanford, Field Officer in Brigade Waiting; HRH The Prince of Wales, Colonel; HM The Queen, Colonel-in-Chief. No. 3 Company is marching past HM. WO2 (CSM) Baldwin is to the front left and to his immediate right is LSgt Soko. *(Welsh Guards Archives)*

The Prince of Wales's Company under enemy fire, Afghanistan, 2009: Gdsm Lodwick (*left*) and LSgt Lawrence. This photograph was taken in Nadi Ali as The Prince of Wales's Company conducted shaping operations prior to the start of Operation Panther's Claw. *(Welsh Guards Archives)*

Lt Col C. K. Antelme, DSO *(left)* commands the Welsh Guards Battle Group from his Tac HQ, Operation Panther's Claw, Afghanistan, 2009. To his right is his Operations Officer, Capt E. N. Launders. *(Welsh Guards Archives)*

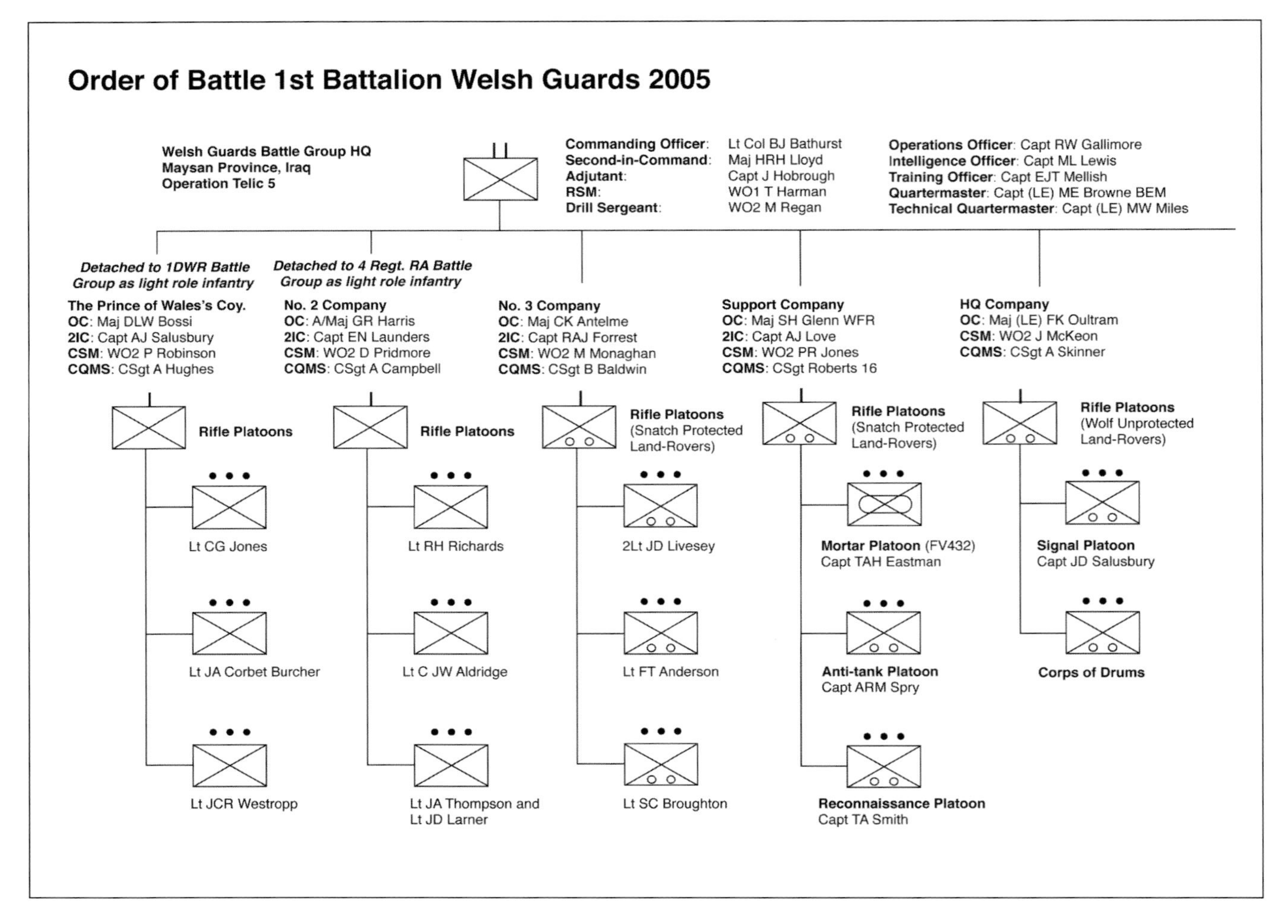

Johnson Beharry. The Welsh Guards adopted a slightly different tempo during their tour, which opened in September 2004 and came to an end in March 2005. Their response was also dictated by prevailing circumstances. 1 PWRR was an armoured infantry battalion, whereas the Welsh Guards were operating in a light infantry role and had 250 fewer soldiers under command. In this respect it helped that the Commanding Officer had a mature understanding of the local security situation, having previously worked with the Coalition Task Force 7, a US three-star headquarters in Baghdad. While serving there he admitted that he had 'stopped trying to compare Iraq with Northern Ireland; that used to irritate US officers so much'.[4]

The difference in pace was also assisted by the proximity of the elections which took place in January 2005 and which gave the Iraqis a belated exposure to democracy and one which the population grasped eagerly. The operation to support these elections was long and nerve-wracking, with The Prince of Wales's Company Group in Basra patrolling for forty-two hours without sleep in an effort to ensure things went smoothly. Security was such that the interpreters arrived for work on election day unaware that they would not be going home for over two days. Once the operation was complete, the Company collapsed exhausted, lying over one another in the communal room and smoking area like dead men. It was very much the main theme of a tour which most Guardsmen agreed was 'not so much enjoyable as rewarding both professionally and morally'.[5] As for the experience itself, much depended on where individual companies were based. For The Prince of Wales's Company, based in the Old State Buildings in Basra, early impressions were formed by the high incidence of violence perpetrated by local insurgents, not least by those loyal to Muqtada al-Sadr, a prominent Shia leader whose father had been executed during the Saddam regime. During the tour the Jam Boys were involved in 56 major incidents: 15 mortar bombs in 5 separate attacks, 2 suicide vehicle bombs, 4 contacts on patrol, 1 very large vehicle bomb and 13 other assorted explosive devices. Added to this were 14 grenade attacks, 2 multi-weapon contacts, 2 sniper shoots, 4 assassinations or attempted assassina-

Order of Battle 1st Battalion Welsh Guards 2005 (Continued)

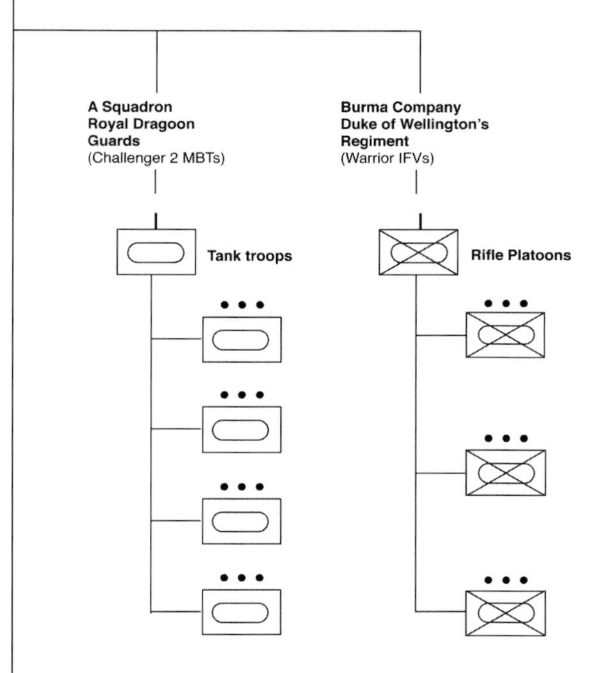

Notes:
(1) 1st Battalion Welsh Guards was periodically reinforced by two Warrior MICV companies (Left Flank and Right Flank) from the 1st Battalion Scots Guards.
(2) The Reconnaissance Platoon was retained by the Commander 4th Armoured Brigade as his Brigade Surveillance Platoon.
(3) The Regimental Medical Officer was Maj A Baker RAMC and the Padre was Capt S Lodwick RAChD.
(4) Additional Battle Group assets included CVR(T) Scimitar reconnaissance vehicles from B Squadron Royal Dragoon Guards; 2 x Chinook/ Merlin Support Helicopters from the Royal Air Force; 2 x Lynx Helicopters from the Army Air Corps; Phoenix Unmanned Aerial Vehicles (UAVs) from 18 Battery Royal Artillery; a Sound Ranging Troop from 5 Regiment Royal Artillery; Explosive Ordnance Disposal (EOD) teams from the Royal Engineers; a Royal Logistic Corps Detachment; A Role 2 Medical Treatment Facility from the Royal Army Medical Corps; and a detachment from the Intelligence Corps.

Col Sir A. J. E. Malcolm, Bt, OBE receives the Regimental and Queen's Colours from Lt Col R. J. A. Stanford, MBE, Commanding Officer 1st Battalion, and Major G. A. G. Lewis at the Laying Up of Old Colours, Bangor Cathedral, June 2007. *(Andrew Chittock)*

tions, and 3 rocket-propelled grenade attacks against the camp or vehicles. The Prince of Wales's Company also identified one device in a polling station amongst the voters, luckily before it could be detonated. That was by no means the sum of their activity. As the Company Commander recorded later, 'We also took part in a controlled explosion on a suspect car bomb, which involved the evacuation of 2,000 bemused Iraqis in their night attire (quite a revelation), at 2 o'clock in the morning. This was against a background of constant small arms fire, explosions and lots of police activity, all while trying to galvanize the Iraqi security forces to function, reform and progress.'[6]

In stark contrast, No. 2 Company operated under the principle of 'no boundaries' because, unlike any other unit in the south of Basra Province, they had free reign to travel unhindered as they went about the business of training and mentoring the Iraqi Police Service (IPS) in the area. They were also responsible for the Permanent Vehicle Checkpoint Police who guarded the perimeter of the city and were a key part of security during the elections. Not that it was a particularly easy task 'considering the very flexible morals

Maj R. E. E. A. Lorriman commanding No. 2 Guard at the Queen's Birthday Parade 2006. *(Welsh Guards Archives)*

and police practice used in the IPS's everyday police work'. As a result, the work, for which the Company had not received any specific training, was extremely challenging and intensive. However, it also proved to be rewarding, although that might not have been the immediate response of a night patrol in the area of the hotel, led by 2nd Lieutenant J. A. Thompson, with Lance Sergeant Furlong and Lance Sergeant Morgan 10, which was engaged by an improvised explosive device (IED) dug in under a railway line and detonated by insurgents at the end of a command wire about 150 yards away. Their vehicle was immobilized by the device, which used a 155 mm artillery shell, losing its right rear wheel. Guardsman Dean Matthews from Swansea, who sat above the seat of explosion, was seriously injured but was safely evacuated back to the UK. Guardsman Miles and Guardsman Atwell were also wounded, the latter incurring spinal injuries as a result of the explosion, but bravely carrying Guardsman Matthews to safety. 'From then on,' noted the Company report, 'the rear right seat of any vehicle was called "the Matthews Seat", and was thereafter allocated, along with a detailed explanation of the name, to several extremely unimpressed journalists when they accompanied a Two Company patrol!'[7] (CSM Pridmore was more succinct, describing the injury as 'a gigantic hole in his [Matthews's] arse'.)

There were other occasions when humour eased tensions. In one incident, No. 2 Company had helped secure the release of some British servicemen who had been taken prisoner by the Iraqi Police Service and were being held in one of the stations covered by its patrols. As luck had it, the company was on the scene when the servicemen were brought in, having been beaten and stripped of all their equipment and personal possessions; over a difficult hour or so, Major Harris managed to persuade the chief of police (who was later sacked for corruption) to release them. It was not the end of the matter. While the men had been returned, some sensitive equipment was still missing and determined efforts were made to recover some of it during the weeks that followed. On a late-night visit to a local police station, an IPS colonel invited Harris,

No. 2 Company advancing along the Shamalan Canal, Operation Panther's Claw, August 2009. *(MoD)*

A Guardsman rests after intense fighting with Taliban insurgents in Helmand during Operation Panther's Claw, Afghanistan, 2009. *(MoD)*

his Intelligence Sergeant (Lance Sergeant Warchol) and the interpreter Ahmed to sit in a room at the back of the building as he had something he wanted to show them. As they sat there waiting, Harris thought how vulnerable they were when, almost on cue, the doors burst open and six or seven men in balaclavas ran into the room and surrounded them. Lance Sergeant Warchol raised his rifle and Harris stood up only to be met by the colonel with a huge grin on his face. 'I would like to present you with my newly formed Quick Reaction Force, and ask you to inspect them.' As Harris tried to compose himself, a last man strode into the room carrying a plump velvet cushion on which sat a 9 mm pistol, one of the stolen weapons. As Harris recalled later, 'It was a typical piece of theatre and no doubt had them in stitches for months. I was less impressed, as you can imagine.'[8]

No. 3 Company also had a varied tour, reflecting both the local security situation and the need to engage with the four main lines of development of the operation – security, governance, economy and communications. Despite the roadside bomb that welcomed them as they made their way to Camp Abu Naji along Route Six, the main north–south road from Basra to Baghdad, they quickly settled into the bustling city of Al-Amarah. Winning local hearts and minds was a priority – a community project patriotically called the 'Red Dragon' eventually employed 750 youngsters – but the main operational task was training a newly formed specialist police unit called the Tactical Support Unit. Patrolling, too, was a priority, especially at night in the run-up to the elections, when it was essential to retain local confidence. Some idea of the problems involved can be discerned in the statistic that their predecessors, 1 PWRR, had been unable to enter Al-Amarah without a contact, nor had they entered the extremely volatile town of Majarr-al-Kabir (location of the murder of six Royal Military Policemen in 2003) which the Battle Group achieved during the tour. Clearly a more aggressive approach was required and one of the first night patrols yielded an important weapons find. As the Officer Commanding No. 3 Company, Major Charlie Antelme, put it in a newspaper interview: 'Maysan is the backwater of Iraq, people respect the tough guy with the gun. It's gangland. It's the Wild West. It's people breaking out of jail. They fish in the river using hand grenades.'[9]

In fact the tone had already been set by the Commanding Officer in his briefing on how to deal with the local population:

Guardsmen fire at suspected enemy positions as they clear a compound in Helmand, 2009. *(MoD)*

A washing line of soldiers' uniforms hung out to dry in a compound occupied by Guardsmen of No. 2 Company, next to an 81 mm mortar of Support Company, Afghanistan, 2009. *(MoD)*

> Avoid making the mistake of thinking that you can turn them into your best mates. You must be robust, impartial and understand that there will be a personal agenda behind everything they ask you to do. Never betray weakness and always be prepared to threaten major violence. You are, after all, the biggest tribe, with the greatest power and influence.[10]

Support Company had also been fully engaged both in dealing with insurgents and in the training and mentoring of Iraqi forces. To the Mortar Platoon fell the distinction of firing operationally for the first time

81 mm Mortar of Support Company during Operation Panther's Claw, Afghanistan, 2009. *(Welsh Guards Archives)*

since the Falklands campaign when Lance Sergeant Evans 62 fired an illumination shot to light up an area thought to be used by insurgents as a launch site for their mortars. An inkling of the conditions faced by the Welsh Guards while on patrol was encapsulated by a visiting US journalist who accompanied the Milan Platoon in the days following the election: 'Jolting along a rutted dirt road cloaked in dust, past a squalid strip of mud huts perched on a canal levee, they had guns and waves at the ready. Either might be needed.'[11]

For visiting journalists covering the election it seemed to be a tough physical assignment for the Battle Group and, to a certain extent, it was – the infrastructure was poor, local conditions were frequently abject, the population was generally hostile, the threat of violence was ever-present and the heat could be enervating – but it has to be said that the Welsh Guards had been well prepared for everything. Pre-deployment training had been sound and the Commanding Officer reckoned that his men were 'probably better prepared for Iraq than other units in 4 Brigade'. It was also clear that they benefited from the residue of previous operational training and experience in Northern Ireland and the Balkans, as well as the exercises in Canada and the USA under former Commanding Officers Sandy Malcolm and Andrew Ford. These had enabled the Battalion to operate effectively as an armoured infantry battle group, as opposed to in its designated light battalion role. Good sense played a major role and the 'Guiding Principles for Operations in Iraq' produced a litany of practical advice which can be summarized in the following main points, listed here in no particular order: never take anything at face value; always operate in strength; avoid the impression of being an occupying power; do not compromise the civil lead in police training; never make promises that cannot be kept; take advantage of all training opportunities; treat elected officials with respect; and at all times work on Iraqi priorities.

In the overall scheme of Operation Telic (the Ministry of Defence codename for the deployment in Iraq) it was always going to be difficult for the Welsh Guards to place their input in a coherent context. Compared to the 1 PWRR deployment, they had a less torrid time because the circumstances were dissimilar and because they reacted differently to them. Their successors, 1st Battalion Coldstream Guards, and 1st Battalion Staffordshire Regiment, both had to contend with a return to violence, which began escalating again later in the year.

In December 2007 the British Army returned

Drapers' Company

The Drapers' Company was established by Royal Charter on 15 July 1364, and celebrated its 650th anniversary in 2014. In 2007, the Regiment was delighted to become affiliated to this venerable body. Not only is the Company one of the most prestigious livery companies in the City, but it supports a very wide range of philanthropic initiatives and endeavours and has very close ties with North Wales.

Despite forming an association relatively recently, the bond between the two institutions has become very close and an excellent relationship has been established. Drapers have managed to visit the Battalion, either on operations or in training, in Bosnia, Belize and Kenya, and many other visits have taken place to the Battalion's base. The Company has been very supportive of the Welsh Guards Afghanistan Appeal and has raised considerable funds for it. The Battalion has also received a generous grant each year.

The Regiment is very proud of its affiliation with the Drapers' Company and therefore it is highly appropriate that the 2015 Welsh Guards Club dinner will be held in the Drapers' Hall.

control of Basra to Iraqi forces and concentrated its much-depleted garrison within the airport perimeter, prior to an official withdrawal the following year. This was known as 'transition with security', but with the Iraq involvement becoming unpopular at home, it increasingly looked like 'cut and run'. Seen from that vantage point, the British withdrawal was projected in much of the media (and also in some sectors of the armed forces) as an avoidable humiliation, but that does not square with the opinion of many in the Welsh Guards Battle Group. Not only did they return to the UK without anyone killed in action, but many of their number, such as Guardsman (later Colour Sergeant) Davies 03 of No. 3 Company, came back with a feeling that they had been involved in a job well done, especially in the period before the elections: 'I

LSgt Braithwaite of Merthyr Tydfil takes part in a joint ANA/No. 3 Company patrol in the Green Zone, near Sangin, Afghanistan, 2009. *(Welsh Guards Archives)*

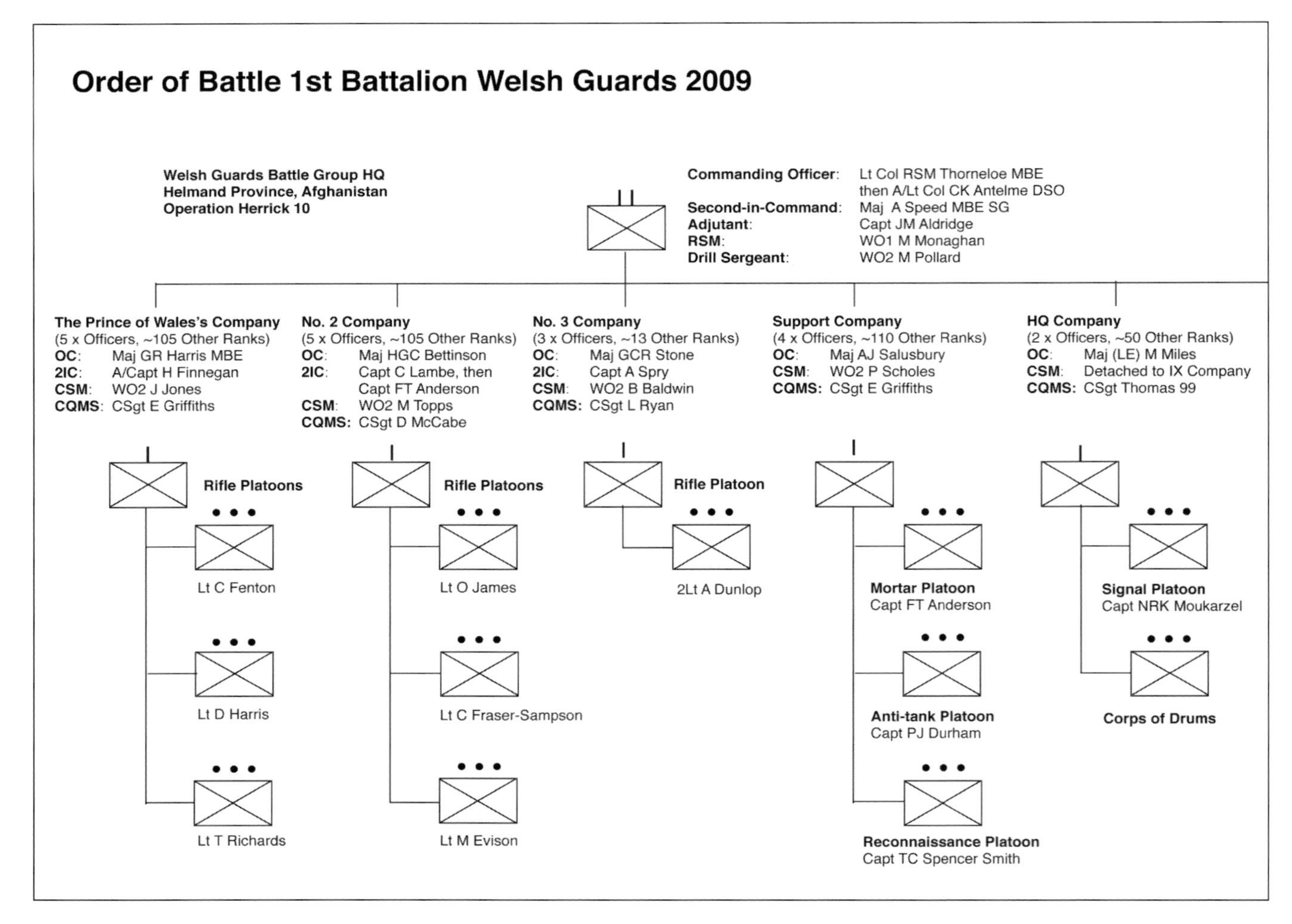

remember that for many of the people it was a feeling of elation as they'd never had free speech or the vote before and they were very responsive to us.' Like many others in the Battalion, he felt that Welsh-speaking North Walians were more capable of dealing with the language differences, being bilingual from birth. And, in common with every other Guardsman, he never accustomed himself to seeing young children running around the streets in ragged clothes and shoeless.[12]

In March 2005 the Battalion returned to St Athan. Already its future movements had been mapped out – a change of station to Wellington Barracks in a year's time, followed by ceremonial duties in London, including Presentation of New Colours and taking part in the Queen's Birthday Parade trooping their new Colour. This would be rounded off by a further six-month operational tour in Bosnia and Kosovo in October 2006. As was noted at the time by the Commanding Officer, they were fully match fit: 'Five years of continual operations have left the Battalion with a real depth of experience and talent which we intend to build on.'[13] And all that professionalism was needed a year later because, when the Battalion re-assembled after Easter leave on 18 April 2006, they only had two weeks left to rehearse before the Presentation of Colours at Windsor Castle on 4 May and 'not a single Guardsman had been issued with a Bearskin or Tunic, let alone worn them in public'. Nevertheless, under the direction of Regimental Sergeant Major Alun Bowen, assisted by Drill Sergeant Pridmore and Drill Sergeant Monaghan, the occasion was considered a great success. It was followed by the Queen's Birthday Parade on 17 June when the new Colour was trooped under the command of Lieutenant Colonel Ben Bathurst. The Prince of Wales's Company provided the Escort to the Colour, the Ensign being 2nd Lieutenant Simon Hillard. No. 2 Company, commanded by Major Rob Lorriman, found Number Two Guard.

Almost immediately it was back to business, with the entire Battalion beginning operational training for the return to the Balkans later in the year under the command of Lieutenant Colonel Richard Stanford, who announced that the deployment would only be counted a success if it allowed the country to 'stand on its own two feet'.[14] The bulk of the Battalion was based

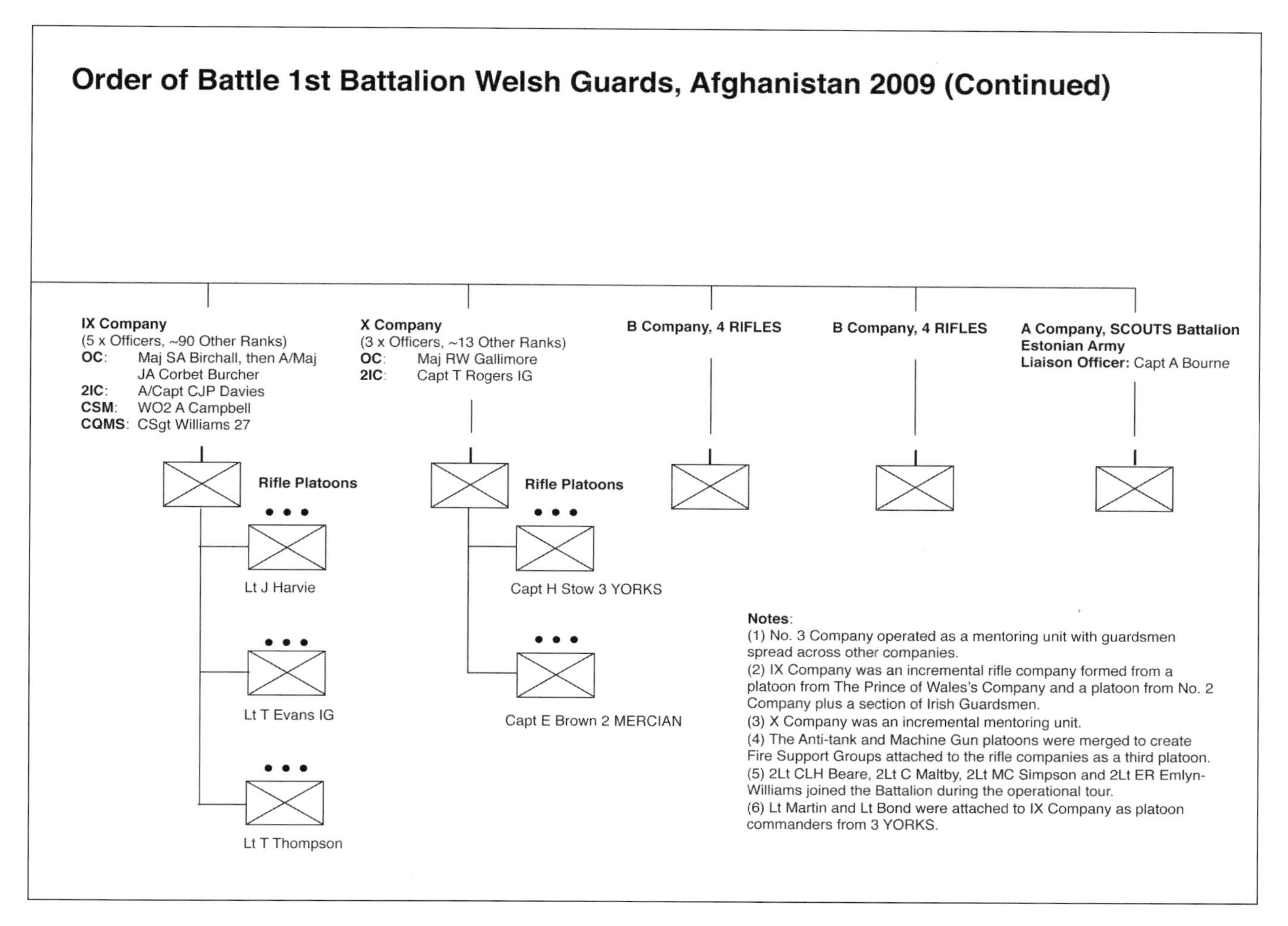

in the familiar surroundings of Banja Luka where their main activity was reconstruction work under the command of EUFOR which had taken over from NATO in 2002. The Battalion had a wide range of nationalities under command, including a detachment from Chile. The Dutch Company under command came from the Princess Irene Guards; it turned out they had fought alongside the Welsh Guards in 1944–5, having been formed in Porthcawl in 1943. No. 2 Company (Major Rupert Pim, CSM Hughes 32) were

1st Battalion Welsh Guards Rugby XV, winners of the Army Rugby Cup, 2011. *(Welsh Guards Archives)*

The 1st Battalion marches past HM The Queen, Colonel-in-Chief, following the Presentation of New Colours, Inner Quadrangle, Windsor Castle, 2006. Painted by Mark Jackson. *(Mark Jackson)*

under NATO command in Kosovo and provided the Force Surveillance Company. They had had to complete an intensive period of training for their role before deploying and had under command an excellent platoon from the Swedish Army. They were kept busy tracking down war criminals and identifying those who were trying to destabilize the fragile peace in Kosovo. In Bosnia, the Battalion had a separate surveillance platoon which was tasked with continuing the work of many previous battalions in finding war criminals from the Bosnian war.

Overall, the tempo of operations in Bosnia was not high and the Battalion spent a great deal of time training, making good use of the huge derelict factory complexes and a very good field firing area. As the British operation was drawing to a close, there was a concerted effort to use up the ammunition rather than ship it back to the UK. Training support was secured from the Bosnian Army with T-55 tanks and there were plenty of helicopter hours to add to the excellent training which was conducted. The Drapers' Company, one of the 'great twelve' City of London livery companies, made their first visit to the Battalion in Bosnia; this was a new affiliation which was cemented with a visit by five members of the Company and set the tone for a connection which has continued to thrive. As a result, the Drapers have been very generous in their support of the Regiment during the testing times in Afghanistan and afterwards.

During the tour, the Ministry of Defence announced that this would be the last major deployment by British forces to the Balkans and this was confirmed on St David's Day in 2007 when the leeks were presented by HRH The Prince of Wales and the Duchess of Cornwall. Their Royal Highnesses also launched the 'Iron Guardsman Challenge' that saw ten Guardsmen cycle 1,300 miles through the Alps, kayak across the English Channel, and then run a series of marathons to London in aid of the Army Benevolent Fund, the South Atlantic Medal Association and Everyman, the testicular cancer research charity. The total raised was £25,000 and the team, organized by Captain James Westropp and Sergeant Mills, made it back to Wellington Barracks on 31 March. On the

Queen's Birthday Parade, 2008. (*L to R*) Maj G. Bartle Jones, Major of the Parade; Capt M. L. Lewis, Adjutant in Brigade Waiting; Field Marshal the Lord Guthrie of Craigiebank, GCB, LVO, OBE, Colonel Life Guards, Gold Stick-in-Waiting; HRH The Prince of Wales, Colonel Welsh Guards; Lt Col R. J. A. Stanford, MBE, Field Officer in Brigade Waiting; Col T. C. S. Bonas, Regimental Adjutant; Lt Col Sir Andrew Ford, KCVO, Comptroller, Royal Household. *(Welsh Guards Archives)*

Battalion's return to London, they were thanked by Defence Minister Adam Ingram at a medals parade in Wellington Barracks on 28 March: 'The fact that the improved security situation has allowed us to end our military operation in Bosnia and bring you home is a huge achievement. It truly marks the end of an era.'

In his address to the Battalion, the Defence Minister accepted that, with 1,400 troops about to be deployed in Afghanistan, there were obvious benefits from the withdrawal from Bosnia but he insisted 'this is not a cut and run, this is a job well done'.[15] He was right to praise the Battalion for its contribution to a deployment which had lasted fifteen years and which had cost the lives of fifty-five British service personnel, but ahead lay difficult times in Afghanistan. In January 2006 the Secretary of State for Defence, Dr John Reid, had announced an increase of 3,500 in British troop numbers in the southern Afghan province of Helmand 'to help and protect the Afghan people construct their own democracy'. As NATO had taken over responsibility for the International Security Assistance Force (ISAF) in the country, it made sense that the UK should shoulder its share but, in words that would come back to haunt him, Dr Reid explained that he hoped it would be a non-combat mission: 'We would be perfectly happy to leave in three years and without firing one shot because our job is to protect the reconstruction.' Within the context of the mission's stated objectives it was a perfectly reasonable comment, but by the end of the year the British Army was engaged in some of the heaviest fighting it had experienced since the Korean War of 1950–53. Between then and the planned withdrawal in 2014, Helmand would become as familiar to many British soldiers as the streets of Belfast or the rolling countryside of South Armagh, with the exception that the opposition was engaged in almost continuous warfare. 1st Battalion Welsh Guards was deployed twice, in 2009 and again in 2012, and many more Welsh Guardsmen would also serve in the country supporting other regiments and battle groups. For example, the Mortar Platoon deployed to Afghanistan on Operation Herrick 6 with 1st Battalion Worcester and Sherwood Foresters. In all, fifty Welsh Guardsmen saw service in Afghanistan before the 1st Battalion deployed under Operation Herrick 10; by way of further example, in 2008 Captain Alex Corbet Burcher served with Inkerman Company Grenadier Guards, and was awarded a Joint Commander's Commendation for his bravery under fire. During Operation Herrick 10, he commanded the force at Patrol Base (PB) Jaker.

CHAPTER FIFTEEN

Fighting the Taliban, 2009–2015

Afghanistan, UK

On the return to London from Bosnia in the spring of 2008, the Battalion enjoyed a year which mixed ceremonial duties with field training and exercises, leading the Regimental Lieutenant Colonel to claim that 'we should not under-estimate how well they have done in switching from red to green and back again'.[1] Earlier in the year the Battalion had been placed on standby for an emergency deployment to Kosovo in the public order role during the period of the elections (Operation Valero), but in the end they were not required. The Regiment Trooped the Colour at the Queen's Birthday Parade on 14 June 2008, with The Prince of Wales's Company providing the Escort to the Colour (Major Ben Ramsay and CSM Pollard) and Lieutenant Henry Finnegan as Ensign. Three weeks earlier the same Company had been reviewed in the gardens of Clarence House by HRH The Prince of Wales, a 'prestigious occasion' which was watched by many families and past members of the Company.

By the autumn the Battalion knew that they would be deployed to Helmand in the following year. Command of the Battalion was due to be changed during the Afghanistan tour on 1 July 2009; however, a new ruling stated that changes of command were not to take place during a deployment and so Lieutenant Colonel Richard Stanford handed over to Lieutenant

Soldiers of No. 2 Company carry a wounded comrade to a Blackhawk medevac helicopter. This followed a prolonged contact with Taliban fighters which saw the enemy attack from various positions on both sides of a river in Helmand during Operation Panther's Claw, Afghanistan, 2009. *(MoD)*

Colonel Rupert Thorneloe in October 2008, the former deployed directly to Iraq on promotion. The new operational deployment entailed a move to Lille Barracks at Aldershot, as part of 19 Light Brigade, under the command of Brigadier Tim Radford, and over the winter months the training cycle took Welsh Guardsmen all over the UK, from Thetford to Otterburn to Salisbury Plain. By then events were moving quickly in Afghanistan. That October the Taliban launched a major offensive in Helmand, mounting an attack on the provincial capital Lashkar Gah which almost succeeded in taking the objective. The incident and the reaction to it would have an impact on the way in which the Welsh Guards were used once they arrived in Helmand for a six-month tour of duty (Operation Herrick 10) in April 2009.[2] During the deployment, the Welsh Guards Battle Group (as it became) was destined to experience some of the most intense fighting of recent times and to lose their Commanding Officer, the first British battalion commander to be killed in action since the Falklands War. The story of that tour has been told in unsparing but sympathetic detail by Toby Harnden, an experienced foreign correspondent and former Royal Navy officer, who had access to operational papers and who also conducted over 300 interviews with the participants both inside and outside the Regiment. In common with the Welsh Guards' two histories of the twentieth-century global conflicts, it should be regarded as essential source material and it is with that in mind that the events of Operation Herrick 10 should be viewed.

Before leaving for Afghanistan, the Battalion was slightly under-strength and the new Commanding Officer was determined to make up numbers by bringing back recently retired Welsh Guardsmen under a scheme known as Full Time Reserve Service, which allowed former soldiers to return to the Army for a specified period to cover manpower shortages. Amongst them was Major Mark Jenkins, a former commander of The Prince of Wales's Company who had left the Army in 2001 but returned to act as Thorneloe's Brigade Liaison Officer. Other replacements came from other Foot Guards regiments but, even so, on deployment the Battalion was still thirty-two men under-strength, a not unnatural circumstance at a time when most of the Army was stretched. As the Battalion was being reconfigured to form a battle group, there was also a need to increase the number of companies to meet the operational requirements of the deployment. Under the original planning requirement,

Guard of Honour found by No. 2 Company in the Inner Quadrangle, Windsor Castle, for Khalifa bin Zayed bin Sultan Al Nahyan, President of the United Arab Emirates, 2013. *(Welsh Guards Archives)*

Thorneloe had been told that he would only have two rifle companies under direct command but, following consultation with senior regimental figures, this was later changed to allow the creation of two new companies for force protection and mentoring and training duties. The solution was both practical and steeped in the Regiment's history. Thorneloe decided to name the first of the new companies IX Company after the lead company in the 2nd Battalion (the Roman numerals reinforced its antiquity). When the

LCpl Nuku of Support Company mentoring the ANP, Afghanistan, 2012. He is carrying an SA80 with UGL. LCpl Nuku is a Commonwealth Soldier from Fiji. On his back we can see the Counter IED Vallon. He has not shaved for medical, rather than operational reasons! *(Welsh Guards Archives)*

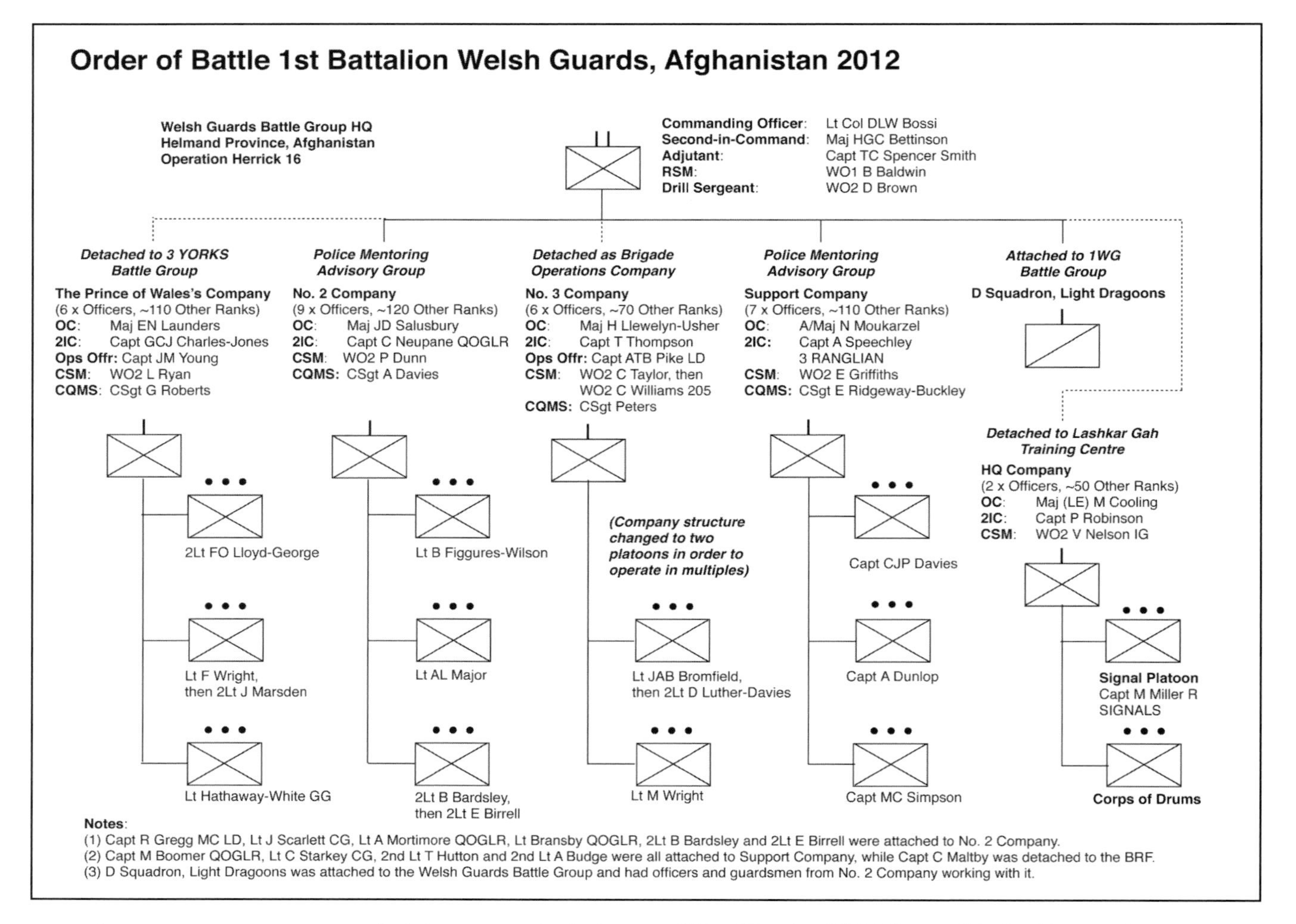

second new company was needed it was a relatively easy matter to term it X Company. Command of the two companies was given, respectively, to Major Sean Birchall and Major Rob Gallimore. Another nod to regimental history came from the honorary appointment of the mother of Major Dai Bevan as 'Company Captain' – she was the daughter of Brigadier James Windsor-Lewis who had served in the 2nd Battalion at Boulogne in 1940 and had escaped from German captivity to command it during the advance into North-West Europe four years later.

By happenstance, No. IX Company was the first to see action while based in the Main Operating Base at Lashkar Gah, supporting the Afghan security forces. This was Operation Zafar, launched on 27 April 2009, to clear the Taliban from several villages around Basharan. Throughout the fighting the Company served alongside troops from the Light Dragoons and 2nd Battalion, Mercian Regiment (2 MERCIAN), and afterwards reported that it had 'acquitted itself well and assisted in establishing three new Check Points: Worcester North and South'.[3] While this was happening, The Prince of Wales's Company (Major Giles Harris, CSM Jones 27) had been given responsibility for Nad-e Ali North based out of Patrol Base (PB) Argyll in the District Centre (DC) and PB Pimon to the north-west. This entailed protecting the DC and engaging the Taliban to keep them away from the main centres of population and facilitating 'shuras', meetings with key personalities in the locality.

No. 2 Company (Major Henry Bettinson, CSM Martin Topps), reinforced by a platoon from No. 3 Company, was tasked with occupying Nad-e Ali South at PB Silab and PB Tanda, but, on arrival in Helmand, they found that they had to man two other locations, Check Point (CP) Paraang and CP Haji Alem, over a mile to the east and one of the most isolated bases in the area of operations. This was a part of the operational area which the Commanding Officer had described as the 'hard shoulder' and which Major Bettinson characterized in the following portrayal: 'It was a predominantly flat area of the Green Zone, interspersed with canals, irrigation ditches, tree lines, fields of poppy and wheat, lone compounds, compound clusters and kalays (villages). As such it afforded good arcs of fire into ISAF locations, as well

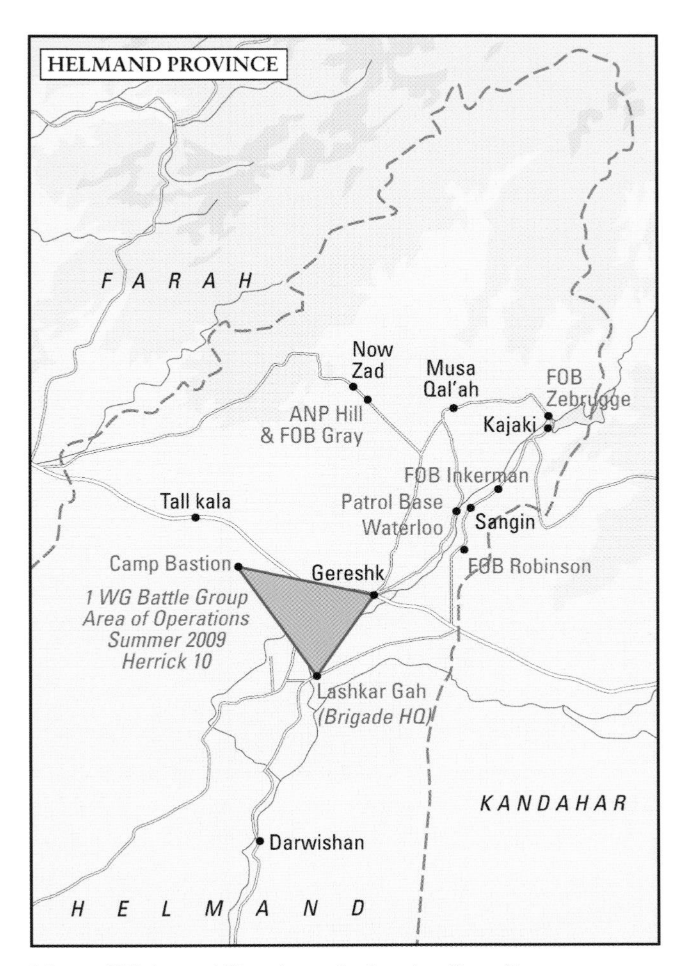

Map of Helmand Province during 1st Battalion deployment, 2009.

as good cover for the insurgents.'[4] Unfortunately that spare description meant that the topography suited the Taliban, who used it to good effect. Soon after the deployment, 7 Platoon in CP Haji Alem became aware of the difficulties facing them – the midday heat of over forty degrees, coming under accurate fire, the problems of re-supply and constant restraints in the communication network. It did not help that 'Every day they were facing several attacks on their bases, as well as Taliban ambushes whenever they ventured out.'[5] On 9 May, while returning from a patrol, Lieutenant Mark Evison, the platoon commander, was hit by Taliban gunfire and later died of his wounds. He was one of the five Welsh Guardsmen (plus two attached) killed during the tour: eleven days earlier, on 28 April, Lance Sergeant Tobie Fasfous, from Pencoed near Bridgend, had been killed by an IED device close to Forward Operating Base (FOB) Keenan.

Support Company (Major Austen Salusbury, CSM Lee Scholes) had its personnel distributed across the Battle Group. To the depleted No. 3 Company (Major Guy Stone, CSM Brian Baldwin) fell the task of mentoring the 2nd Kandak (Battalion) of the Afghan National Army (ANA) at Sangin, which had been branded 'the deadliest area in Afghanistan'. This brought it under operational command of 2 MERCIAN within the 2nd Battalion, The Rifles' (2 RIFLES) area of operations, but despite significant challenges, including the threat of roadside bombs, No. 3 Company completed its tour without suffering any casualties. Elsewhere, the Battle Group's Headquarters was based at Camp Bastion, but the Commanding Officer was keen to move to a new headquarters at Nad-e Ali town so that he could be closer to his men and to local Afghan officials.

At the time, Brigadier Radford had seven battle groups under his command, each in a different location and each of a different size with different tasks to fulfil. Broadly speaking, the British plan for the reconstruction and pacification of Helmand identified the area around Lashkar Gah and Gereshk as crucial for improving conditions, which could then be exploited by the Foreign and Commonwealth Office and the Department for International Development and their associated agencies. With that in mind, the Helmand Task Force had been ordered to establish a British centre of operations at Camp Bastion and to secure a triangle of territory between that base, Lashkar Gah and Gereshk. Successive brigades had arrived for six-month tours and each had their own objectives, often leading to a dispersal of effort. As Professor Anthony King, a policy analyst, saw the situation:

Sharpshooter from No. 3 Company with the 7.62 mm L129A1 rifle, Afghanistan, 2012. *(Welsh Guards Archives)*

During Operation Panther's Claw, Guardsmen of No. 7 Platoon, together with members of the Mortar Platoon, respond to enemy fire as they occupy a compound near Checkpoint Yellow 7 on the Shamalan Canal in Helmand, Afghanistan. *(MoD)*

A Guardsman of No. 1 Platoon, based at Patrol Base Argyll, Nad e Ali, is pictured returning from a patrol around the village of Zarghun Kalay, Afghanistan. *(MoD)*

1st Battalion Battle Group prepares to deploy on Operation Panther's Claw, Afghanistan, 2009. *(Welsh Guards Archives)*

Viking vehicles carrying Welsh Guardsmen and driven by men of the Royal Tank Regiment engage enemy positions after coming under fire from compounds surrounding Checkpoint Yellow 7 on the Shamalan Canal. *(MoD)*

Moments before a contact with the enemy, a column of Viking armoured vehicles, led by a Ridgback Armoured Fighting Vehicle, rolls up to Checkpoint Yellow 7 on the Shamalan Canal in Helmand Province, Afghanistan. *(MoD)*

Welsh Guardsmen training in Kenya during Exercise Askari Storm, September 2014. *(MoD)*

Queen's Birthday Parade, 15 June 2013. Left, Maj H. G. C. Bettinson, Major of the Parade and, right, the Field Officer in Brigade Waiting, Lt Col D. W. L. Bossi, who was also Commanding Officer of the Battalion at the time. *(Welsh Guards Archives)*

> It is noticeable that each brigade tour of Helmand has sought to define itself by a major operation: 16 Brigade 'broke in', 3 Commando Brigade retook Sangin, 12 Brigade 'mowed the grass', 52 Brigade retook Musa Qala, 16 Brigade transported a huge turbine to the Kajaki dam project, 3 Commando Brigade seized Nad-e-Ali ... there was very little continuity between the tours, as brigade commanders defined their own objectives.[6]

Although King admired the professionalism of those involved, he counselled that, while there had been gains, 'They have contributed less to the overall development of the campaign than the resources and efforts invested in them might demand.'

All that was supposed to change with the main offensive for the summer of 2009 – Operation Panchai Palang, or Panther's Claw – which was aimed at securing a swathe of the Helmand river valley roughly the size of the Isle of Wight, with a population of around 80,000. With casualties mounting, the British government was anxious to see a decisive offensive ahead of the forthcoming elections. However, the operation was stymied before it began by a shortage of resources and a growing belief that 19 Brigade would not be able to hold ground it had secured, leaving the Taliban free to return once the fighting had died down. Having been given the responsibility for securing Route Cornwall, which ran alongside the Shamalan Canal on the eastern side of the operational area, the Welsh Guards Battle Group began to fight its way up the track on 25 June. It was a well-made plan, but even before it began, Thorneloe had expressed worries about its purpose and the capacity of the Brigade to undertake it successfully. He was particularly exercised by the shortage of available helicopters to counter the growing tendency of the Taliban to place land mines or IEDs on routes used by British forces. Those concerns were increased on 19 June when Major Sean Birchall was killed after his Jackal armoured vehicle hit an IED on the track between CP South and CP North. The driver, Lance Corporal Jamie Evans 15, was badly wounded in the incident.

LSgt Jones during Foxhound familiarization training, May 2014, Warminster, Wiltshire. *(MoD)*

Earlier, after Operation Zafar, Birchall had insisted that his men should assist in reopening the local school at Basharan as this would represent 'a significant step in bringing the local population onside'.[7]

In the initial stages of the advance, the Welsh Guards made good progress. Achieving total surprise by driving into the heart of the town and clearing from the inside out, The Prince of Wales's Company moved up to take Chah-e Anjir, the 'prize in the north', without a shot being fired. That night the Taliban counter-attacked and continued to do so every day thereafter in a failed attempt to disrupt the Company's counter-insurgency efforts. At one stage Sergeant Parry 700 and his men were holding the eastern position of the town with no cover other than ditches in the fields. This bold and daring move allowed No. 2 Company to echelon through them as they moved north in a mixture of Mastiff, Ridgeback and Viking vehicles towards the northern tip of the canal where they were due to link up with B Company, 3rd Battalion Royal Regiment of Scotland (3 SCOTS). It was at this point that it became clear that Route Cornwall was too narrow to allow the advance to continue unimpeded. Matters were not helped on 29 June when a Viking vehicle carrying the Reconnaissance Platoon slipped into the canal and overturned, almost killing all of those on board. In the midst of the battle the incident shook those involved and, according to Harnden who interviewed many of those involved, 'The Sharmalan Canal was already taking on an aura of almost mystical foreboding.'[8] Aware that his men had been shaken and keen to get forward to the front, Thorneloe set out along Route Cornwall on 1 July, his intention being to pass through PB Shahzad where he could be briefed before continuing north with the supply convoy consisting of ten Vikings, seven trucks and a Panther.[9]

Typically, Thorneloe rode in the lead vehicle, occupying the top cover position and taking a full part in the operation to check the road ahead of the convoy – known as 'barma-ing', using Vallon metal detectors to find suspect objects dug into the road. At 15.18 there was a massive explosion which killed Thorneloe as well as Trooper Josh Hammond of the Royal Tank Regiment. Coming after the earlier casualties, this could have been a body blow to the Battalion but the men held firm, largely as a result of their own training and discipline and also because of a desire to take the fight back to the Taliban. The mood was summed up later by Martin Topps, at the time CSM of No. 2 Company and later Regimental Sergeant Major: 'The

Welsh Guards recruiting poster, 2014. *(MoD)*

hardest thing I found was that whenever there was a fatality I'd do the vigil service and the act of remembrance and then I'd say to the guys to stay focused or they would become a statistic. Forget about it now and we'll give him a send-off when we get back. It sounded heartless but we had to stay focused.'[10]

Thorneloe was the second Welsh Guards Commanding Officer to be killed in action, the first being Johnny Fass at Cheux in Normandy in June 1944. Two days later Lieutenant Colonel Doug Chalmers, PWRR, took over interim command of the Welsh Guards Battle Group after flying to Afghanistan from Cyprus. Having been Thorneloe's predecessor in Afghanistan, he was a sound replacement, but a new Commanding Officer had to be found for 1st Battalion Welsh Guards and the choice fell on Major Charlie Antelme, who was in a staff appointment in

Lt Col G. R. Harris, DSO, MBE, Commanding Officer, with Capt C. J. P. Davies, Adjutant (*right*), and W01 (RSM) M. Topps (*left*), Cavalry Barracks, Hounslow, May 2014. *(MoD)*

London and who had served as a Company Commander in Iraq where he had been awarded a DSO 'for leading an attack on an al-Qaeda safe house … armed only with a pistol after his rifle had jammed'.[11] But it was not the end of the setbacks. Four days later, on 5 July, Lance Corporal Dane Elson was killed by an IED while on foot patrol. Despite the losses and the fact that the Battle Group was involved in repelling multiple attacks from the insurgents on what seemed to be a daily basis, the mood remained positive as the Guardsmen moved north along the Sharmalan Canal, clearing the area of insurgents and securing fourteen crossing points. Operation Panther's Claw came to an end on 31 July and the official view was that, despite the twenty-three killed in action, it had been a success. Brigadier Radford told CNN: 'We've had a significant impact on the Taliban in this area – both in terms of their capability and their morale.'[12] Overlapping with the British offensive, 4,000 US Marines from the 2nd Marine Expeditionary Brigade, plus 650 Afghan soldiers, had launched Operation Strike of the Sword (Operation Khanjar), a major offensive around the town of Nawa-

Welsh Guardsmen showing uniform and badges of rank for home service clothing. (*L to R*) Gdsm Stevens, LCpl Selby, LSgt Shapland, Sgt Deren, CSgt Lias, Drum Maj De-Wit, and WO2 (CSM) Bowen. Cavalry Barracks, Hounslow, May 2014. *(MoD)*

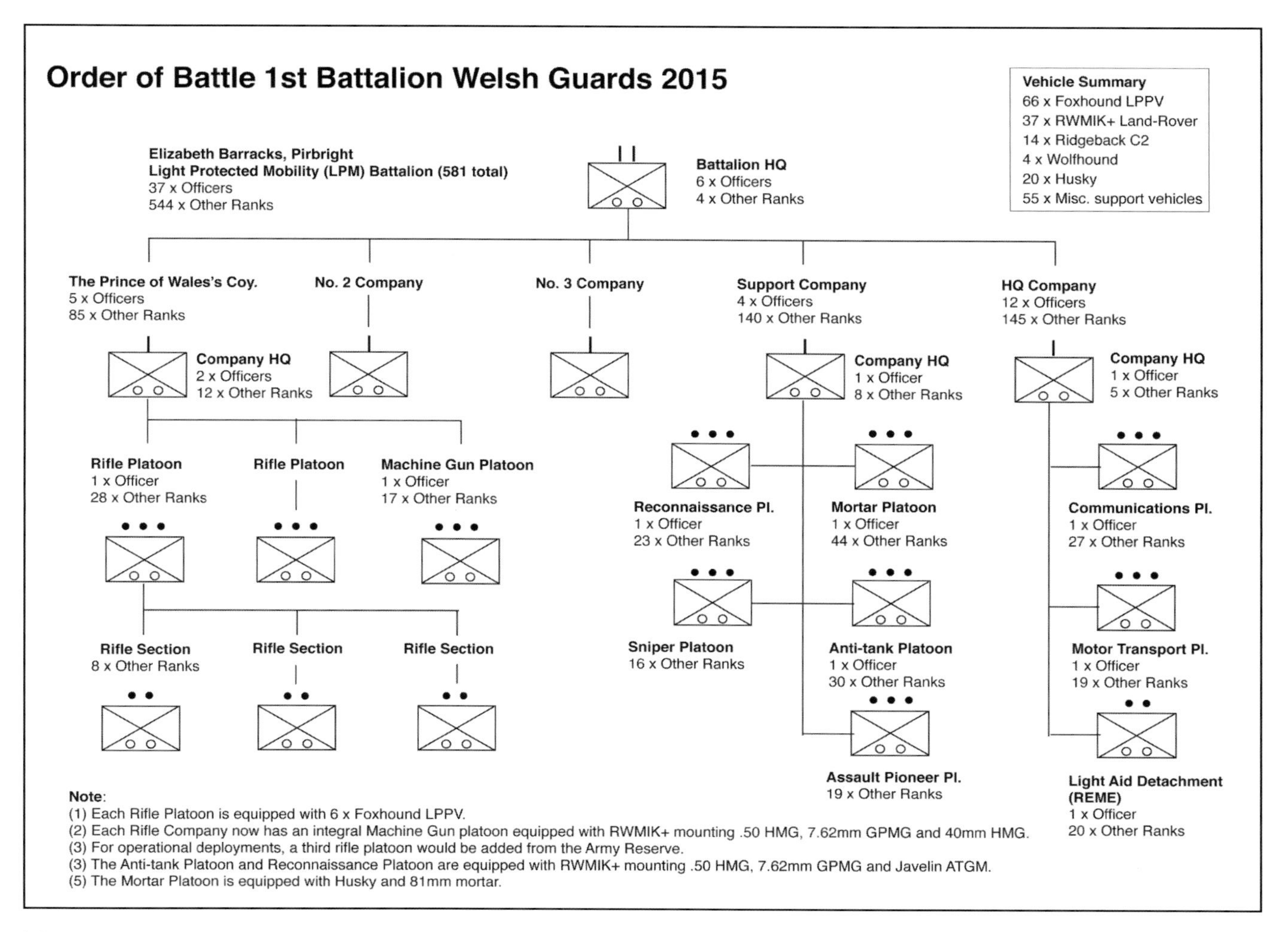

l-Barakzayi, south of Lashkar Gah. The operation began when units moved into the Helmand River valley in the early hours of 2 July 2009; it was the biggest offensive airlift by the US Marine Corps since the Vietnam War. In their mentoring role with the ANA, Major Gallimore's X Company was attached to this operation, serving alongside Golf and Echo Companies of 2nd Battalion 8th Marine Regiment, an involvement which he described as 'both frenetic and critical'.[13]

In the period after Panther's Claw, IEDs continued to present a problem across the 19 Brigade area of operations but, despite the continuing violence, the presidential elections went ahead as planned. Across the Brigade area, polling stations remained open, although the numbers of voters varied according to local circumstances and the capacity of Taliban fighters to prevent voters reaching polling stations. According to the new Commanding Officer, this had been prompted by 'a potent cocktail of fear and apathy', but other factors also intruded, notably corrupt electoral practices and the unpopularity of the Kabul government. Although it was something of a miracle that the elections had taken place at all, the turnout of 30 per cent overall was disappointing. However, there were grounds for optimism in Chah-e Anjir, the largest town in the Nad-e Ali district, which was under the control of The Prince of Wales's Company whose Guardsmen showed what could be done in terms of counter-insurgency doctrine by partnering Afghan security forces in joint patrols in a successful attempt to persuade 'local nationals that we were the good guys'. By the end of the tour, the local school had been reopened – something previously unthinkable before The Prince of Wales's Company fought to unseat the Taliban from the town.

On 3 October the Battalion handed over to 1st Battalion Grenadier Guards, and on their return to the UK settled back in at Lille Barracks in Aldershot. Inevitably the casualties made it seem that the tour had been a downbeat experience, but the reality was that the efforts of the Welsh Guards Battle Group had laid the foundations for securing the operational area, not just through their own efforts but with the help of others, including an Estonian company, one of whose soldiers had already served in Afghanistan with the

Afghanistan Appeal

During the Welsh Guards' first tour of Afghanistan in the summer of 2009, the significant number of Welsh Guardsmen killed in action or suffering serious injuries, many of them life-changing, led the Regimental trustees to launch an Afghanistan Appeal in September of that year, recognizing that there would be increased welfare needs – both physical and psychological – for those affected. Given the high number of casualties received in Afghanistan, the Regiment knew that existing Regimental charitable funds would be insufficient to meet expected needs.

The response to the Welsh Guards Afghanistan Appeal has been magnificent. The Regiment prides itself in being a close-knit family regiment and the support received from both serving and retired Welsh Guardsmen has made a huge difference. Support from the general public, particularly in Wales, has also been impressive and humbling. The Regiment is incredibly grateful to all those who have raised funds for the appeal. Special mention must be made of the Peterson family, Ryan Jones and the Walk on Wales initiative. The former Welsh Rugby Captain and renowned international player chose the Appeal as his charity for his Testimonial Year and has since donated more than £70,000 to it. The Walk on Wales initiative, which took place between August and November 2013, became a pilgrimage for thousands of Welsh Guardsmen who walked around the coastline of Wales and in doing so raised over £300,000, divided between the Welsh Guards Appeal and Combat Stress. The Drapers' Company has also been particularly generous.

The importance of the Appeal was again brought into sharp focus when the Battalion returned to Afghanistan for a second tour in 2012. As in 2009, the Battalion performed with distinction, but the deployment was not without cost, with three Welsh Guardsmen killed in action and many others wounded. The funds raised by the Appeal have been used to support the families of those killed in action on both tours, as well as the injured, often in ways that other charities cannot. The Appeal has also allowed the Regiment to employ a former Welsh Guards Warrant Officer as a full-time Regimental Casualties Officer, providing a more direct means of identifying and assisting the bereaved, the wounded and those affected by combat operations.

Looking to the future, the evidence of past conflicts warns us that many stress-related psychological injuries do not surface for ten to fifteen years afterwards. Our vision is that the Welsh Guards Afghanistan Appeal will provide a fighting fund to ensure that when veterans need help in years to come – when the memory of the campaign has faded from the public consciousness – the fund will be ready to respond quickly to their needs.

Soviet Army. Unlike previous homecomings from operational duty, there had been a change in the public's perception of the Army. Throughout 2009 backing for the troops fighting in Afghanistan had been expressed by displays of public support, most notably in the town of Wootton Bassett, close to RAF Lyneham, where hearses carrying coffins draped in the Union flag were greeted by large crowds and colour parties from the Royal British Legion. Returning battalions and regiments were also greeted by 'homecoming parades' as local communities welcomed back men and women who had served in Afghanistan by showing their support as they marched through the streets. It was no different for the Welsh Guards. The first inkling of what lay ahead came in October when over 300 Guardsmen and their families were the guests of Cardiff City Football Club: a crowd of 25,000 cheered eighty members of the Battalion, including some of the wounded, as they marched onto the pitch before the game against Crystal Palace. This was followed in March 2010 by a month-long event which saw a number of parades, some of them through Freedom Towns, with the Battalion divided into three areas – The Prince of Wales's Company in North Wales (Donington Barracks), No. 2 Company in the south-west (Penally Camp) and No. 3 Company in Cardiff (Maindy Barracks). On the back of this widespread enthusiasm, the Regiment's Afghanistan Appeal was launched with a variety of fund-raising events which were characterized by the chairman, Brigadier Peter Williams, DL, as 'skydiving, running against a horse, desert marathon, climbing every conceivable peak in Wales as well as the Three Peaks Challenge, and even someone tattooed in the cause'.[14]

The launch of the Appeal and the season of homecoming parades were a welcome fillip to the Regiment, especially in Wales where the Battalion's tour of duty in Afghanistan enjoyed a high profile across the Principality. In 2010 Brigadier Robert Talbot Rice took over as Regimental Lieutenant Colonel from Colonel Sandy Malcolm, who had been responsible for leading the Regiment during the turbulent and frequently stressful time of the operational tours in Iraq and Afghanistan and was the guiding light behind the creation of the Afghanistan Appeal. With recruiting holding up and with the Companies involved in a wide range of training exercises, morale was high and remained so throughout the year – one of the highlights being the presence of seventy-six Guardsmen from No. 2 Company (Major Dai Bevan, CSM Paul Dunn) at Russia's Victory Parade in Moscow on 9 May to mark the sixty-fifth anniversary of the end of the Second World War. They were the first British troops to be granted this honour, taking part in a lavish multi-million-pound parade, which included a 1,000-strong military band, a variety of tanks and missiles and 127 aircraft.

Lessons were learned from Afghanistan, not least in Support Company with the adoption of fire support groups which could operate in support of the rifle companies. During the year they began to receive the new .50 Heavy Machine Gun as well as the 40 mm Grenade Machine Gun. Towards the end of the year the Battalion deployed to Kenya on Exercise Askari Thunder and, on return to the UK, moved into Cavalry Barracks at Hounslow beneath the flight path to Heathrow Airport. Ahead lay the training cycle which would take the Battalion back to Helmand on Operation Herrick 16 in April 2012, on this occasion acting in a training role as the Police Mentoring and Advisory Group (PMAG).

All this was happening against a background of further change in defence policy, with the Strategic Defence and Security Review of 2010, which planned to reduce the Regular Army to 90,000 soldiers by 2015 and to 82,000 by 2020. Despite the turbulence, the Regimental Lieutenant Colonel remained bullish in his report for the year, telling his fellow Guardsmen that it was not all doom and gloom: 'I can say with absolute honesty that at no time in my military service has the Battalion enjoyed better organized training, better personal equipment or better soldier living accommodation than it has now.'[15] Several layers of icing were added throughout 2011. After twenty-nine years of disappointment, the rugby team won the Army Premiership Cup (as the Army Cup had become in the season 2006–7) beating 2nd Battalion

Guardsmen deploying from the Foxhound LPPV during familiarization training, Salisbury Plain, May 2014.

The Royal Welsh 28–9 in an exciting final. It was the Regiment's eleventh triumph in the competition and their twenty-third appearance in the final. An unexpected replay took place on 22 September when the teams repeated the match at the Millennium Stadium in Cardiff, but this time the Royal Welsh won 19–12. The Welsh Guards team captain, Lance Corporal Lewis 23, was selected to play for the Barbarians (as was Lance Sergeant Dwyer) and he also captained the Army XV in their win against the Royal Navy at Twickenham. Another highlight came on 29 April when the Welsh Guards played a pivotal role in the wedding of the Duke and Duchess of Cambridge, providing a Guard of Honour, formed by No. 2 Company, in the forecourt of Buckingham Palace. In May the Regiment received the Freedoms of the County of Powys and the County Borough of Bridgend, both occasions being marked by large and enthusiastic crowds, with No. 2 Company forming the Guard of Honour. It was also decided to re-launch the Afghanistan Appeal to meet the continuing needs of those affected by service on Operation Herrick.

In advance of the next deployment to Helmand, the Battalion set about training in earnest for its new role. In the middle of September a cadre was held at Cavalry Barracks which included background briefings on this quite different role, a presentation by 5th Battalion Royal Regiment of Scotland (5 SCOTS) on their experience of the PMAG role, and a realistic demonstration by Welsh-speaking Guardsmen, under the command of Lance Sergeant Jones 88, playing the role of an Afghan National Police commander and insisting that everyone at the checkpoint was called Jones. By the time the Battalion deployed on Herrick 16 the security situation had changed radically in Afghanistan. In the previous year, 2011, the al-Qaeda

Guardsmen wearing the Army's new No. 2 dress at Cavalry Barracks, Hounslow, preparing for public duties during the summer of 2014. *(MoD)*

leader Osama bin Laden had been killed by US special forces during an operation against his hideaway in Abbottabad in Pakistan in May and, in the following month, President Barack Obama announced that 10,000 US troops would be withdrawn from Afghanistan by the end of 2011 and an additional 23,000 troops would leave the country by the summer of 2012. Following suit, other NATO countries, including the UK, also announced reductions in troop numbers, with a final withdrawal planned for the end of 2014; in December 2012, Prime Minister David Cameron announced that the UK's conventional force levels in Afghanistan would draw down to around 5,200 by the end of 2013. In addition to profiting from Afghanistan's improving security situation, the reduction brought economic benefits by improving UK public finances. It meant that the Treasury could plan to cut spending on Operation Herrick: £3.7bn in 2012–13, £3bn in 2013–14 and £2bn in 2014–15.[16]

These changes made an impact on the Battalion's tour, which began in March 2012 when they formed part of 12 Mechanised Brigade under the command of Brigadier Doug Chalmers, an old friend of the Battalion. Quite apart from the fact that the action was less kinetic than it had been three years previously, the main thrust of the Battalion's work centred on its PMAG responsibilities across the British-controlled zone in Helmand, which were to develop the Afghan uniformed police there so that they could provide security for ordinary Helmandis to go about their day-to-day lives. This was a key part of the UK's withdrawal plan – if indigenous forces could be trained to provide their own security, UK troops would no longer be needed to fend off insurgent attacks. Battalion Headquarters with No. 2, Headquarters and Support Companies formed the nucleus of PMAG and were reinforced by two company-strength groups from the Queen's Own Gurkha Logistic Regiment and the Light Dragoons, plus a further company from the Danish Army. (Headquarters Company formed the mentoring team in the Lashkar Gah Police Training Centre – a key location that was visited by more ISAF senior officers than anywhere else.) In addition to contingents from the Royal Military Police, Royal Engineers and others, the Commanding Officer, Lieutenant Colonel Dino Bossi, estimated that twenty-three cap badges were represented in his 'rainbow nation'. Outside those mentoring and training duties, The Prince Wales's Company (Major Ed Launders, CSM Ryan) occupied the remote Patrol Base 5 in the Nahr-e Saraj district where they operated as the main deployable force for the 3rd Battalion Yorkshire Regiment (3 YORKS)

A Foxhound LPPV shows its cross-country performance. It carries a driver, vehicle commander and four soldiers in the rear compartment. *(MoD)*

Lieutenant Colonel Rupert Thorneloe, MBE

On Wednesday, 1 July 2009, at approximately 15.15 hours Afghan local time, Lieutenant Colonel Rupert Thorneloe, MBE, Commanding Officer, 1st Battalion Welsh Guards, was killed in action as a result of a Taliban improvised explosive device detonating underneath his Viking armoured vehicle north of Lashkar Gah, Helmand Province. Trooper Joshua Hammond of the 2nd Royal Tank Regiment was also killed in the same incident. The Welsh Guards Battle Group under Colonel Thorneloe's command was in the opening stages of Operation Panther's Claw, designed to extend security force control in the north of the region in advance of the Afghan presidential elections in August. Colonel Thorneloe was one of five Welsh Guardsmen killed in action on Operation Herrick 10 in 2009.

Thorneloe was an outstanding Welsh Guardsman and Commanding Officer, who was at the leading edge of his generation and universally liked and respected by all ranks in the Regiment and across the Army. Educated at Radley and passing out of Sandhurst in 1992, he later became Adjutant and commanded No. 2 Company. He held a number of key staff appointments and was awarded an MBE for his considerable personal contribution to the work of HQ 1 (UK) Armoured Division on their seven-month deployment to south-east Iraq in 2005–6. He was also Military Assistant to the Assistant Chief of Defence Staff (Policy) and, as a lieutenant colonel, served two Defence Secretaries of State. His intellectual ability, charm and complete reliability stood him in good stead for this critically important appointment. He took over command of the Battalion in October 2008.

Perhaps no better epitaph on Colonel Thorneloe can be given than the one by Brigadier Tim Radford, his brigade commander in Afghanistan in 2009, when he said, 'He died as he had lived his life, leading from the front. I valued his leadership, his honesty and his enormous moral and physical courage. He was destined for greatness in the Army.'

Colonel Thorneloe is survived by his wife, Sally, and their two daughters.

Lt Col R. Thorneloe, MBE, takes a break during Operation Panther's Claw, Afghanistan, July 2009. *(Welsh Guards Archives)*

Battle Group. During their tour the Company conducted some thirty-five operations resulting in the successful interdiction of insurgents and the seizure weapons and equipment. Despite losing several vehicles to IEDs, this was achieved with no serious injuries or loss of life as a result of enemy action.

On arrival in Helmand, No. 3 Company (Major Henry Llewelyn-Usher, CSM C. Williams) occupied PB Pimon in the north-west of the British zone where they were surrounded by significant insurgent activity. Worryingly, the opposition fighters were found to be well-disciplined, decent shots and able to execute well-planned attacks, making good use of ground. In the first week two Guardsmen were badly wounded, providing proof of the Taliban capacity 'to manoeuvre at ease in their own surroundings'. In June the Company was re-roled as the Brigade Operations Company, tasked with disrupting insurgent activity in areas uncontrolled or unvisited by ISAF forces. This produced a total of 260 foot patrols aimed at keeping the opposition off-balance and it was achieved 'by either conducting long covert night patrols, then cordoning compounds of interest prior to first light or alternatively flying by either UK or USMC [United States Marine Corps] aviation into areas of interest'.[17] While Herrick 16 was quieter than its predecessor in 2009, it was not without drama. There were eleven Guardsmen wounded in action, and thirty-three non-battlefield injuries, but the most serious incident occurred when No. 2 Company (Major Julian Salusbury, CSM Paul Dunn) suffered two 'green-on-blue' attacks involving Guardsmen being attacked and killed by rogue Afghan policemen. The first took place near PB Attal when Lance Corporal Lee Davies of Carmarthen was shot dead on 12 May, together with Corporal John McCarthy of the RAF. The second incident took place when Guardsman Craig Roderick from Pencoed and Guardsman Apete Tuisovurua were killed near Forward Operating Base Oullette on 1 July. Also killed in the attack was WO2 Leonard Perran

Major General Robert Talbot Rice

Robert Harry Talbot Rice was educated at Eton College and commissioned into the Welsh Guards in 1983. His father was a Welsh Guards National Service officer and his grandfather was one of the Regiment's founding officers who served in both world wars and commanded the Prince of Wales Company.

Having started in the Battalion as a platoon commander in No. 3 Company, Talbot Rice completed an in-service degree at Durham University and returned to the Battalion as Adjutant. After completing Staff College, he commanded The Prince of Wales's Company, and then the 1st Battalion from 2002 to 2004. His operational service included four tours in Northern Ireland (including command in Londonderry); command of the British Battle Group in Bosnia; and tours on the staff in Kosovo (in Headquarters KFOR, J5 Campaign Plans) and Iraq (in General Petraeus's headquarters as the chief liaison officer to the Iraqi Ministry of Defence).

As a staff officer, he served in a range of planning and equipment acquisition posts. As a Brigadier, he was Director Equipment in Army Headquarters, responsible for managing all the vehicles and land equipment used by the three services. He was then appointed Head of Armoured Vehicles in the Ministry of Defence's procurement organization to manage the Army's principal capital equipment projects, including the new Scout reconnaissance vehicle, in which role he was promoted to Major General in 2014. He was appointed Regimental Lieutenant Colonel Welsh Guards in 2010.

Major General R. H. Talbot Rice, Regimental Lieutenant Colonel, 2010–present. *(Welsh Guards Archives)*

Thomas of 37 Signals Regiment (Volunteers), who had served previously with 1st Battalion Welsh Guards, reaching the rank of colour sergeant in the Recce Platoon before leaving the Regular Army in 2000 and later rejoining the Royal Signals as a reservist and volunteering to deploy to Afghanistan with his former Regiment. While these attacks were serious setbacks which could have damaged the PMAG concept, the Company Commander reported later that 'Everyone continued to work as hard as possible to maintain relations with the Afghan Police and ensure that all the good work was not lost.'

For Support Company (Major Naim Moukarzel, CSM E. Griffiths), the primary task was to provide the District Advisory Team for the Afghan Police in Nad-e Ali. During the tour the Company was reinforced by multiples from the Queen's Own Gurkha Logistic Regiment and the Light Dragoons. They also took under command two police advisory teams (PATs) from 3rd Battalion The Rifles (3 RIFLES), and 1st Battalion Royal Anglian Regiment (1 R ANGLIAN). For those who had served in the previous tour it was a strange though welcome experience to be operating over ground that was so familiar to the Battalion yet so dramatically changed, having been transformed into an Afghan government sphere of influence. Instead of the mayhem of the previous tour, Nadi-e Ali had been secured from south to north and as a result, was a relatively quiet place where the police were 'clearly functioning and we were really focusing on developing and improving existing structures, rather than starting from scratch'.[18] The Company also contributed to the Brigade Reconnaissance Force (BRF), with Captain Chris Fenton as second-in-command and Lance Sergeant Evans 88 in charge of the snipers, while Captain Charlie Maltby was one of the troop commanders. During the tour the BRF enjoyed a number of successes, conducting over sixty deliberate operations including removing illegal weapons and ammunition and disrupting nine facilities for making IEDs.

Roll of Honour: Afghanistan

Lance Sergeant Tobie 'Fas' Fasfous, 28 April 2009
Lieutenant Mark Evison, 12 May 2009
Major Sean Birchall, 19 June 2009
Lieutenant Colonel Rupert Thorneloe, MBE, 1 July 2009
Lance Corporal Dane Elson, 5 July 2009
Private John Brackpool, formerly The Princess of Wales's Royal Regiment, 9 July 2009
Guardsman Christopher King, 1st Battalion Coldstream Guards, attached 1st Battalion Welsh Guards, 22 July 2009
Lance Corporal Lee Thomas Davies, 12 May 2012
Guardsman Craig Andrew Roderick, 1 July 2012
Guardsman Apete Saunikalou Ratumaiyale Tuisovurua, 1 July 2012

The tour ended in October and, while the Battalion had suffered casualties, the overall feeling was that Operation Herrick 16 had been a success. Levels of violence had decreased and, most importantly, progress had been made in mentoring and training the Afghan security forces. On the return to the UK, Brigadier Chalmers described the tour as a period in which the emphasis had been on enabling the Afghan National Security Forces to take over responsibility for security, while at the same time driving away the insurgents from the main centres of population. All this boded well for the country's future and he hoped that it would allow a smooth transition, telling a debrief at the Ministry of Defence: 'It felt that this tour, perhaps more than any other, brought together all the elements that had been achieved in previous Herricks and made sense of all the work that has been done over the years.'[19] Following the traditional well-earned leave and the by now familiar homecoming parades, including one mounted in Hounslow, the Battalion settled back into a refurbished cavalry barracks where the Commanding Officer promised a varied programme of public duties spiced by sport and adventure training. To the great pleasure of the Regiment, in June 2012 one of its most distinguished officers, Charles Guthrie, was appointed to the rank of field marshal, having retired as Chief of the Defence Staff in 2001.

Major General Ben Bathurst, OBE

Ben Bathurst was commissioned into the Welsh Guards after attending Bristol University. He joined No. 2 Company in Hohne in 1987 before becoming Aide-de-Camp to the Commander of 1st (British) Corps in Germany, Lieutenant General Sir Charles Guthrie, followed by operational tours in Belfast, Ballykelly and Bosnia. He commanded No. 3 Company in 1997 for the deployment to South Armagh and to Belize in 1998. As an SO2, he worked in the Ministry of Defence's equipment capability area where he defined requirements for new anti-armour weapons including JAVELIN and NLAW.

As a staff officer, he was involved in the strategic planning for operations in Iraq, before being posted to the US Headquarters in Baghdad as the insurgency developed. Promoted to Lieutenant Colonel, he commanded the Welsh Guards Battle Group in Maysan, Iraq, 2004–5. This was a highly successful operational tour and the Battalion returned to St Athan without losing a single Guardsman. Promoted to Colonel, he directed the Army Public Relations team from 2006 to 2008 and was responsible for managing the media during HRH Prince Harry's first deployment to Afghanistan. On promotion to Brigadier in 2008, he returned to Baghdad to work in the US Headquarters. In 2010, he commanded the Initial Training Group in Pirbright. In 2011, he became Director Training at Army Headquarters as well as Commandant of the Royal Army Physical Training Corps. Promoted to Major General in 2014, he was posted to HQ ISAF in Kabul, Afghanistan, to lead the MoD's Ministerial Advisory Group.

Major General B. J. Bathurst, OBE, Ops Offr 1WG, 1991–2, OC No. 3 Company, 1st Battalion, 1993–4 and 1996–8, CO 1 1st Battalion, 2005–7. *(Welsh Guards Archives)*

The Battalion entered 2013 in good heart with a busy schedule of public duties ahead of it. It was not an easy transition from the rigours of operational service but, as the Commanding Officer noted in his report for the year, 'In the process the Quartermaster, Sergeant Major, Drill Sergeants and Master Tailor have sustained numerous grey hairs and worry lines – but we got there despite the loss of voices (and not a few tempers) en route.'[20] On the Queen's Birthday on 15 June it provided three Guards for Trooping the Colour. In addition, the Escort to the Colour was The Prince of Wales's Company, the Field Officer was Lieutenant Colonel Dino Bossi, the Regimental Sergeant Major Martin Topps and the Ensign 2nd Lieutenant Joe Dinwiddie. Another highlight of the year was the mounting of the Walk on Wales (WOW), an ambitious and successful attempt to walk the 870 miles of the Welsh coastal path to honour the fifty Welsh Guardsmen who had been killed in action since the end of the Second World War. The aim was to raise £1 million to be shared by the Welsh Guards Afghanistan Appeal and the charity Combat Stress, a veterans' mental health charity. Walk on Wales was dreamed up by two Falklands veterans, Jan Koops and David (Dai) Graham, and the first of eleven relay teams set off on 25 August carrying a specially commissioned silver baton inscribed with the names of each of the fifty Guardsmen. It reached a triumphant conclusion on 2 November when the walkers arrived in Cardiff to a tumultuous reception, some of them having walked the entire route in sixty-one days. The appeal was also helped separately and significantly by the generosity of the Welsh international rugby player Ryan Jones, who chose the Welsh Guards Afghanistan Appeal as the beneficiary charity of his testimonial year in 2013. Together with the Walk on Wales, the linking of Wales's most capped captain of all time with the Welsh Guards produced a huge amount of well-merited publicity for the Regiment. It helped greatly that the regimental rugby team repeated its feat of two years earlier by winning the Army Premiership Cup, beating 39 Engineer Regiment 11–3 on 20 March.

As the Regiment's first hundred years came to an end, the Army, too, was changing. In response to the government's Strategic Defence and Security Review of 2010, the Chief of the General Staff instituted a top-to-bottom reassessment of the Army which was designed to bring 'a generational change in its vision, structure, composition and capability to ensure that it can meet the challenges of 2020 and beyond'.[21] Set against the background of the coming withdrawal

The Welsh Guards Association

Within six months of the end of the 1914–18 War, senior Welsh Guardsmen met under the chairmanship of the Regimental Lieutenant Colonel, Colonel Murray-Threipland, to consider the formation of a Welsh Guards Comrades Association 'to maintain connection between the past and serving members of the Welsh Guards and thereby promote their mutual interest and welfare of the Regiment generally'.

There were other objectives, which included 'the circulation of information concerning the Regiment and to encourage desirable candidates to join' and 'to arrange social gatherings in suitable places'. At a meeting in Cardiff in November 1919, it was agreed to form local Branches. In 1965, 'Comrades' was omitted from the Association title to encourage serving soldiers to be active members and the links between past and present have continued and are greatly enjoyed.

Until 1992 the Regimental Lieutenant Colonel was President, and the Superintending Clerk at Regimental Headquarters was the General Secretary. In that year a regimental office was opened at Maindy Barracks, Cardiff, and a past-serving Guardsman from within the Association became General Secretary, a position later changed to Secretary-General. Newsletters were introduced in the 1970s and at the same time the Association Report became the *Welsh Guards Magazine*.

Association events have included a biennial dinner, darts and shooting competitions, battlefield tours and a number of other occasions which bring past and present Welsh Guardsmen together, none more so than on St David's Day. The branches form the heart of the Association and are administered by enthusiastic volunteers who arrange many social events and uphold the standards and traditions of the Regiment in their areas.

The current President of the Association is Colonel (Retd) T. C. S. Bonas; there are two Vice-Presidents, Brigadier J. F. Rickett, CBE, and Roy Lewis; the Secretary-General is Brian Keane; and there are seventeen branches which are listed here with the date of their formation: Aberdare (1951), Cardiff (1919), Cardiganshire (1947), East Glamorgan (1920), Llanelli (1974), London (1926), Merthyr Tydfil (1953), Midlands (1953), Monmouthshire (1920), Montgomeryshire & Shropshire (1950), North America (1986), North of England (1946), North Wales (1936), Ogmore (1982), Pembrokeshire (1957), Swansea & West Glamorgan (1927) and Welsh Guards Reunited (2006).

from Afghanistan at the end of 2014 and an earlier decision to withdraw the garrison from Germany, the review was predicated on the fact that the bulk of the nation's armed forces would be primarily UK-based for the first time in many years and that this new configuration brought challenges and opportunities to embrace change. There would also be a new reliance on the integration of reserve forces. For the Welsh Guards this meant that it would become part of the planned Adaptable Force, as part of 11 Infantry Brigade South-East in the light protected mobility role, equipped with the Foxhound armoured patrol vehicle. Its paired reserve light role infantry battalion would be 3rd Battalion The Royal Welsh, based in Cardiff. Transition to the new Army 2020 structure was planned to take place between mid-2014 and mid-2015. During that period 1st Battalion Welsh Guards relocated to Pirbright towards the end of November 2014. That same year, in January, Lieutenant Colonel Giles Harris, DSO, MBE, had taken over command of the Battalion and, with deployments planned for 2014 to include the Falklands, Belize and Kenya, the Regimental Lieutenant Colonel was surely right to claim that 'as ever, it was a great time to be a Welsh Guardsman!'[22]

* * *

One hundred years on from their beginnings to fight in the First World War and to provide Wales with a regiment of Foot Guards, the Welsh Guards have come a long way. They received a baptism of fire at Loos and spent the bulk of their early years in action on the Western Front, becoming an integral part of the Guards Division and in so doing upholding the history and traditions of the regiments which made up the Brigade of Guards. In the Second World War they maintained and supported three active-service battalions, learned the mysteries of armoured warfare and emerged from the battlefronts in North Africa and Europe with a deservedly high reputation. Peace brought little respite as the old empire started contracting, forcing the Army to hold the line in dismal sweaty places while the politicians attempted to find workable solutions to seemingly intractable problems. The long war in Northern Ireland consumed huge resources, awakened ancient enmities and cost lives; the short sharp campaign in the Falklands ran close to catastrophe but ended triumphantly; the insurgencies in Iraq and Afghanistan tested ingenuity and capabilities but proved that they could be mastered by the unchanging standards of British infantry soldiering. In all those varied campaigns, Welsh Guardsmen played a prominent role, proving their worth and adding to an already illustrious history. As Lieutenant Colonel Charlie Antelme, Commanding Officer in 2009, told the Battalion on their return from Afghanistan, they were the equal of their forebears: 'The threads of Agincourt, Rorke's Drift and Monte Picolo remain unbroken; Welshmen shining amongst their comrades in the defence of the nation.'[23]

With good reason the centenary year, 2015, will be one of celebration. A number of events have been planned, the highlights being the presentation of new Colours by Her Majesty The Queen at Windsor Castle and Trooping the Colour on the Queen's Birthday Parade. The Battalion will also surge into Wales, marching through many of the Freedom Towns, and the Regimental Band will be busy giving a number of concerts in Cardiff and elsewhere. The Freedom of the City of Newport will also be granted to the Regiment in 2015. But the Centenary will also be a time for remembrance and reflection on the many who paid the ultimate sacrifice in the service of their country in both world wars as well as the many other operations and conflicts in which the Regiment has been involved, not forgetting also the many wounded, both physically and mentally. Throughout 2015, the Battalion will also be involved in state ceremonial and public duties in London and Windsor and training in preparation for its new role with 11 Infantry Brigade, starting in January 2016. The Battalion is likely to remain at Pirbright in the light protected mobility role for six years before re-roling back into public and ceremonial duties. As part of the Adaptable Force, it is very probable that the Battalion, or elements of it, will see itself deployed on operations should they arise, and in doing so, will maintain the excellence the Regiment's forebears achieved in the last hundred years.

In the midst of all those changes, most of them rapid enough to bewilder even the most involved observer, some things remain satisfyingly familiar. Between 1992 and 2012 the Battalion enjoyed one of the busiest periods in its history, seeing operational service in Northern Ireland, Bosnia, Iraq and Afghanistan. During that time its reputation grew exponentially and it quickly cemented its name as a reliable and thoroughly professional outfit, the kind that any senior officer would want to have under his command. During that time, too, Welsh Guardsmen proved over and over again that they could switch between operational and ceremonial duties, able to serve the Sovereign equally well in bearskins or body armour, being as professional in the sands of Helmand as they are in the streets of London. Despite the introduction of modern equipment and the arrival of new technologies, the Regiment retained a sense of itself and its history and traditions that was both heartening and strengthening. It also remained solidly rooted. The vast majority of the Guardsmen still hail from Wales, Welsh is still spoken by many Guardsmen and the links with the Principality have been strengthened and not weakened by the experience of otherwise unpopular operational tours in Iraq and Afghanistan. Above all, Army numbers are still as important as names. In a Regiment of Joneses, Evanses and Thomases, the last two digits of an Army number remain an important means of identification and nobody thinks any the worse of maintaining this long-standing custom. As the Regiment looks forward to the second hundred years of its existence, ageless virtues of that kind are essential certainties in an ever-changing world. Long may they continue.

The Welsh Guards Collection

Stan Evans served in the Regiment from 1960 to 1968 and started collecting Welsh Guards photographs in 1994, initially by placing advertisements in every Welsh newspaper. After an overwhelming response, he suggested to Regimental Headquarters that a database of those photographs should be created. In 1999, a committee was set up at Wellington Barracks to manage and protect what had become a growing collection of Regimental memorabilia, comprising not only photographs but also an extensive range of other items, including uniforms and badges. Brigadier Johnny Rickett became President and Lieutenant Colonel Brian Morgan became Treasurer. The Collection was initially displayed and stored in a separate building adjacent to Stan Evans' home, but with more and more pieces being added, a new location was needed. In 2001, a Victorian coach house at Park Hall Farm, Oswestry, was offered by the Directors of the Park Hall Countryside Experience, Martin Hughes and Richard Powell. Park Hall is a few hundred metres from the former Junior Leaders barracks where many Welsh Guardsmen previously served. Geographically, the site is situated in the middle of all seventeen Welsh Guards Association branches. Lord Harlech and Sir Francis Lloyd, both instrumental in the formation of the Welsh Guards, also lived locally. In February 2014, the Collection was given full accreditation as a military museum by the Army Museums Ogilvy Trust. Today, Park Hall attracts approximately 30,000 visitors annually. The Collection is now home to more than 4,000 artefacts and 2,500 photographs, making it a must for any visiting Welsh Guardsman. The Regiment is hugely appreciative of Park Hall Farm's support. It is also grateful to the curator, Stan Evans, and his band of volunteers (including the late Brian Morgan) whose efforts have made the Welsh Guards Collection an exceptionally successful archive and an important part of our Regimental heritage.

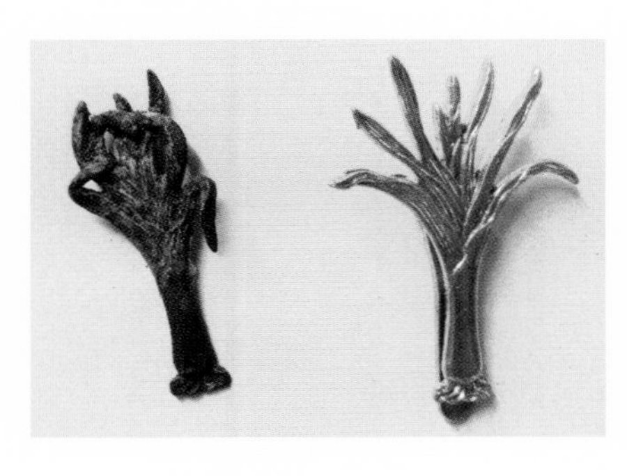

LEFT: A Welsh Guards brass leek from 1916 which was found on a battlefield dig at Ginchy, one of the Regiment's Battle Honours, next to a present-day leek. *(Welsh Guards Collection)*

RIGHT: A Guardsman's 1902 Pattern green khaki Service Dress tunic with 1908 Pattern Webbing Equipment and Service Dress Cap. *(Welsh Guards Collection)*

Appendices

Appendix 1

AFFILIATIONS

The Regiment was formerly affiliated with the 5th Battalion Australian Regiment. The affiliation started in 1965 and finished in March 1973 when the 5th Battalion joined with the 7th Battalion, whereupon the affiliation became 5th/7th Battalion Royal Australian Regiment.

Against the background of the government of Australia wishing to expand the Australian Army, 5th/7th Battalion was formally unlinked on 3 December 2006, thus marking the return of 5th Battalion Royal Australian Regiment, to the Army's order of battle. The Welsh Guards are, therefore, re-affiliated with the 5th Battalion Royal Australian Regiment.

In 1986, the Regiment was affiliated to HMS *Andromeda*. This ship was laid up in 1993. From 1994 the Regiment was affiliated to HMS *Campbeltown* until that vessel was decommissioned under defence cuts in 2011.

Appendix 2

BATTLE HONOURS INCLUDING THOSE NOT BORNE ON THE COLOURS

The Battle Honours, granted by royal authority in commemoration of war service, are shown below. Those printed in capital letters are borne on the colours:

The Great War

'LOOS'
'BAPAUME, 1918'
'Somme, 1916–18'
'Arras, 1918'
'GINCHY'
'Albert, 1918'
'FLERS-COURCELETTE'
'Drocourt-Queant'
'MORVAL'
'Hindenburg Line'
'Ypres, 1917'
'Havrincourt'
'PILCKEM'
'CANAL DU NORD'
'POELCAPPELLE'
'Celle'
'Passchendaele'
'SAMBRE'
'CAMBRAI, 1917, 18'
'France and Flanders, 1915, 18'

The Second World War

'DEFENCE OF ARRAS'
'Djebel el Rhorab'
'BOULOGNE, 1940'
'Tunis'
'St Omer–La Bassee'
'HAMMAM LIF'
'Bourguebus Ridge'
'North Africa, 1943'
'Cagney'
'MONTE ORNITO'
'MONT PINCON'
'Liri Valley'
'BRUSSELS'
'MONTE PICCOLO'
'HECHTEL'
'Capture of Perugia'
'Nederrijn'
'Arezzo'
'Lingen'
'Advance to Florence'
'Rhineland'
'Gothic Line'
'North West Europe 1940, 44–45'
'BATTAGLIA'
'FONDOUK'
'Italy 1944–45'

Post-1945

'FALKLAND ISLANDS 1982'

Appendix 3

LOCATIONS OF BATTALIONS, 1915–2015

1st Battalion

February–August 1915: London District
August 1915–November 1918: Western Front, France and Flanders
December 1918–March 1919: Germany, army of occupation
March 1919–April 1929: London District
April 1929–December 1930: British Troops Egypt, Cairo
January 1931–April 1939: London District
April 1939–November 1939: Gibraltar Command, Gibraltar
November 1939–May 1940: British Expeditionary Force, Arras, France
June 1940–June 1944: England
June 1944–April 1945: France and North-West Europe
May 1945–October 1945: 22 Guards Brigade, Stobo Camp, Hawick; London District
October 1945–March 1948, 1 Guards Brigade, Palestine
March 1948–March 1950: London District
March 1950–February 1952: 4 Guards Brigade, Anglesey Barracks, Wuppertal, West Germany
March 1952–March 1953 Berlin Brigade, Wavell Barracks, West Berlin
March 1953–September 1953: London District.
September 1953–March 1956: Suez Canal Zone, Egypt
March 1956–December 1960: 19 Infantry Brigade, Colchester
December 1960–November 1969: 4 Guards Brigade Group, Gort Barracks, Hubblerath, West Germany
November 1963–October 1965: London District, Chelsea Barracks
October 1965–October 1966: 24 Infantry Brigade, Salerno Lines, Aden
October 1966–March 1970: London District, Victoria Barracks, Windsor
March 1970–December 1974: 4 Armoured Brigade, Waterloo Barracks, Munster, West Germany
(25 March–28 July 1971: Northern Ireland: Carnmoney, Belfast).
(17 June–25 October 1972: Northern Ireland: Belfast City) .
December 1972–1974: London District
(November 1973–March 1974: Northern Ireland, Bessbrook, two Coys in Belfast)
January 1975–January 1977: London District, Caterham
October 1976–May 1977: United Nations, Cyprus
January 1977–July 1979: Berlin Brigade, Wavell Barracks, West Berlin
July 1979–February 1984: London District
(October 1979–February1980: Northern Ireland, Bessbrook)
(April–June 1982: 5 Infantry Brigade, Falkland Islands)
February 1984–November 1988: 22 Armoured Brigade, Campbell Barracks, Hohne, West Germany
(March–July 1986: Northern Ireland, Belfast)
November 1988–April 1992: London District, Pirbright
(April–October 1989: Belize)
April 1992–April 1994: 8 Infantry Brigade, Shackleton Barracks, Ballykelly, Northern Ireland
April 1994–February 1996: 143 Brigade, Clive Barracks, Tern Hill
March 1996–March 2000: London District, Wellington Barracks
April 2000–April 2003: 12 Mechanised Brigade, Bruneval Barracks, Aldershot
(March–September 2002: Bosnia, Operation Palatine)
May 2003–March 2006: Wales: West Camp, RAF St Athan
(May–December 2003: Northern Ireland: Ebrington Barracks, Londonderry)
(September 2004–March 2005: 4 Armoured Brigade, al-Amarah, Iraq, Operation Telic 5)
April 2006–July 2008: London District, Wellington Barracks
(October 2006–March 2007: Bosnia)
(November–December 2007: Belize)
August 2008–2009: Lille Barracks, Aldershot
(April–October 2009: 19 Light Brigade, Afghanistan: Helmand Province, Operation Herrick 10)
2009–2014: London District, Cavalry Barracks, Hounslow
(April–October 2012: 12 Mechanised Brigade. Afghanistan, Operation Herrick 16)
November 2014 onwards: Elizabeth Barracks, Pirbright, London District, and from January 2016 11 Brigade

2nd Battalion

1939: Tower of London
(May, June 1940: Hook of Holland. Boulogne)
July 1940–June 1944: England
June 1944–May 1945: France, North-West Europe
May 1945–July 1947: North Germany

3rd Battalion
June 1942: Uxbridge, Hampstead
February 1943: 1st Guards Brigade, Tunisia, North Africa
February 1944–April 1945, 1st Guards Brigade, Italy, Austria
May 1945–April 1946, Selkirk, Great Missenden

Appendix 4

MOTTO, BADGE, MARCHES AND THE REGIMENTAL COLLECT

The motto of the Regiment is 'Cymru am Byth' ('Wales for Ever').

The leek, which is the national emblem of Wales, was approved by King George V on the formation of the Regiment as the principal symbol on the regimental cap badge. Sealed patterns of the form in which it appears on the regimental uniform are kept at Regimental Headquarters. No variation from these patterns is allowed without the approval of the Colonel-in-Chief, the Colonel, the Major General and the Ministry of Defence.

In June 2009 a new silver design was approved for the warrant officers' cap badge. It has the same specification of the brass cap badge with eight leaves on the leek.

REGIMENTAL MARCHES

The Regimental Quick March is 'The Rising of the Lark' and the Slow March is 'Men of Harlech'. These marches are invariably used when the Regiment or a detachment of it is marching past in slow or quick time.

Through continual usage, certain quick marches have become associated with each company of the 1st Battalion as follows:

The Prince of Wales's Company
'God Bless The Prince of Wales'
No. 2 Company
'Deep and Wide'
No. 3 Company
'Happy Wanderer'
Headquarter Company
'Old Comrade'
Support Company
'The Old Grey Mare'

WELSH GUARDS COLLECT

O Lord God, who has given us the Land of Our Fathers for our inheritance, help thy servants, the Welsh Guards, to keep thy laws as our heritage for ever until we come to that better and heavenly Country which thou hast prepared for us: through Jesus Christ our Lord. Amen

O Arglwdd Dduw, a roddaist I. ni wlad ein tadau yn dreftadaeth, Cynorthwya dy weision, y Gwarchodlu Cymreig, I. gadw Dy Ddeddfau yn etifeddiaeth inni dros fyth, nes y doen I'r wlad well A. nefol a baratoaist I. ni, trwy Iesu Grist ein Harglwydd.

Appendix 5

REGIMENTAL APPOINTMENTS

Colonels-in-Chief
His Majesty King George V – August 1915
His Majesty King Edward VIII – January 1936
His Majesty King George VI – December 1936
Her Majesty Queen Elizabeth II – February 1952

Colonels
His Royal Highness Edward, Prince of Wales – June 1919
Colonel W. Murray-Threipland, DSO – March 1937
Brigadier the Earl of Gowrie, VC – June 1942
His Royal Highness The Prince Philip, Duke of Edinburgh – July 1953
His Royal Highness Prince Charles, The Prince of Wales – March 1975

Regimental Lieutenant Colonels
Colonel W. Murray-Threipland, DSO – February 1915
Colonel Lord Harlech, TD – June 1915
Colonel W. Murray-Threipland, DSO – October 1917
Colonel the Hon. A. G. A. Hore-Ruthven, VC, CB, CMG, DSO – December 1920
Colonel T. R. C. Price, CMG, DSO – October 1924
Colonel R. E. K. Leatham, DSO – October 1928
Colonel M. B. Beckwith-Smith – October 1934
Colonel W. A. F. L. Fox-Pitt, MVO, MC – January 1938
Colonel R. E. K. Leatham, DSO – September 1939
Colonel A. M. Bankier, DSO, OBE, MC – February 1942
Colonel Sir Alexander Stanier, Bt, DSO, MC – March 1954

Colonel G. W. Browning, OBE – March 1948
Colonel J. C. Windsor-Lewis, DSO, MC – May 1951
Colonel D. G. Davies-Scourfield – July 1954
Colonel H. C. L. Dimsdale, OBE, MC – July 1957
Colonel C. A. la T. Leatham – July 1960
Colonel M. C. Thursby-Pelham – November 1964
Colonel V. G. Wallace – June 1967
Colonel J. W. T. A. Malcolm – September 1972
Colonel M. R. Lee, OBE – September 1976
Colonel S. C. C. Gaussen – June 1978
Colonel D. R. P. Lewis – November 1982
Lieutenant Colonel C. J. Dawnay – March 1987
Brigadier J. F. Rickett, OBE – January 1989
Brigadier C. F. Drewry, CBE – December 1994
Major General C. R. Watt, CBE – January 2000
Colonel A. J. E. Malcolm, OBE – November 2005
Major General R. H. Talbot Rice – June 2010

Regimental Adjutants
Captain M. O. Roberts – January 1916
Captain B. T. V. Hambrough – February 1916
Captain R. W. Lewis, MC – May 1916
Captain H. E. Allen – July 1916
Captain P. L. M. Battye – October 1916
Major R. G. Williams-Bulkeley, MC – March 1917
Captain J. W. L. Crawshay – February 1918
Captain B. T. V. Hambrough – May 1918
Captain The Earl of Lisburne – June 1918
Captain J. J. P. Evans, MC – May 1919
Captain H. Talbot Rice – November 1920
Major R. T. K. Auld – February 1923
Captain F. A. V. Copland-Griffiths, MC – October 1924
Major R. W. Lewis, MC – October 1927
Captain G. D. Young – June 1931
Captain A. W. A. Malcolm – January 1934
Captain W. D. C. Greenacre, MVO – January 1937
Major C. Dudley-Ward, DSO, MC – September 1939
Major The Earl of Lisburne – January 1944
Major H. G. Moore-Gwyn – March 1945
Major A. A. Duncan – July 1947
Captain R. V. J. Evans – August 1950
Major C. A. la T. Leatham – October 1950
Major J. M. Miller, DSO, MC – June 1953
Major M. C. Thursby-Pelham – July 1956
Major V. G. Wallace – January 1958
Major R. P. Hedley-Dent – February 1960
Major N. Webb-Bowen – February 1962
Major J. W. T. A. Malcolm – March 1964
Major S. C. C. Gaussen – May 1966
Major D. C. MacDonald-Milner – January 1969
Major J. B. B. Cockcroft – January 1971
Major D. R. P. Lewis – December 1972
Major J. F. Rickett, MBE – January 1975
Major C. F. B. Stephens – December 1976
Major R. F. Powell – December 1978
Major J. L. Goodridge – February 1980
Major M. R. Senior – December 1981
Major A. D. I. Wall, MBE – December 1984
Major C. R. Watt – March 1987
Major J. D. G. Sayers – October 1987
Lieutenant Colonel C. F. B. Stephens – July 1990
Colonel T. C. S. Bonas – October 2007

Superintending Clerks
2730093 C. E. Woods, MM – February 1915
2730003 I. M. Smith – January 1927
2730090 J. C. Buckland – September 1934
2730376 J. Copping, MBE – August 1937
2732789 J. W. Edwards – July 1946
2733669 H. W. Humphries – December 1947
2733903 D. J. Griffiths – November 1951
2735387 R. C. Williams – September 1956
2739311 K. D. Lewis – January 1961
2739697 D. W. Wilcox – September 1965
22217175 D. J. L. Jones – September 1968
22217718 I. G. Jones – June 1971
22217895 G. L. Evans – July 1974
23523319 G. F. Taylor – July 1976
23877373 L. Ellson – June 1979
23908674 C. F. G. Owen – January 1984
24125178 P. J. Richardson – March 1986
24125065 T. D. J. Thorne – August 1989
24385093 K. W. Stacey – May 1992
24173601 A. J. Powell – May 1993

Following a Defence Review, the historical post of WO1 Superintending Clerk was disestablished and the WO2 Regimental Quartermaster Sergeant became the senior Non-Commissioned Officer at Regimental Headquarters.

Regimental Quartermaster Sergeants
24220120 A. Bennett – April 1994
24378054 J. B. Williams – March 1997
24498669 K. J. Sincock – October 1999
24631207 G. W. Lloyd – November 2001
24682012 G. Lewis – August 2003
24788725 W. J. Williams – September 2005
24797497 A. Campbell – February 2010
24738054 L. T. Scholes – April 2011
24815189 D. P. Brown – November 2012
25028703 S. J. Boika – February 2014

Directors of Music

Major A. Harris, MVO, LRAM (Bandmaster from Oct 1915–Mar 1919) – March 1919
Major T. S. Chandler, MVO, LRAM, ARCM – January 1938
Major F. L. Statham, MBE, LRAM, ARCM – March 1948
Major A. Kenney, LRAM, ARCM – June 1962
Major D. K. Walker, ARCM – October 1969
Major D. N. Taylor, ARCM, LTCL – September 1974
Lieutenant Colonel P. Hannam, BEM, psm – February 1986
Lieutenant Colonel C. J. Ross, ARCM, MISM, psm – January 1993
Lieutenant Colonel S. A. Watts, ARAM, psm – June 1994
Major T. S. Davies, FTCL, ARCM, psm – May 1997
Captain P. D. Shannon, MBE, ARAM, LRAM, psm – May 1998
Major D. W. Creswell, BBCM, psm – June 2005
Lieutenant Colonel S. C. Barnwell, BBCM, psm – November 2008
Major K. F. N. Roberts, MMus, FLCM – November 2013

1st BATTALION, WELSH GUARDS

Commanding Officers

Lieutenant Colonel W. Murray-Threipland, DSO – February 1915
Lieutenant Colonel G. C. D. Gordon, DSO – December 1916
Lieutenant Colonel H. Dene, DSO – February 1918
Captain W. B. L. Bonn, MC – August 1918
Lieutenant Colonel R. E. C. Luxmoore-Ball, DSO, DCM – August 1918
Major C. H. Dudley-Ward, DSO, MC – October 1918
Lieutenant Colonel R. E. C. Luxmoore-Ball, DSO, DCM – November 1918
Lieutenant Colonel G. C. D. Gordon, DSO – March 1919
Lieutenant Colonel The Hon. A. G. A. Hore-Ruthven, VC, CB, CMG, DSO – May 1919
Lieutenant Colonel T. R. C. Price, CMG, DSO – December 1920
Lieutenant Colonel R. E. K. Leatham, DSO – August 1924
Lieutenant Colonel R. T. K. Auld – October 1928
Lieutenant Colonel M. B. Beckwith-Smith, DSO, MC – October 1932
Lieutenant Colonel W. A. F. L. Fox-Pitt, MVO, MC – October 1934
Lieutenant Colonel F. A. V. Copland-Griffiths, MC – January 1938
Lieutenant Colonel J. Jefferson – July 1940
Lieutenant Colonel G. St V. J. Vigor – September 1941
Lieutenant Colonel G. W. Browning – March 1944
Lieutenant Colonel J. E. Fass – June 1944
Lieutenant Colonel C. H. R. Heber-Percy, DSO, MC – July 1944
Lieutenant Colonel J. F. Gresham, DSO – August 1944
Lieutenant Colonel C. H. R. Heber-Percy, DSO, MC – October 1944
Lieutenant Colonel R. B. Hodgkinson – August 1945
Lieutenant Colonel A. W. A. Malcolm – February 1947
Lieutenant Colonel D. G. Davies-Scourfield, MC – July 1949
Lieutenant Colonel R. C. Rose-Price – August 1952
Lieutenant Colonel A. C. W. Noel, MC – November 1952
Lieutenant Colonel C. A. la T. Leatham – November 1955
Lieutenant Colonel J. M. Miller, DSO, MC – November 1958
Lieutenant Colonel V. G. Wallace – March 1961
Major J. D. N. Rettallack – June 1963
Lieutenant Colonel P. R. Leuchars – August 1963
Lieutenant Colonel P. J. N. Ward – September 1965
Lieutenant Colonel N. Webb-Bowen – January 1968
Lieutenant Colonel J. W. T. A. Malcolm – March 1970
Lieutenant Colonel M. R. Lee, OBE – July 1972
Lieutenant Colonel P. R. G. Williams – December 1974
Lieutenant Colonel C. R. L. Guthrie – September 1977
Lieutenant Colonel J. F. Rickett, MBE – March 1980
Lieutenant Colonel R. F. Powell – October 1982
Lieutenant Colonel C. F. Drewry – March 1985
Lieutenant Colonel D. P. Belcher – August 1987
Lieutenant Colonel C. R. Watt – March 1990
Lieutenant Colonel T. C. R. B. Purdon, MBE – June 1992
Lieutenant Colonel R. L. Traherne – November 1994
Lieutenant Colonel A. J. E. Malcolm – November 1996
Lieutenant Colonel A. C. Ford – August 1999
Lieutenant Colonel R. H. Talbot Rice – January 2002
Lieutenant Colonel B. J. Bathurst – July 2004
Lieutenant Colonel R. J. A. Stanford, MBE – July 2006
Lieutenant Colonel R. S. M. Thorneloe, MBE – November 2008
Lieutenant Colonel C. K. Antelme, DSO – July 2009
Lieutenant Colonel D. L. Bossi – July 2011
Lieutenant Colonel G. R. Harris, DSO, MBE – January 2014

Adjutants

Captain G. C. D. Gordon – February 1915
Captain J. A. D. Perrins – July 1915
Captain G. C. Devas – August 1917
Captain J. W. L. Crawshay – November 1918
Captain W. A. F. L. Fox-Pitt, MC – May 1919
Captain A. M. Bankier, DSO, MC – May 1922
Lieutenant Sir Alexander Stanier, Bt, MC – September 1923
Lieutenant W. D. C. Greenacre, MVO – October 1926
Captain J. Jefferson – October 1929
Lieutenant J. C. Windsor-Lewis – October 1932
Lieutenant H. M. C. Jones-Mortimer – October 1935
Captain J. E. Gurney – September 1938
Captain H. J. Moore-Gwyn – June 1940
Captain J. M. Miller – August 1942
Captain J. M. Spencer-Smith, MC – May 1944
Captain P. R. Leuchars – April 1946
Major N. S. Kearsley – March 1949
Captain V. G. Wallace – December 1950
Captain N. Webb-Bowen – December 1952
Captain G. S. Lort Phillips – February 1955
Major W. J. Burchell – January 1957
Major D. M. St G. Saunders – October 1959
Captain M. R. Lee – July 1960
Captain N. Van Moppes – June 1962
Captain C. R. L. Guthrie – June 1964
Captain G. F. Richmond Brown – April 1966
Captain J. F. Rickett, MBE – April 1968
Captain W. J. H. Moss – August 1970
Captain G. N. R. Sayle – August 1972
Captain R. E. H. David – June 1974
Captain C. F. Drewry – May 1976
Major C. R. Watt – July 1978
Captain A. C. Richards – July 1980
Captain R. S. Mason – December 1981
Captain W. de B. Prichard – January 1983
Captain P. G. de Zulueta – August 1983
Captain A. J. E. Malcolm – July 1985
Captain P. Owen-Edmunds – December 1986
Captain R. H. W. St G. Bodington – June 1988
Captain R. H. Talbot Rice – January 1991
Captain M. B. D. Jenkins – December 1992
Captain R. J. Williams, MC – December 1995
Captain R. S. M. Thorneloe – January 1998
Captain D. L. W. Bossi – July 1999
Captain G. R. Harris – December 2000
Captain D. W. N. Bevan – May 2002
Captain J. M. Hobrough – August 2003
Captain D. H. Basson – July 2005
Captain M. L. Lewis – March 2007
Captain J. W. Aldridge – July 2008
Captain N. R. K. Moukarzel – February 2010
Captain T. C. Spencer-Smith – May 2011
Captain C. J. P. Davies – March 2013
Captain C. H. L. Beare – January 2015

Quartermasters

Captain W. B. Dabell, MBE, MC – February 1915
Major W. L. Stevenson, MBE, DCM, MM – June 1928
Lieutenant J. C. Buckland, MBE – August 1937
Captain W. L. Bray, DCM, MM – September 1941
Captain I. Roberts, MBE – October 1945
Captain H. W. Humphries, MBE – December 1951
Captain A. Rees, MBE – March 1956
Captain W. S. Phelps – February 1961
Captain R. C. Williams, MBE – June 1963
Captain A. P. Joyce, MM, BEM – December 1968
Captain R. E. Fletcher – January 1972
Captain I. A. James – April 1974
Captain B. D. Morgan – February 1978
Captain G. White – October 1982
Captain E. L. Pridham, MBE – July 1985
Captain A. J. Davies, MBE – April 1988
Major A. O. Bowen – March 1991
Major D. R. Evans – March 1993
Captain N. Harvey – December 1996
Major K. W. Stacey – April 2000
Captain F. K. Oultram – November 2001
Captain M. E. Browne, BEM – March 2004
Captain M. W. Miles – September 2005
Captain M. Cooling – July 2007
Captain N. P. Mott – August 2008
Major A. F. Bowen – December 2010
Major K. Dawson – December 2013
Major D Pridmore – June 2014

Regimental Sergeant Majors

1 (2730001 SG) W. Stevenson – February 1915
1898 E. Barnes – August 1915
2 (11285 GG) W. Bland, MC – October 1915
1 (2730001 SG) W. Stevenson, MBE, DCM, MM – May 1918
2604434 L. Pownall – June 1928
2730040 J. O. Hughes – February 1934
2605066 W. L. Bray, DCM, MM – September 1934
2730511 E. L. G. Richards, MM – March 1938
2731403 A. R. Baker, MBE – June 1940
2733690 A. Rees, MBE – August 1945
2733893 B. F. Hillier, DCM – September 1951
2735209 W. S. Phelps – September 1956
2733914 E. K. L. Edwards – March 1958

2739213 C. S. Payton – March 1960
2740446 D. Rowden – October 1961
2741094 D. I. Williams – February 1963
14076196 I. A. James – March 1966
22217856 R. J. H. Coe – January 1970
22831416 B. D. Morgan – July 1973
22856003 G. White – July 1975
23456550 E. L. Pridham, MBE – August 1977
23876522 A. J. Davies – August 1980
23876787 D. R. Evans – June 1983
23908628 M. L. Hughes – April 1985
23929784 M. E. Evans – June 1988
24184337 N. Harvey – March 1990
24203114 I. Dyas – August 1992
24263839 J. W. Harford – March 1994
24263896 F. K. Oultram – November 1996
24471311 M. W. Miles – June 1998
24464663 M. Cooling – April 2000
24596186 A. Davies – January 2002
24714346 T. Harman – January 2004
24713091 A. F. Bowen – August 2005
24772309 D. W. Pridmore – March 2007
24797524 M. Monaghan – April 2008
25788299 A. L. Roberts – March 2010
24823463 B. J. Baldwin – September 2011
25014288 M. Topps – March 2013
25015807 P. J. Dunn – March 2015

Regimental Quartermaster Sergeants
4 (2730003 GG) I. Smith – February 1915
2730004 H. J. Pursey – August 1915
2730090 J. C. Buckland – January 1927
2730044 A. H. Evans – May 1931
2730580 A. V. Wenban – February 1935
6190690 H. H. Webb – February 1939
2731283 L. C. Welsh – November 1939
2732789 J. Edwards – April 1941
2733801 D. C. Jenkins – October 1943
2732864 C. S. E. Blackmore – September 1945
2735209 W. S. Phelps – March 1952
2734410 C. H. Dewar – August 1954
2737357 E. Davies, DCM – February 1955
4547748 A. Joyce, MM, BEM – January 1956
2738972 R. E. Fletcher – December 1961
2740109 S. G. Richards – November 1966
2741826 W. G. Northwood – July 1968
22217134 J. E. Jones – January 1971
22831465 D. C. Hearne – December 1973
2344644 M. E. Smith – June 1975
23523021 D. M. Davies – May 1977
23323201 J. M. Jones – July 1979
23879381 R. J. Parry – October 1980
23908633 W. E. Evans – October 1982
23908628 M. L. Hughes – January 1984
23929677 J. M. F. Everett – April 1985
23929784 M. E. Evans – November 1986
24125035 A. L. Denman – June 1988
24184337 N. Harvey – March 1989
24141818 D. A. Williams – March 1990
24274049 S. Cox – September 1991
24263850 J. W. Harford – August 1993
24242690 M. Frost – March 1994
24263896 F. K. Oultram – March 1995
24471311 M. W. Miles – March 1997
24464663 M. Cooling – June 1998
24596486 A. Davies – April 2000
24593324 M. Evans – January 2002
24714346 T. J. Harman – November 2002
24736038 J. P. Mott – January 2004
24683200 P. L. Robinson – July 2005
24772309 D. W. Pridmore – August 2006
24797524 M. Monaghan – April 2007
24772832 A. J. Hughes – June 2008
25788299 A. L. Roberts – April 2009
24830440 M. Pollard – November 2010
25014288 M. Topps – September 2011
25015807 P. J. Dunn – March 2013
24921250 E. Griffiths – March 2014

2nd BATTALION, WELSH GUARDS

Commanding Officers
Lieutenant Colonel Sir Alexander Stanier, Bt, DSO, MC – May 1939
Lieutenant Colonel G. St V. J. Vigor – October 1940
Lieutenant Colonel J. Jefferson – September 1941
Lieutenant Colonel W. D. C. Greenacre, MVO – October 1941
Lieutenant Colonel J. C. Windsor-Lewis, DSO, MC – December 1943
Lieutenant Colonel A. W. A. Malcolm – May 1946
Lieutenant Colonel Sir William V. Makins, Bt – January 1947

Adjutants
Captain R. C. Rose Price – May 1939
Captain A. A. Duncan – November 1940
Captain R. J. A. Watt – February 1942
Captain S. O. F. Bateman – January 1944
Captain M. A. L. F. Pitt-Rivers – December 1944
Captain D. A. Gibbs – August 1945
Captain R. J. V. Evans – May 1946

Quartermasters
Lieutenant W. L. Bray, DCM, MM – May 1939
Captain J. C. Buckland, MBE – September 1941
Captain H. H. Webb, MBE – January 1945

Regimental Sergeant Majors
2730700 K. W. Grant – May 1939
2731644 A. P. Maskell – February 1942
2732331 I. Roberts – March 1942
2731556 T. Rees – September 1945
2732389 S. Webb – July 1946
2733395 T. J. H. John, MM – May 1947

Regimental Quartermaster Sergeants
2732115 F. T. Jones, DCM – May 1939
2731968 T. Curtis – February 1943
2733669 H. W. Humphries – July 1945
4073926 H. G. Maisey – June 1946

HOLDING AND 3rd BATTALION, WELSH GUARDS

Commanding Officers
Lieutenant Colonel W. D. C. Greenacre, MVO – April 1941
Lieutenant Colonel A. M. Bankier, DSO, OBE, MC – September 1941
Lieutenant Colonel G. W. Browning – February 1942
Lieutenant Colonel D. E. P. Hodgson – December 1942
Lieutenant Colonel Sir William V. Makins, Bt – July 1943
Lieutenant Colonel D. G. Davies-Scourfield, MC – May 1944
Lieutenant Colonel J. E. Gurney, DSO, MC – June 1944
Lieutenant Colonel R. C. Rose Price, DSO – April 1945
Lieutenant Colonel A. W. A. Malcolm – October 1945

Adjutants
Captain K. A. S. Morrice (Holding Battalion) – April 1941
Captain K. A. S. Morrice – October 1941
Captain G. D. Rhys-Williams – January 1943
Captain A. W. Stephenson – April 1943
Captain J. D. Gibson-Watt, MC – July 1943
Captain B. R. T. Greer – April 1944
Captain F. L. Egerton – June 1944
Captain M. C. Thursby-Pelham – September 1944
Captain F. L. Egerton – February 1945
Captain J. D. N. Retallack – October 1945

Quartermasters
Lieutenant F. Starnes – May 1941
Lieutenant K. W. Grant – February 1942

Regimental Sergeant Majors
2731283 I. Welsh – April 1941
3951122 M. Davies – August 1941
2559572 A. Barter – December 1942
2731624 H. Dunn, MBE – July 1944
2732791 F. Dodd – June 1945
2731403 A. Baker, MBE – August 1945

Regimental Quartermaster Sergeants
2731624 H. Dunn, MBE – April 1941
2733673 C. Jones – June 1944

TRAINING BATTALION

Commanding Officers
Lieutenant Colonel Lord Glanusk, DSO – September 1939
Lieutenant Colonel T. A. Oakshott – February 1942
Lieutenant Colonel D. G. Davies-Scourfield, MC – July 1945

Adjutants
Captain J. C. Windsor-Lewis – September 1939
Captain C. A. la T. Leatham – March 1940
Captain Lord Delamere – January 1940
Captain B. R. T. Greer – August 1942
Captain D. T. Llewellyn – March 1944
Captain B. R. T. Greer – February 1945
Captain P. R. H. Hastings – May 1945

Quartermasters
Lieutenant L. Pownall – September 1939
Captain J. O. Hughes – March 1941
Lieutenant F. Starnes – January 1944
Lieutenant H. H. Webb, MBE – May 1944
Captain J. C. Buckland, MBE – January 1945

Regimental Sergeant Majors
2730248 C. Sayers – September 1939
6190690 H. H. Webb – November 1939
3951122 M. Davies – May 1944
299370 P. Dunne, MBE – July 1944

Regimental Quartermaster Sergeant
2730580 A. V. Wenban – September 1939

HIGHER GALLANTRY AWARDS 1915–2015

VC – Victoria Cross
DSO – Distinguished Service Order
MC – Military Cross
MM – Military Medal
DCM – Distinguished Conduct Medal

FIRST WORLD WAR, 1915–1918

Victoria Cross
939 Sergeant R. Bye

DSO & Bar
Lieutenant Colonel The Hon. A. G. A. Hore-Ruthven, VC

DSO, MC
Captain W. B. L. Bonn
A/Major C. H. Dudley-Ward
A/Captain L. F. Ellis

DSO
Captain H. G. G. Ashton
A/Lieutenant Colonel H. Dene
Major G. C. D. Gordon
2/Lieutenant R. R. Jones
A/Lieutenant Colonel R. E. C. Luxmoore-Ball
Lieutenant Colonel W. Murray-Threipland
Major T. R. C. Price
Captain Sir R. Williams, Bt

MC & Bar
Captain J. J. P. Evans
Captain J. A. D. Perrins

MC
Lieutenant W. Arthur
A/Captain F. L. T. Barlow
Captain P. L. M. Battye
Lieutenant O. Bird
RSM A. Bland
Lieutenant R. C. Bonsor
Lieutenant J. P. T. Burchell
A/Captain F. A. V. Copland-Griffiths
Lieutenant J. W. L. Crawshay
Captain & Quartermaster W. B. Dabell
Lieutenant D. B. Davies
A/Captain G. C. Devas
Captain C. C. L. Fitzwilliams
Captain W. A. F. L. Fox-Pitt
A/Captain A. Gibbs
A/Captain R. E. O. Goetz
A/Captain B. T. M. Hebert
2/Lieutenant H. C. N. Hill
T/Captain G. C. L. Insole
Captain R. W. Lewis
Lieutenant P. Llewellyn
Captain K. G. Menzies
2/Lieutenant R. R. D. Paton
Lieutenant H. A. St G. Saunders
Lieutenant R. C. R. Shand
2/Lieutenant A. B. G. Stanier
A/Captain H. S. Stokes
Major. R. G. W. Williams-Bulkeley
Captain B. C. Williams-Ellis
Lieutenant R. L. Wreford-Brown
Lieutenant F. B. Wynne-Williams

DCM & Bar, MM
2259 L/Corporal O. F. Waddington

DCM, MM & Bar
872 L/Corporal E. W. Gordon

DCM, MM
1 RSM W. Stevenson
1663 Sergeant A. G. Ham

DCM & Bar
114 Sergeant R. Mathias

DCM
24 Sergeant O. Ashford
408 Sergeant S. E. Davies
668 Guardsman J. Duffy
1229 Sergeant A. H. Evans
1228 Sergeant C. L. Glover
48 Sergeant G. C. Grant
1185 L/Sergeant F. Hall
1209 Guardsman W. Hughes
23 CQMS L. Hunter
823 Sergeant E. Jones
6 CSM A. Pearce
2361 Guardsman J. O. Pritchard
395 L/Sergeant W. Roberts
3093 Guardsman A. Thomas
1529 L/Corporal G. Thomas
2541 Guardsman S. T. Thomas
858 Guardsman T. Thomas

MM & Bar

1037 Sergeant F. Aspinall
2763 Guardsman D. T. Clancy
1010 L/Corporal F. E. H. Cosford
1392 Guardsman J. Hammond
1972 L/Corporal W. A. Harries
1063 L/Sergeant H. F. Hutchings
756 L/Sergeant W. M. Jones
162 Corporal D. J. Luker

MM

2783 Guardsman J. Airey
2717 Guardsman G. E. Allen
3292 Guardsman J. Anderson
1078 L/Sergeant C. Attfield
2328 Guardsman D. I. Aubrey
1254 Guardsman S. T. Baldwin
128 Guardsman A. J. Barlow
1439 Guardsman H. J. Barther
364 Sergeant C. W. F. Bartlett
193 Sergeant W. C. Beazer
761 Sergeant C. A. Bonar
280 Guardsman W. Botcher
403 Sergeant C. O. Bowles
2478 Guardsman E. Boyle
1869 Corporal J. Broom
2220 Guardsman D. Brown
882 L/Sergeant G. W. Burman
589 Guardsman W. C. Burton
3275 Guardsman R. F. Charnley
4016 Guardsman J. H. Cornelius
4153 Guardsman J. H. Crebbin
648 Corporal J. T. Crumb
1938 Corporal J. H. Cummings
1573 Guardsman T. A. Daniels
2885 Guardsman C. Davies
2169 Guardsman E. S. Davies
4020 Guardsman J. T. Davies
216 Guardsman R. Davies
2333 Guardsman L. Edwards
2533 Guardsman J. Ellis
1802 Guardsman J. Emanuel
1360 Corporal D. J. Evans
423 Guardsman J. R. Evans
2851 Guardsman T. Evans
311 L/Corporal E. Fairbanks
2782 Guardsman J. Feely
2164 Guardsman E. Fitzgerald
834 L/Sergeant W. C. Gardiner
3850 Guardsman W. Garnett
32 Sergeant E. J. Gibbs
2144 Guardsman T. Gibbs
1876 L/Corporal H. Gilbert
420 Sergeant J. Gough
2759 L/Sergeant T. Griffiths
492 Guardsman T. Harris
1900 Sergeant T. H. Haylock
744 L/Sergeant A. Hicks
3683 Guardsman A. S. Hill
832 Guardsman H. Holbrook
130 A/CSM G. H. Holme
273 Sergeant T. A. Hughes
29 Sergeant A. J. O. Humphreys
613 L/Corporal D. James
141 Guardsman A. John
1795 Sergeant A. E. Johnson
1568 Guardsman W. T. Johnson
468 Guardsman A. Jones
3015 Guardsman D. E. Jones
2251 Guardsman D. O. Jones
1523 Guardsman D. R. Jones
2627 Guardsman E. L. Jones
2256 Guardsman G. R. Jones
723 Guardsman J. Jones
4068 Guardsman J. G. Jones
82 Guardsman M. Jones
153 Corporal P. E. Jones
2708 Guardsman T. M. Jones
1189 Guardsman W. Jones
3558 Guardsman W. D. Jones
25 L/Sergeant R. Lawson
38 Sergeant E. Lewis
2661 Guardsman J. Lewis
529 Guardsman W. J. Lewis
202 Guardsman J. Lloyd
1631 Guardsman T. G. Lucas
1050 Sergeant W. Manuel
263 Guardsman H. J. Matthews
2380 Corporal R. Merrett
252 Guardsman G. Messer
3080 L/Corporal W. H. Mills
2760 Sergeant G. H. Moore
954 Sergeant E. Morgan
290 L/Corporal M. J. Morgan
623 Guardsman J. Morrisey
2111 L/Corporal H. G. Neale
487 Corporal R. Needs
3180 Guardsman J. Newport
1889 Guardsman T. Norgate
3417 Guardsman E. R. Owen
3555 Guardsman I. H. Owen
21 A/CQMS S. Owen

1317 Guardsman J. C. Palmer
188 Guardsman L. Phillips
1320 Guardsman W. Phillips
495 Corporal S. Pinkham
265 Guardsman S. G. Powell
706 Guardsman W. Powell
2523 Guardsman B. T. Pritchard
624 L/Corporal A. Raisey
2721 Guardsman J. T. Richards
1522 Guardsman E. Roberts
2138 Guardsman J. L. Roberts
2954 Guardsman E. D. Rowlands
3069 Guardsman C. G. Sendy
1845 Guardsman H. Sheppard
3861 Guardsman W. Snell
144 Guardsman S. G. Spencer
92 Guardsman A. G. Sully
145 Guardsman E. Tanner
798 L/Sergeant A. G. Thomas
974 Guardsman C. Thomas
1276 Guardsman G. Thomas
1125 Guardsman T. H. Thomas
3208 Guardsman W. J. Thomas
3931 Guardsman A. J. Turley
1043 Guardsman J. Ulyatt
2881 Guardsman J. E. Vaughan
371 Guardsman F. J. Vowles
2854 L/Sergeant T. Walker
859 Guardsman J. R. Wallace
336 Musn. J. Walters
3366 Guardsman E. G. Ward
3580 L/Corporal A. J. Wellings
758 Guardsman A. A. West
235 L/Sergeant E. J. Wheatley
1465 Corporal S. White
3548 Guardsman C. H. Willett
794 L/Corporal G. T. Williams
2163 Guardsman J. Williams
485 Sergeant L. Williams
776 L/Sergeant G. Winter
496 Sergeant F. Winter
4092 Guardsman G. E. Wood
43 Sergeant C. H. Wren

SECOND WORLD WAR 1939–1945

Victoria Cross
Lieutenant the Hon. C. Furness

DSO & Bar, MC
Lieutenant Colonel J. Windsor-Lewis

DSO, MC & Bar
Lieutenant Colonel J. E. Gurney

DSO & Bar
Brigadier Sir Alexander Stanier, Bt, MC

DSO, MC
Lieutenant Colonel C. H. R. Heber-Percy
Major J. M. Miller

DSO
Brigadier F. A. V. Copland-Griffiths, MC
Brigadier W. A. F. L. Fox-Pitt, MC
Brigadier W. D. C. Greenacre
Lieutenant Colonel J. F. Gresham
Lieutenant Colonel R. M. V. Ponsonby
Lieutenant Colonel R. C. R. Rose Price
Lieutenant Colonel C. P. Vaughan

MC & Two Bars
Major J. D. Gibson-Watt

MC & Bar
Major N. M. Daniel
Major Sir R. G. D. Powell, Bt

MC
Captain D. A. N. Allen
Captain Sir M. G. Beckett, Bt
Major F. B. Bolton
Captain D. N. Brinson
Captain W. H. Carter
Major W. L. Consett
Major A. H. S. Coombe-Tennant
Lieutenant D. H. Cottom
Lieutenant Colonel D. G. Davies-Scourfield
Captain M. C. Devas
Major H. C. L. Dimsdale
Captain F. M. Eastwood
Captain F. L. Egerton
Major W. D. D. Evans
Captain R. P. Farrer
Major N. T. L. Fisher
Major H. S. Forbes
Major B. P. R. Goff
Captain W. H. Griffiths
Lieutenant Colonel R. B. Hodgkinson
Major H. E. Lister
Major W. H. R. Llewellyn
Major D. C. M. Mather
Major A. C. W. Noel

Captain B. B. Pugh
Major R. E. W. Sale
Major R. C. Sharples
Major J. R. Martin-Smith
Captain J. M. Spencer-Smith
Captain D. J. C. Stevenson
Captain A. G. Stewart
Captain A. W. S. Wheatley
Major W. G. M. Worrall

DCM
2734613 CSM J. A. Barham
2733197 D/Sergeant D. Davies
2737357 D/Sergeant E. Davies
2733504 CSM U. Davies
2736473 CQMS E. Hart
2733893 CQMS B. F. Hillier
2734818 Sergeant L. A. Hodgson
2733460 Sergeant D. J. Jones
2732412 L/Sergeant E. Jones
2732115 RSM F. T. Jones
4194198 Guardsman J. E. Jones
2733681 L/Sergeant J. King
2733864 Sergeant P. G. Matthews
2735425 Guardsman F. B. Morris
2737136 Sergeant W. H. Parsons
2734495 Sergeant A. G. H. Phillips
2733752 CSM H. L. Richardson
2734436 CQMS A. A. Semark
2733239 RSM S. J. Slack
2734077 Sergeant A. J. Webb

MM
2734602 Sergeant O. F. Abrams
2656525 L/Sergeant R. A. Appleby
2737253 L/Corporal T. J. L. Arnold
2737129 L/Corporal D. W. Beynon
2732548 Guardsman B. E. Booker
2737190 L/Corporal D. F. Burgess
2734798 L/Corporal J. W. Catherall
2736675 L/Sergeant M. G. E. Chatwin
2733705 Guardsman I. C. Davage
2733126 Corporal B. J. Davies
2734842 Sergeant K. Davies
2733442 L/Corporal T. Easter
2736347 L/Corporal G. Evans
2734193 Sergeant D. H. Griffiths
2733475 Sergeant S. Gough
2734347 Sergeant L. J. Gower
2737293 Sergeant C. H. Hansford
2733395 D/Sergeant T. H. John
2738430 Guardsman H. R. Jones
2735182 Guardsman T. Jones
4547748 L/Sergeant A. Joyce
2738611 Sergeant T. L. Kennedy
2738850 L/Corporal C. J. Keogh
2731773 CSM R. Larcombe
2736303 Sergeant C. G. Lawrie
2734794 CSM W. Mairs
2737222 L/Sergeant J. H. McGhan
2734526 CSM E. Morgan
2737677 L/Corporal R. D. Porter
2733284 Guardsman T. F. Potter
2734220 Sergeant N. Rees
2730511 RSM E. L. G. Richards
4191702 L/Sergeant J. L. Roberts
2735833 Sergeant R. L. Roberts
552159 MQMS F. Roughton
2735101 L/Corporal D. G. Ruddle
2738159 L/Corporal C. N. Sparrow
2732706 Guardsman C. R. Stone
2734039 Sergeant F. Townsend
2737061 CQMS J. Tumelty
2733185 Sergeant K. Vaughan
2735523 L/Sergeant L. A. R. Webb
2736203 Sergeant J. Westrop
2734166 CSM E. Williams
2735799 L/Sergeant F. E. Williams

POST-1945

Distinguished Service Order
Major C. K. Antelme (Iraq)
Major G. R. Harris (Afghanistan)

Military Cross
Captain A. J. G. Wight (Falklands)
Captain R. J. W. Williams (Cambodia)
Captain J. Stenner (Iraq)

Military Medal
Sergeant T. Edwards (Aden)
Guardsman H. Holland (Aden)
Colour Sergeant A. Richards (Northern Ireland)
Company Sergeant Major B. Neck (Falklands)
L/Corporal D. Loveridge (Falklands)
Guardsman S. Chapman (Falklands)

Queen's Gallantry Medal
Company Sergeant Major P. John (Northern Ireland)

Notes

(Full details of abbreviated titles are given in the Bibliography.)

Chapter 1

1 Edmonds, *Official History*, 1914, I, 11.
2 David Lloyd George, quoted in *The Times*, 20 September 1914.
3 Michael and Eleanor Brock, eds., *H. H. Asquith: Letters to Venetia Stanley*, Oxford, 1983.
4 Morris, *Man who Ran London*, 82–5.
5 RHQ, RHQ Diary, February 1915.
6 RA PS/PSO/GV/PS/ARMY 15246/5, Stamfordham to Kitchener.
7 Ibid.
8 RA PS/PSO/GV/PS/ARMY 15246/3, Stamfordham to Maj Gen Sir F. S. Robb, Military Secretary, War Office, 11 Feb. 1915.
9 RHQ, Registers of Enlistments, File 3, 4207 to 6306.
10 RHQ, RHQ Diary, 22 February 1915.
11 'Recruiting in the Great War', *Welsh Guards Magazine*, 2008, 70.
12 RHQ, 1st Battalion Archive, 7, Letter from Grenadier officer turning down offer to join the Welsh Guards.
13 Army Order 338, 1920.
14 'Formation', *Welsh Guards Magazine*, 2004–5, 39.
15 RHQ, Welsh Guards Photograph Albums and Scrap Books, Album 1.
16 RHQ, Regimental Standing Orders, Welsh Guards, 1993, 7.
17 Dudley-Ward, *History*, 6.
18 RA PS/PSO/GV/PS/ARMY 15934, Stamfordham to Murray-Threipland.
19 Gary Sheffield and John Bourne, eds. *Douglas Haig: War Diaries and Letters 1914–1918*, London, 2005, 128.
20 RHQ, Welsh Guards Photograph Albums and Scrap Books, Album 1.
21 Dudley-Ward, *History*, 30.
22 RHQ, Box 100, Perrins to Leatham, 25 April 1964.
23 RHQ, Welsh Guards Photograph Albums and Scrap Books, Album 1.
24 Ibid.
25 RHQ, Box 100 Murray-Threipland to Lord Harlech.
26 RHQ, Welsh Guards Photograph Albums and Scrap Books, Album 1.
27 Dudley-Ward, *History*, 42.
28 Ibid., 67.

Chapter 2

1 Denis Winter, *Haig's Command: A Reassessment*, London, 1991, 45.
2 RHQ, Box 100, Perrins to Leatham, 25 April 1964.
3 Ibid., Murray-Threipland to Lord Harlech.
4 Dudley-Ward, *History*, 133.
5 Keeling, *Arthur Gibbs: Letters Home*, 30.
6 Dudley-Ward, *History*, 194.
7 Cardiff School of History, Archaeology and Religion, *Welsh Voices of the Great War*, World War 1 material collected from families across Wales.
8 RHQ, Box 100, Perrins to Leatham, 25 April 1964.
9 Dudley-Ward, *History*, 156.
10 *London Gazette* (Supplement) no. 30272, p. 9260, 4 September 1917.
11 Dudley-Ward, *History*, 167.
12 IWM Document 6374, Papers of C. H. Dudley-Ward, 1915–19.
13 NA WO 95, War Diary, 1st Battalion Welsh Guards.
14 IWM Document 6374, Papers of C. H. Dudley-Ward, 1915–19.
15 IWM Sound Archive 10487, Charles Evans.
16 Edmonds, *Official History*, 1918, I, 258.
17 RHQ, Box 100.
18 IWM Document 6374, Papers of C. H. Dudley-Ward, 1915–19.
19 Ibid., 8 April 1918.
20 Dudley-Ward, *History*, 227.
21 Stanier, *Sammy's Wars*, 8.
22 IWM Document 6374, Papers of C. H. Dudley-Ward, diary, 8 August 1918.
23 Ibid., Dudley-Ward, diary, 27 September 1918.
24 Dudley-Ward, *History*, 282.
25 NA WO 85, War Diary, 1st Battalion Welsh Guards, 11 November 1918.

Chapter 3

1 Stanier, *Sammy's Wars*, 22.
2 RHQ, RHQ Diary, Murray-Threipland to London District, 9 April 1919.
3 *The Times*, 5 June 1920.
4 Lt Gen Sir Francis Lloyd, letter to the editor, *The Times*, 8 June 1920.
5 Leading article, *Daily Telegraph*, 7 June 1920.
6 Lord Chandos, *The Memoirs of Lord Chandos: An Unexpected View from the Summit*, London, 1962, 50.
7 *Western Mail*, 9 June 1920.
8 *The Times*, 16 June 1920.
9 *Western Mail*, 1 March 1926.
10 His nickname 'Chicot' derives from Jean-Antoine d'Anglere, or Chicot, the jester of King Henry III of France and later Henry IV, who was renowned for being sharp-tongued and for his willingness to speak to the king without formalities. He appears as a character in the novel *La Dame de Monsoreau* (1846) by Alexander Dumas.
11 Graham, *Recollections*, 6.
12 'A Proud Day for the Welsh Guards', *The Times*, 24 June 1925.
13 Barnett, *Britain and her Army*, 411.
14 RHQ, private letter from Roddy Sale, 1 August 2012.
15 Gibson-Watt, *Undistinguished Life*, 115–16.
16 Holmes, *Tommy*, 618.
17 RHQ, Archives 1st Battalion.
18 'In the Beginning was the Word and the Word was Depot', *Welsh Guards Magazine* 1993–4, 18.
19 David Niven, *The Moon's a Balloon*, London, 1971, 54.
20 Stanier, *Sammy's Wars*, 23.
21 Royle, *Anatomy of a Regiment*, 42–3.
22 Watcyn Watcyns, 'How I became a Singer', in *Musical Masterpieces*, Percy Pitt, ed., London,1925.
23 Dudley-Ward, *History*, 392–3.
24 'Flour convoyed from the docks', *The Times*, 10 May 1926.
25 Keith Laybourn, *The General Strike of 1926*, London, 1994, 64–5.

26 HRH The Duke of Windsor, *A King's Story*, London, 1951, 190.
27 Brigadier J. C. Windsor-Lewis, 'The Regiment Between the Wars', in Lewis, *Welsh Guards 1915–1965*, 40.
28 RHQ, Archives, 1st Battalion, 16, 'Message from the King'.
29 RHQ, Regimental Headquarters Diary, April 1929.
30 Mrs R. M. Stevenson, 'An Old Wife's Tale', in Lewis, *Welsh Guards 1915–1965*, 34.
31 Stanier, *Sammy's Wars*, 26.

Chapter 4

1 Michael Davie, ed., *The Diaries of Evelyn Waugh*, London, 1976, 447.
2 Graham, *Recollections*, 7.
3 Graham, *Recollections*, 9; Brutton, *Ensign in Italy*, 7; private information.
4 Spencer-Smith, *Rex Whistler's War*, 25–7.
5 NA WO 171, War Diary, 1st Battalion, Welsh Guards, 9 September 1944.
6 RHQ, Ellis, Book of Anecdotes, 80.
7 Anthony Kemp, *The SAS at War 1941–1945*, London, 1991, 1.
8 RHQ, 1st Battalion Archive, 'Arma Virosque'.
9 RHQ, 1st Battalion Archive, 25, Eric Coles, 'The Hell where Youth and Laughter Go', 1.
10 RHQ, 1st Battalion Archive, 28, Guardsman Bill Williams, 'Welsh Guards: Tough at the Bottom'.
11 RHQ, 1st Battalion Archive, 25, Coles, 1.
12 BBC WW2 People's War, Guardsman Harold Harrison's Story, ID A7589244, contributed on 7 December 2005.
13 'The Late Field Marshal Viscount Gort', House of Lords Debate, 2 April 1946 Hansard vol 140 cc473-7.
14 RHQ, 1st Battalion Archive, 29, 2Lt H. E. Lister, Carrier Platoon War Diary, Arras, May 1940.
15 Ellis, *Welsh Guards at War*, 107.
16 RHQ, 1st Battalion Archive, 28, Guardsman Bill Williams, 'Welsh Guards: Tough at the Bottom', 21.
17 Ibid., 23.
18 NA WO 166/4112, War Diary, 1st Battalion Welsh Guards, July 1940–December 1941.
19 RHQ, 1st Battalion Archive, 24, 'A Dunkirk Incident'.
20 NA WO 166/4113, War Diary, 2nd Battalion, Welsh Guards, September 1939–April 1940.
21 BBC WW2 People's War, Captain P. R. J. Tilley's Story, ID A4678121, contributed on 3 August 2005.
22 IWM Sound Archives, Fox-Pitt interview.
23 NA CAB 106/228, Report on operations of 2nd Battalion Welsh Guards, at Boulogne, 21–24 May 1940, by Major J. C. Windsor-Lewis.
24 Stanier, *Sammy's Wars*, 32.
25 NA WO 166/4113, War Diary, 2nd Battalion, Welsh Guards, September 1939–December 1941.
26 Major L. F. Ellis, *The War in France and Flanders 1939–1940, History of the Second World War (United Kingdom Military Series)*, London, 1954, 157.
27 NA CAB 106/228, Report on operations of 2nd Battalion Welsh Guards, at Boulogne, 21–24 May 1940, by Major J. C. Windsor-Lewis.
28 NA WO 166/4105, War Diary, 2nd Battalion Irish Guards, September 1939 – April 1940, Appendix D.

Chapter 5

1 RHQ, Meiron Ellis, *Iron Men*, 31.
2 RHQ, 1st Battalion Archive, 28, Guardsman Bill Williams, 'Welsh Guards: Tough at the Bottom', 29.
3 Ellis, *Welsh Guards at War*, 19–20.
4 Rosse and Hill, *Guards Armoured Division*.
5 NA WO 166/4113, War Diary, 2nd Battalion Welsh Guards, September 1941; Ellis, *Welsh Guards at War*, 21.
6 Spencer-Smith, *Rex Whistler's War*, 75.
7 Ellis, *Welsh Guards at War*, 38.
8 RHQ, 1st Battalion Archive, 28, Guardsman Bill Williams, 'Welsh Guards: Tough at the Bottom', 32.
9 RHQ, Meiron Ellis, *Iron Men*, 34.
10 RHQ, 1st Battalion Archive, Williams, 32.
11 NA WO 166/4112, War Diary, 1st Battalion Welsh Guards, October–December 1941.
12 Major General Peter Leuchars, interview with author, summer 1988.
13 RHQ, 1st Battalion Archives, Williams, 38.
14 RHQ, Ellis, *Iron Men*, 59.
15 NA WO 373/60, Recommendations for Honours and Awards for Gallant and Distinguished Service (Army) 29 November 1940–15 July 1941.
16 RHQ, Ellis, Book of Anecdotes, 89.
17 Ibid., 81.
18 NA WO 175/490, War Diary, 3rd Battalion Welsh Guards, January–June 1943.
19 RHQ, Box 118, Welsh Guards Battlefield Tour, Tunisia, 28–30 April 1993, W. B. Davies, 'A Sergeant's Story', 19.
20 RHQ, 3rd Battalion Archives, 4, W. K. Buckley, 'Memories of the 3rd Battalion, Welsh Guards, January to May 1943'.
21 Atkinson, *Army at Dawn*, 473.
22 RHQ, Box 118, Davies, 40.
23 Ibid., 42.
24 RHQ, 3rd Battalion Archives, Address at funeral of Leatham, 15 February 2001.
25 Quoted in Ellis, *Welsh Guards at War*.
26 RHQ, 3rd Battalion Archives, Buckley, 26.
27 Alan Moorehead, *The End in Africa*, London, 1943, 519.
28 Quoted in Atkinson, *Army at Dawn*, 475.
29 W. B. Davies (4034108), 'The 3rd Battalion's Farewell to Arms', *Welsh Guards Magazine*, 1998–9, 37.
30 RHQ, 3rd Battalion Archives, Buckley, 18.
31 Ellis, *Welsh Guards at War*.
32 NA WO 35/1211, 'The Guards in Tunisia', 4 June 1943.
33 NA WO 175/490, War Diary, 3rd Battalion Welsh Guards, January–June 1943.
34 Samuel W. Witcham Jr, 'Arnim: General of Panzer Troops Hans Jurgen von Arnim', in Correlli Barnett, ed., *Hitler's Generals*, London, 1989, 353.
35 Ellis, *Welsh Guards at War*, 35.
36 RHQ, 3rd Battalion Archives, Buckley, 24.

Chapter 6

1 Murrell, *Dunkirk to the Rhineland*, 44. The original drawings are held in the Welsh Guards Collection.
2 RHQ, Ellis, *Iron Men*, 61.
3 Spencer-Smith, *Rex Whistler's War*, 158.
4 Murrell, *Dunkirk to the Rhineland*, 93.
5 NA WO 171/1259, War Diary, 1st Battalion Welsh Guards, January–December 1944.
6 RHQ, Archives, 1st Battalion, 38, Major General P. R. Leuchars, 'Recollections of Fighting in France and Belgium, August–September 1944', 1.
7 Ellis, *Welsh Guards at War*, 204.
8 NA WO 171/1259, War Diary, 1st Battalion Welsh Guards, January–December 1944.
9 RHQ, Ellis, Book of Anecdotes, 36.
10 Quoted in Ellis, *Welsh Guards at War*, 208.
11 Murrell, *Dunkirk to the Rhineland*, 147.

12 RHQ, Archives, 3rd Battalion, 1, Major J. K. Cull, 'Cerasola', 1.
13 RHQ, Archives, 3rd Battalion, 'Monte Cerasola, 10–20 February 1944', 2.
14 RHQ, Archives, 3rd Battalion, Cull, 5.
15 RHQ, Archives, 3rd Battalion, 3 WG Medical Records, Battalion Sick Book.
16 NA WO 170/1355, War Diar, 3rd Battalion Welsh Guards, January–December 1944.
17 National Library of Wales, *From Warfare to Welfare*, Online resource.
18 NA WO 170/1355, War Diary, 3rd Battalion Welsh Guards, January–December 1944.
19 Brutton, *Ensign in Italy*.
20 Retallack, *Welsh Guards*, 105–6.
21 NA WO 170/1355, War Diary, 3rd Battalion Welsh Guards, January–December 1944.
22 *Western Mail*, 30 August 2010.
23 Gibson-Watt, *Undistinguished Life*, 115.
24 Ibid., 144.
25 RHQ, Archives, 3rd Battalion, 10, Brigadier M. C. Thursby-Pelham, 'Lorna', 2.

Chapter 7

1 RHQ, Meiron Ellis, Iron Men, 91.
2 NA WO 170/1259, War Diary, 1st Battalion Welsh Guards, 7 September 1944.
3 Ellis, Iron Men, 93.
4 NA WO 171/1259, War Diary, 1st Battalion Welsh Guards, 7 September 1944.
5 RHQ, 1st Battalion Archives.
6 Syrett, *Letters*, 20.
7 RHQ, Ellis, Iron Men, 98.
8 RHQ, Ellis, Book of Anecdotes, 36.
9 Major General Peter Leuchars, interview with author, summer 1988.
10 NA WO 171/1259, War Diary, 1st Battalion Welsh Guards, 12 September 1944.
11 Ellis, *Victory in the West*, Vol. II, p. 237.
12 Ellis, *Welsh Guards at War*, 227.
13 Murrell, *Dunkirk to the Rhineland*, 166.
14 Ibid., 185.
15 NA WO 171/1259, War Diary, 1st Battalion Welsh Guards, 9 November 1944.
16 Ibid., War Diary, 1 December 1944.
17 Ibid., War Diary, 17 December 1944.
18 Ellis, *Welsh Guards at War*, 253.
19 Ibid., 234.
20 RHQ, Major General P. R. Leuchars, 'Recollections of Italy 1945', 1.
21 Ibid., 253.
22 NA WO 170/1355, War Diary, 3rd Battalion Welsh Guards, March 1945.
23 Ibid., 26 April 1945.
24 Ibid., 2 May 1945.
25 Ibid., 4 May 1945.
26 Ben Shepherd, *The Long Way Home: The Aftermath of the Second World War*, London, 2010, 78–82.
27 NA WO 170/1355, War Diary, 3rd Battalion Welsh Guards, 19 May 1945.
28 IWM 10968/2/2, 15, Nigel Nicolson.
29 Ibid., 21 May 1945.
30 Shepherd, *Long Way Home*, 79.
31 NA WO 170, War Diary, 3rd Battalion Welsh Guards, 8 May 1945.
32 Sir Frederic Bolton, MC, '3rd Battalion, Welsh Guards', *Welsh Guards Magazine*, 1985–6, 9.
33 Dick Fletcher, 'Imber Court', *Welsh Guards Magazine*, 2001–2, 44–5.

Chapter 8

1 Nigel Hamilton, *Monty: The Field Marshal 1944–1976*, London, 1986, 633–721.
2 Author's interview with Arthur Rees, summer 1989, quoted in Royle, *Anatomy of a Regiment*, 47.
3 Roy Lewis, 'For Them the War was Over', *Welsh Guards Magazine*, 2004–5, 35.
4 *Household Division Magazine*, Winter 1946–7.
5 NA WO 170/1355, War Diary, 3rd Battalion Welsh Guards, 22 June 1945.
6 W. B. Davies, 'The 3rd Battalion's Farewell to Arms', *Welsh Guards Magazine*, 1998–9, 37.
7 RHQ, Ellis, Book of Anecdotes, 82.
8 Avi Shlaim, *The Iron Wall: Israel and the Arab World*, London, 2000, 13.
9 'The 1st Battalion in 1945 – A Year of Contrasts', *Welsh Guards Magazine*, 1996–7, 39–40.
10 Barnett, *Lost Victory*, 61.
11 RHQ, Box 109, War Diary, 1st Battalion, 27 October 1945.
12 RHQ, 1st Battalion Archive, Major General Peter Leuchars, 'Palestine 1945–1948', 1.
13 RHQ, Box 109, 1st Battalion Orders, 4 December 1945.
14 RHQ, 1st Battalion Archive, 'Palestine 1945–1948, 2'.
15 Ibid.
16 Ibid., 3.
17 RHQ, Box 109, Orders, 7 January 1947.
18 RHQ, Box 46, Major General P. R. Leuchars, '1WG Palestine 1945-49'.
19 RHQ, Box 48, Major General P. R. Leuchars, '1WG Palestine 1947/48'.
20 Trevor Royle, *Glubb Pasha: The Life and Times of Lieutenant-General Sir John Bagot Glubb, Commander of the Arab Legion*, London, 1991, 350.
21 RHQ, Major General P. R. Leuchars, CBE, '1WG in Palestine 1947–48 – The Last Year', *Welsh Guards Magazine*, 1999–2000, 36.
22 Major General John Marriott, *Scots Guards Magazine*, 1962.
23 Royle, *Anatomy of a Regiment*, 173; further information from Glyn Davies.
24 RQMS Richard Fletcher, 'The Regiment Today', *Welsh Guards 1915–1965*, 108–9.
25 Hansard, HC Deb, 15 April 1947, vol. 436 c8.
26 Royle, *Best Years*, 144.
27 Fletcher, 'Regiment Today', 110.
28 Ibid., 113.
29 RHQ, 1st Battalion Archive, Major General P. R. Leuchars CBE, 'England, Germany and Egypt 1948–1956', 4.
30 RHQ, 1st Battalion Archive, P. R. Leuchars CBE, 'Aqaba', 6.

Chapter 9

1 Retallack, *Welsh Guards*, 158.
2 Strawson, *Gentlemen in Khaki*, 255.
3 RHQ, Archives, Personal Memories, 54, Personal Account of Service of 23523004 Alan Acreman.
4 RHQ, Archives, 3rd Battalion, 3, Brigadier Thursby-Pelham, 'Colonel C. A. la T. Leatham, Address at Memorial Service', 3.
5 Author's interview with Arthur Rees, summer 1989, quoted in Royle, *Anatomy of a Regiment*, 49.
6 Author's interview with Peter Leuchars, summer 1989.
7 S. Tapper-Jones, Town Clerk of Cardiff, 'Freedoms', in Lewis, *Welsh Guards 1915–1965*, 96–102.
8 Paradoxically, less than ten years after forcing the rest of

NATO to adopt 7.62 mm ammunition, the USA switched to the 5.56 mm round.

9 When No. 1 (Guards) Independent Parachute Company was eventually disbanded on 24 October 1975, most members returned to their parent regiments, although some Guards soldiers went to Hereford to join the Guards Squadron, 22 SAS Regiment, which had been formed in 1966. There is still a platoon of Guards parachute soldiers attached to 3rd Battalion, Parachute Regiment, which has served with distinction in Afghanistan.

10 Paradata website, http://www.paradata.org.uk/people/elwyn-glyndwr-roberts.

11 Lieutenant General Sir Charles [later Field Marshal the Lord] Guthrie, KCB, LVO, OBE, 'The Last Forty Years', *Welsh Guards Magazine*, 1990–1, 18.

12 N. J. Lambert and Umit Candan, 'Analysis & Evaluation of the Immediate Reaction Task Force (Land) Command and Control Concept: Applying the COBP', Paper presented at the RTO SAS Symposium on 'Analysis of the Military Effectiveness of Future C2 Concepts and Systems', held at NC3A, The Hague, Netherlands, 23–25 April 2002, and published in RTO-MP-117.

13 BBC1 Television News, 6 September 2005.

14 Richard Fletcher, 'The Regiment Today', in Lewis *Welsh Guards 1915–1965*, 119–22.

15 Gow, *Trooping the Colour*, 92.

16 Lieutenant Colonel Paddy Stevens, 45 Commando, RM, quoted in Neillands, *A Fighting Retreat*, 337.

17 J. M. Jones, 'Aden … 40 Years On', *Welsh Guards Magazine*, 2005–6, 46.

18 Captain Dick Fletcher, 'Aden and the Radfan – 1965–1966', *Welsh Guards Magazine*, 1995–6, 32.

19 Quoted in Allen, *Savage Wars of Peace*, 165.

20 Lieutenant Colonel Philip Ward, *All the Queen's Men*, produced and directed by Kevin Millington, ATV, 23 November 1966.

Chapter 10

1 Retallack, *Welsh Guards*, 165.

2 IWM Department of Sound Records, 21566, Interview with Ivar Froystad, 20 April 2001.

3 Retallack, *Welsh Guards*, 166.

4 Captain C. N. Black, 'Soldiering in Northern Ireland – a Guardsman's View', *Welsh Guards Magazine*, 1986–7, 5.

5 Robert Barlett, ed, *The Working Life of the Surrey Constabulary, 1851–1992*, International Centre for the History of Crime, Policing and Justice, Part Three, Open University, 2013.

6 Retallack, *Welsh Guards*, 170.

7 1st Battalion Notes, *Welsh Guards Magazine*, 1977–8, 11.

8 Mike Jackson, *Soldier: The Autobiography*, London, 2010, 132.

9 RHQ, information from Lieutenant General Sir Christopher Drewry, 2012.

10 RHQ, *The Leek*, Issue 1, November 1979, 3.

11 Ibid., Issue 3, January 1980, 10.

12 Ibid., Issue 2, December 1979, 3.

13 Ibid.

14 Ibid., Issue 3, January 1980, 12.

15 Ibid., Issue 4, February 1980, 2.

16 RHQ, '1st Battalion, Welsh Guards', *Welsh Guards Magazine*, 1979–80, 14.

17 Colonel J. W. T. A. Malcolm, 'Cavalry Unseated: Polo History is made', *Guards Magazine*, Summer 1976, 61.

18 '1st Battalion, Welsh Guards', *Welsh Guards Magazine*, 1980–1, 12.

19 Hansard, HC Deb, 12 December 1991 vol. 200 c514W.

20 '1st Battalion, Welsh Guards', *Welsh Guards Magazine*, 1980–1, 14.

21 Ibid., 14.

22 '1st Battalion, Welsh Guards', *Welsh Guards Magazine*, 1981–2, 10.

Chapter 11

1 '1st Battalion, Welsh Guards', *Welsh Guards Magazine* 1982–3, 10.

2 Weston, *Walking Tall*, 74.

3 Jackson and Bramall, *The Chiefs*, 402–3.

4 Freedman, *Official History of the Falklands Campaign*, II, 21–2.

5 Major General Julian Thompson, 'War in the South Atlantic', *Imperial War Museum Book of Modern Warfare*, 294.

6 Freedman and Gamba-Stonehouse, *Signals of War*, 324.

7 Bramall and Jackson, *The Chiefs*, 412.

8 Hastings and Jenkins, *Battle for the Falklands*, 269–70.

9 '1st Battalion, Welsh Guards', *Welsh Guards Magazine*, 1982–3, 12.

10 Captain R. S. Mason, 'The Falklands Campaign: Personal Account', *Welsh Guards Magazine*, 1982–3, 16.

11 RHQ, 'Battle Account of the 1st Battalion, Welsh Guards', Falkland Islands, 2 June–14 June 1982'.

12 RHQ, Battle Account, 3; Rickett interview, Royle, *Anatomy of a Regiment*, 225.

13 RHQ, Box File 3DI, Commander's Diary 1WG, 8 June 1982.

14 Geoff Puddefoot, *No Sea too Rough, The Royal Fleet Auxiliary in the Falklands War: The Untold Story*, London, 2007, 148.

15 Sayle interview, Royle, *Anatomy of a Regiment*, 228.

16 Robert S. Bolia, 'The Falklands War: The Bluff Cove Disaster', *Military Review*, November–December 2004, 69.

17 Hastings and Jenkins, *Battle for the Falklands*, 350.

18 Sayle, interview, Royle, *Anatomy of a Regiment.*

19 Hilarian Roberts, *Guards Magazine.*

20 Guardsman Wayne Trigg, in Bilton and Kosminsky, *Speaking Out*, 174.

21 Charles Bremner, quoted in Arthur, *Above All, Courage*, 132.

22 Crispin Black, 'Titanic and Sir Galahad: What links the two tragedies', *The Week*, April 2012.

23 Rickett, interview, Royle, *Anatomy of a Regiment*, 235.

24 Captain R. S. Mason, 'The Falklands Campaign – Personal Account', *Welsh Guards Magazine*, 1982–3, 17.

25 Ibid.

26 '1st Battalion, Welsh Guards', *Welsh Guards Magazine*, 1982–3, 14.

27 Major C. F. Drewry, 'Sealink Sets You Free', *Welsh Guards Magazine*, 1982–3, 18–21.

28 Harnden, *Dead Men Risen*, 22.

29 NA ADM 330, Board of Inquiry into the loss of RFAs *Sir Tristram* & *Sir Galahad*, June 1982.

Chapter 12

1 '1st Battalion, Welsh Guards', *Welsh Guards Magazine*, 1984–5, 10.

2 RHQ, Ellis, *Iron Men*, 123.

3 '1st Battalion, Welsh Guards', *Welsh Guards Magazine*, 1986–7, 10.

4 Captain C. N. Black, 'Soldiering in Northern Ireland – A Guardsman's View', *Welsh Guards Magazine*, 1986–7, 5.

5 Kevin Myers, 'Two extremes in North like spoilt brats on a day out', *Irish Independent*, 15 January 2013.

6 Arthur, *Northern Ireland: Soldiers Talking*, 246.
7 Royle, *Anatomy of a Regiment*, 193.
8 Foreword by the Regimental Lieutenant Colonel, *Welsh Guards Magazine*, 1991–2, 4.
9 Jane Richards, 'Boys Own Epic', *The Independent*, 24 November 1995.
10 Headquarter Company Notes, *Welsh Guards Magazine*, 1991–2, 18–20.
11 '1st Battalion, Welsh Guards', *Welsh Guards Magazine*, 1992–3, 6.
12 Post-Operational Interview with Lieutenant Colonel B. J. Bathurst, conducted by Brigadier (Retd) I. A. Johnstone, SO1 Mission Support Group, Land Warfare, 20 July 2005.
13 'UK troops Angola-bound', *The Independent*, 14 April 1995.
14 Headquarter Company, *Welsh Guards Magazine*, 1995–6, 18.
15 '1st Battalion, Welsh Guards', *Welsh Guards Magazine*, 1996–7, 4.

Chapter 13
1 The Prince of Wales's Company Group, *Welsh Guards Magazine*, 2002, 6–8.
2 Hansard, HC Debate, vol 318, 3 November 1998, c444.
3 Battalion Tactical Headquarters, *Welsh Guards Magazine*, 1997–8, 4.
4 Ibid.
5 'Report into a Complaint from Rita and John Restorick Regarding the Circumstances of the Murder of their son Lance Bombardier Stephen Restorick on 12 February 1997', Police Ombudsman for Northern Ireland, 13 December 2006.
6 Number Three Company Group, *Welsh Guards Magazine*, 1997–8, 13–15.
7 Battalion Headquarters, *Welsh Guards Magazine*, 1998–9, 8.
8 Support Company, Corps of Drums, *Welsh Guards Magazine*, 1999–2000, 19.
9 Hansard, HC 138, 1997–1998, The Strategic Defence Review, 3 September 1998, 39.
10 Foreword, *Welsh Guards Magazine*, 2000–1, 3.
11 Major D. L. W. Bossi, 'Number Eight Company', *Welsh Guards Magazine*, 2002–3, 29–31.
12 Misha Glenny, *The Balkans 1804–1999*, London, 1999, 80.
13 Number Two Company, *Welsh Guards Magazine*, 2002–3, 9.
14 Report to HM The Queen, January 2003.
15 Ibid.
16 Report to HM The Queen, July 2004.
17 '1st Battalion, Welsh Guards, Rugby Tour to South Africa 25 August–8 September 2003', *Welsh Guards Magazine*, 2003–4, 25–7.
18 Report to HM The Queen, January 2004.
19 Borrow, *Wild Wales*, 366.

Chapter 14
1 Foreword, *Welsh Guards Magazine*, 2004–5, 3.
2 'Fact Sheet on Iraq's Major Shia Political Parties and Militia Groups', Institute for the Study of War, Washington, April 2008.
3 Holmes, *Dusty Warriors*, 108.
4 Post-Operational Interview with Lieutenant Colonel B. J. Bathurst, conducted by Brigadier (Retd) I. A. Johnstone, SO1 Mission Support Group, Land Warfare, 20 July 2005.
5 Headquarter Company, *Welsh Guards Magazine*, 2004–5, 20.
6 The Prince of Wales's Company, ibid., 8.
7 Number Two Company, ibid.
8 Communication from Lieutenant Colonel Giles Harris, 24 February 2015.
9 Toby Harnden, 'Welsh Guards Seek Justice for Redcap Killings', *Daily Telegraph*, 12 December 2004.
10 Post-Operational Interview with Lieutenant Colonel B. J. Bathurst, conducted by Brigadier (Retd) I. A. Johnstone, SO1 Mission Support Group, Land Warfare, 20 July 2005, 5.
11 Doug Struck, 'Exit for British in Poor Corner of Iraq', *Washington Post*, 12 February 2005.
12 Interview with author, Cavalry Barracks, Hounslow, 30 September 2013.
13 Foreword, *Welsh Guards Magazine*, 2006–7, 5.
14 BBC News, 26 March 2007.
15 'Praise for Bosnia "job well done"', *Birmingham Post*, 30 March 2007.

Chapter 15
1 Colonel A. J. E. Malcolm, OBE, Foreword, *Welsh Guards Magazine*, 2008, 3.
2 19 Light Brigade to replace 3 Commando Brigade, Royal Marines, in Afghanistan, *Defence News*, 16 December 2008.
3 Number IX Company, *Welsh Guards Magazine*, 2009, 41–2.
4 Number Two Company, ibid., 16.
5 Harnden, *Dead Men Risen*, 148.
6 Anthony King, 'Understanding the Helmand Campaign: British Military Operations in Afghanistan', *International Affairs* 86: 2 (2010), 311–32.
7 Number IX Company, *Welsh Guards Magazine*, 2009, 42.
8 Harnden, *Dead Men Risen*, 316.
9 Ibid., 209–10.
10 Interview with author, Cavalry Barracks, Hounslow, 30 September 2013.
11 Harnden, *Dead Men Risen*, 347.
12 'Britain hails "success" of anti-Taliban push', CNN, 27 July 2009.
13 Number X Company, *Welsh Guards Magazine*, 2006, 41–7.
14 Brigadier P. R. G. Williams, 'Welsh Guards Afghanistan Appeal', *Welsh Guards Magazine*, 2010, 34–5.
15 Brigadier R. H. Talbot Rice, Foreword, *Welsh Guards Magazine*, 2011, 3.
16 Hansard, HC Deb, 14 May 2013, c509.
17 Number Three Company, *Welsh Guards Magazine*, 2012, 19.
18 Support Company, ibid., 34–5.
19 'Commander Task Force Helmand reports on HERRICK 16 progress', Ministry of Defence, 23 October 2012.
20 Lieutenant Colonel D. L. W. Bossi, Foreword, *Welsh Guards Magazine*, 2013, 7.
21 *Transforming the British Army: Modernising to Face an Unpredictable Future*, Ministry of Defence, July 2012, 2.
22 Brigadier R. H. Talbot Rice, Foreword, *Welsh Guards Magazine*, 2013, 5.
23 Harnden, *Dead Men Risen*, 496.

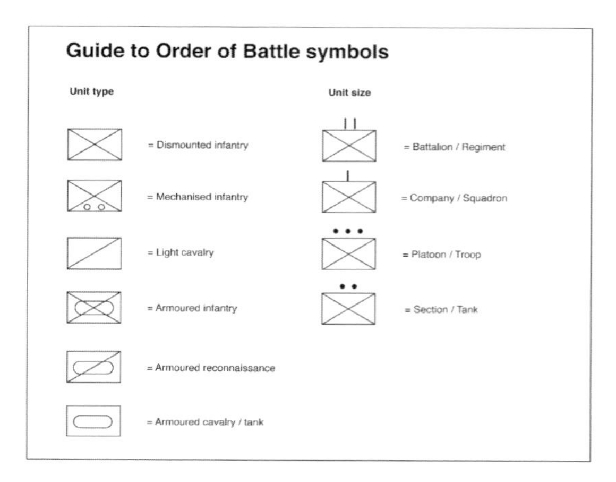

Bibliography and Official Papers

PRIMARY SOURCES

Regimental Archives (RHQ)
The bulk of the archives are held by Regimental Headquarters, Welsh Guards, Wellington Barracks, London, publishers of *The Welsh Guards Magazine*.

National Archives (NA)
War diaries and other government papers are deposited in the National Archives, Public Record Office, Kew.

Royal Archives (RA)
Papers relating to the foundation of the regiment are deposited in the Royal Archives at Windsor Castle.

Imperial War Museum (IWM)
Papers of C. H. Dudley-Ward, 1915–1919, Document 6374.
Department of Sound Archives, interviews with various Welsh Guardsmen.

British Broadcasting Corporation (BBC)
WW2 People's War Project, contributions by various Welsh Guardsmen, 1939–1945.

Welsh Guards Collection, Park Hall, Oswestry
A unique collection of artefacts and archives depicting and honouring the history of life in the Welsh Guards since the formation in February 1915.

Welsh Guards Reunited website
http://www.welshguardsreunited.co.uk

SECONDARY SOURCES

Books about the Welsh Guards or books written by Welsh Guardsmen

Brutton, Philip, *Ensign in Italy*, London, 1992

——, *A Captain's Mandate*, London, 1996

Cook, Denys, *Missing in Action, Or My War as a Prisoner of War*, Trafford, 2013

Dudley-Ward, C. H., *History of the Welsh Guards*, London, 1920

——, *The Welsh Regiment of Footguards*, London, 1936

Ellis, L. F., *Welsh Guards at War*, Aldershot, 1946

——, *Victory in the West: The Defeat of Germany*, Uckfield, 1968

Evison, Margaret, *Death of a Son: A Mother's Story*, London, 2012

Gibson-Watt, Andrew, *An Undistinguished Life*, Lewes, 1990

Graham, Andrew, ed., *Recollections of Colonel R. E. K. Leatham, DSO*, Hinckley, n.d.

Harnden, Toby, *Dead Men Risen: The Welsh Guards and the Real Story of Britain's War in Afghanistan*, London, 2011

Keeling, Jenifer, ed., *Captain Arthur Gibbs MC: Letters Home 1914–1918*, privately published, Sedlescombe, 2010

Lewis, K. D., ed., *Welsh Guards 1915–1965: An Informal Account of the First Fifty Years*, Devonport, 1965

Lloyd George, Owen, *A Tale of Two Grandfathers*, London, 1999

Mather, Sir Carol, *Aftermath of War: Everyone Must Go Home*, London, 1992

——, *When the Grass Stops Growing*, London, 1999

Murrell, Nick, ed., *Dunkirk to the Rhineland: Diaries and Sketches of Sergeant C. S. Murrell, Welsh Guards*, Barnsley, 2011

Pritchard, Sydney, *My Life in the Welsh Guards, 1939–1946*, Talybont, 2007

Rees, Tim, *In Sights: The Story of a Welsh Guardsman*, Stroud, 2013

Retallack, John, *The Welsh Guards*, London, 1981

Rosse, Captain, the Earl of, and Colonel E. R. Hill, *The Story of the Guards Armoured Division*, London, 1956

Royle, Trevor, *Anatomy of a Regiment: Ceremony and Soldiering in the Welsh Guards*, London, 1990

Spencer-Smith, Jenny, *Rex Whistler's War, 1939–July 1944: Artist into Tank Commander*, London, 1994

Stanford, Fiona, *Don't Say Goodbye: Our Heroes and the Families They Leave Behind*, London, 2011

Stanier, Sir Beville, Bt., ed., *Sammy's Wars: Recollections of War in France and Other Occasions*, privately published, 1998

Syrett, John David Alfred, *Letters*, privately printed, 1945

Warburton-Lee, John, *Roof of Africa*, Shrewsbury, 1992
——, *Roof of the Americas*, Shrewsbury, 1996
Weston, Simon, *Walking Tall: An Autobiography*, London, 1989

Other Books Consulted

Allen, Charles, *The Savage Wars of Peace: Soldiers' Voices 1945–1989*, London, 1990
Arthur, Max, *Above All, Courage: Personal Stories from the Falklands War*, London, 1985
——, *Northern Ireland: Soldiers Talking*, London, 1987
Ascoli, David, *A Companion to the British Army, 1660–1983*, London, 1983
Atkinson, Rick, *An Army at Dawn: The War in North Africa, 1942–1943*, London, 2003
Barnett, Correlli, *Britain and her Army, 1509–1970*, London, 1970
——, *The Lost Victory: British Dreams, British Realities, 1945–1950*, London, 1995
Beevor, Antony, *D-Day: The Battle for Normandy*, London, 2009
Bilton, Michael, and Peter Kosminsky, *Speaking Out: Untold Stories from the Falklands War*, London, 1989
Borrow, George, *Wild Wales*, London, 1862
Brereton, J. M., *The British Army: A Social History of the British Army from 1661 to the Present Day*, London, 1986
Chandler, David, and Ian Beckett, eds, *The Oxford Illustrated History of the British Army*, Oxford, 1994
Dear, I. C. B., ed., *The Oxford Companion to the Second World War*, Oxford, 1995
Edmonds, James E., general ed., *History of the Great War, Based on Official Documents: Military Operations, France and Belgium*, 14 vols, London, 1922–49
Ford, Ken, *Mailed Fist: 6th Armoured Division at War*, Stroud, 2005
Freedman, Lawrence, *Official History of the Falklands Campaign*, 2 vols, London, 2005
Freedman, Lawrence, and Virginia Gamba-Stonehouse, *Signals of War: The Falklands Conflict, 1982*, London, 1990
Gander, Terry, ed., *Encyclopedia of the Modern British Army*, London, 1986
Gow, General Sir Michael, *Trooping the Colour*, rev. ed., London, 1988
Hammond, Bryn, *Cambrai 1917: The Myth of the First Great Tank Battle*, London, 2008
Hastings, Max, *Overlord: D-Day and the Battle for Normandy*, London, 1984
Hastings, Max, and Simon Jenkins, *The Battle for the Falklands*, London, 1983
Headlam, Cuthbert, *The Guards Division in the Great War*, 2 vols, London, 1924
Holmes, Richard, ed., *The Oxford Companion to Military History*, Oxford, 2001
——, *Tommy: The British Soldier on the Western Front*, London, 2004
——, *Dusty Warriors: Modern Soldiers at War*, London, 2006
Jackson, Bill, and Edwin Bramall, *The Chiefs: The Story of the United Kingdom Chiefs of Staff*, London, 1992
Mallinson, Allan, *The Making of the British Army: From the English Civil War to the War on Terror*, London, 2009
Mead, Gary, *The Doughboys: America and the First World War*, London, 2000
Middlebrook, Martin, *Operation Corporate: The Falklands War*, London, 1985
Morris, Richard, ed., *The Man Who Ran London during the Great War: The Diaries and Letters of Lieutenant General Sir Francis Lloyd, GCVO, KCB, DSO, 1853–1926*, Barnsley, 2009
Neillands, Robin, *A Fighting Retreat: The British Empire, 1947–1997*, London, 1996
Royle, Trevor, *The Best Years of Their Lives: The National Service Experience, 1945–1963*, London, 1986
Strawson, John, *Gentlemen in Khaki: The British Army, 1890–1990*, London, 1989
Thompson, Major General Julian, ed., *No Picnic: 3 Commando Brigade in the South Atlantic, 1982*, London, 1985
——, *The Imperial War Museum Book of Modern Warfare*, London, 2002
Van der Bijl, Nick, and David Aldea, *5th Infantry Brigade in the Falklands War*, Barnsley, 2003

Film and Television

They Were Not Divided, written and directed by Terence Young, Two Cities Films, 1950
All the Queen's Men, produced and directed by Kevin Millington, ATV, 1966
In the Company of Men, written and directed by Molly Dineen, BBC, 1995
The Lost Platoon, narrated by Shaun Dooley, BBC3, September 2012

Index